# International Conference on Geospatial Science for Digital Earth Observation (GSDEO-2021)



IOP Conference Series: Earth and Environmental Science
Volume 1164

Online
26-27 March 2021

**Editors:**

**Tuhin Bhadra**
**Kasturi Mukherjee**
**Anu Rai**
**Rajib Sarkar**
**Pratik Dash**
**Debaroti Das**

ISBN: 978-1-7138-7285-6
ISSN: 1755-1307

**Printed from e-media with permission by:**

Curran Associates, Inc.
57 Morehouse Lane
Red Hook, NY 12571

**Some format issues inherent in the e-media version may also appear in this print version.**



Institute of Physics
Dirac House, Temple Back
Bristol BS1 6BE UK

Phone: 44 1 17 929 7481
Fax: 44 1 17 920 0979

techtracking@iop.org

**Additional copies of this publication are available from:**

Curran Associates, Inc.
57 Morehouse Lane
Red Hook, NY 12571 USA
Phone: 845-758-0400
Fax: 845-758-2633
Email: curran@proceedings.com
Web: www.proceedings.com

# TABLE OF CONTENTS

## GEOMORPHOLOGY, CLIMATE AND ENVIRONMENT



## SOCIETY, CITY AND SUSTAINABILITY

IOP Conf. Series: Earth and Environmental Science 1164 (2023) 011001 doi:10.1088/1755-1315/1164/1/011001

# PREFACE

## International Conference on Geospatial Science for Digital Earth Observation (GSDEO 2021)

**T Bhadra[1], K Mukherjee[1], A Rai[1], R Sarkar[1], P Dash[2], D Das[3]**
[1]Department of Geography, Adamas University, Kolkata
[2]Department of Geography, Khejuri College, Purba Medinipur
[3]Department of Geography, P. N. Das College, Kolkata

The international conference on "Geospatial Science for Digital Earth Observation" (GSDEO) 2021 was successfully held on a virtual platform of Zoom on March 26$^{th}$ and 27$^{th}$, 2021. The conference was jointly organized by the Indian Society of Remote Sensing (ISRS), Kolkata chapter, and the Department of Geography, School of Basic and Applied Sciences, Adamas University. Due to the non-predictable behaviour of the COVID-19 second wave, which imposed restrictions on organizing offline events, the GSDEO (2021) organizing committee decided to organize the conference online, instead of postponing the event.

Remotely sensed data and geographic information systems have been increasingly used together for a vast range of applications, which include land use/land cover mapping, water resource management, weather forecasting, environmental monitoring, agriculture, disaster management, etc. Currently, intensive research is being carried out using remotely sensed data on the geoinformatics platform. New developments have led to dynamic advances in recent years. The objective of the international conference on Geospatial Science for Digital Earth Observation (GSDEO 2021) was to bring the scientists, academicians, and researchers, in the field of geo-environmental sciences on a common platform to exchange ideas and their recent findings related to the latest advances and applications of geospatial science. The call for papers received an enthusiastic response from the academic community, and over 100+ participants from 50+ colleges, universities, and institutions participated in the conference. In total 50+ research papers had been presented through the virtual Zoom conference platform in GSDEO 2021.

The conference witnessed the presentation of research papers from diverse applied fields of geospatial sciences, which include the application of geoinformatics in geomorphology, hydrology, urban science, land use planning, climate, and environmental studies. There were four sessions namely, TS 1: Geomorphology and Hydrology, TS 2: Urban Science, TS 3: Social Sustainability and Land Use Planning, and TS 4: Climate and Environment. Each session was further subdivided, into two parts, namely Technical Session 1-A and 1-B. Each sub-session had been designed with one keynote speech and 5 oral presentations. Oral sessions were organized in two parts and offered through live and pre-recorded components based on the preference of the presenters. The presentation session was followed by a live Q&A session. The session chairs moderated the discussions. Similarly, poster sessions were organized in three parts and offered e-poster, live, and pre-recorded components. The best presenter of each sub-session received the best paper award.

Dr. Prithvish Nag, Ex-Director of NATMO & Ex Surveyor General of India delivered the inaugural speech, and Dr. P. Chakrabarti, Former Chief Scientist of the DST&B, Govt. of West Bengal delivered a special lecture after the inaugural session. Eight eminent keynote speakers, Prof. S.P. Agarwal from the Indian Institute of Remote Sensing, Prof. Ashis Kumar Paul from Vidyasagar University, Prof. Soumya Kanti Ghosh from the Indian Institute of Technology, Kharagpur, Prof. L. N. Satpati from the University of Calcutta, Prof. R.B. Singh from the University of Delhi, Dr. A.K. Raha, IFS (Retd), Prof. Gerald Mills from the University College Dublin and Prof. Sugata Hazra from Jadavpur University enriched the knowledge of participants in the field of geoinformatics by their informative lectures. The presentations and discussions widely covered the various spectrums of geoinformatics and its application in monitoring natural resources like vegetation mapping, agricultural resource monitoring, forest health assessment, water, and ocean resource management, disaster management, land resource management, water and climate studies, drought vulnerability assessment, groundwater quality monitoring, accretion mapping and the use of geospatial science in studying morphological, hydrological, and other biophysical characteristics of a region etc. Application of geoinformatics in predicting urban expansion, urban climate, disaster management, healthcare accessibility, anthropogenic resource monitoring, spatial-interaction mapping, and, sustainable regional planning were well-discussed topics of the conference.

**List of Committees:**

**Patrons**
- Dr. Prithvish Nag, Former Director, NATMO, Former Surveyor General, Survey of India
- Prof. Samit Ray, Hon'ble Chancellor, Adamas University, Kolkata
- Prof. (Dr.) Deependra Kumar Jha, Vice Chancellor, Adamas University, Kolkata

**General Chairpersons**
- Er. S. Halder, Chairman, ISRS – Kolkata Chapter
- Prof. (Dr.) Jitendra K. Pandey, Pro Vice-Chancellor and Dean, Adamas University, Kolkata

**Organizing Chairpersons**
- Dr. A.K. Mishra, Vice Chairman, ISRS – Kolkata Chapter
- Dr. Kasturi Mukherjee, HOD, Geography, Adamas University, Kolkata

**Conference Conveners**
- Dr. A. Lahiri, Life Member, ISRS – Kolkata Chapter
- Dr. Tuhin Bhadra, Assistant Professor, Dept. of Geography, Adamas University, Kolkata

**Organizing Secretaries**
- Ms. D. Das, Joint Secretary, ISRS – Kolkata Chapter
- Dr. Rajib Sarkar, Assistant Professor, Dept. of Geography, Adamas University, Kolkata

**Joint Organizing Secretaries**
- Mr. S. Maity, Executive Committee Member, ISRS – Kolkata Chapter
- Dr. Anu Rai, Assistant Professor, Dept. of Geography, Adamas University, Kolkata

**Technical Chairpersons**
- Mr. P. Das, Executive Committee Member, ISRS – Kolkata Chapter
- Dr. Rajib Sarkar, Assistant Professor, Dept. of Geography, Adamas University, Kolkata

**Treasurer**
- Ms. D. Biswas, Treasurer, ISRS – Kolkata Chapter
- Dr. Anu Rai, Assistant Professor, Dept. of Geography, Adamas University, Kolkata

**Editorial Board (Conference Abstract Volume)**
- Prof. P. K. Sikdar, Vice Chairman, ISRS – Kolkata Chapter
- Prof. B. Biswas, Executive Committee Member, ISRS – Kolkata Chapter
- Dr. Kasturi Mukherjee, HOD, Geography, Adamas University, Kolkata
- Dr. Anu Rai, Assistant Professor, Dept. of Geography, Adamas University, Kolkata
- Dr. Rajib Sarkar, Assistant Professor, Dept. of Geography, Adamas University, Kolkata
- Dr. Tuhin Bhadra, Assistant Professor, Dept. of Geography, Adamas University, Kolkata
- Dr. Pratik Dash, Khejuri College, Vidyasagar University, Purba Medinipur

IOP Conf. Series: Earth and Environmental Science 1164 (2023) 011001 doi:10.1088/1755-1315/1164/1/011001



Screenshots of GSDEO 2021 |26th to 27th March 2021|Virtual| 9.00 HRS – 18.00 HRS (IST)

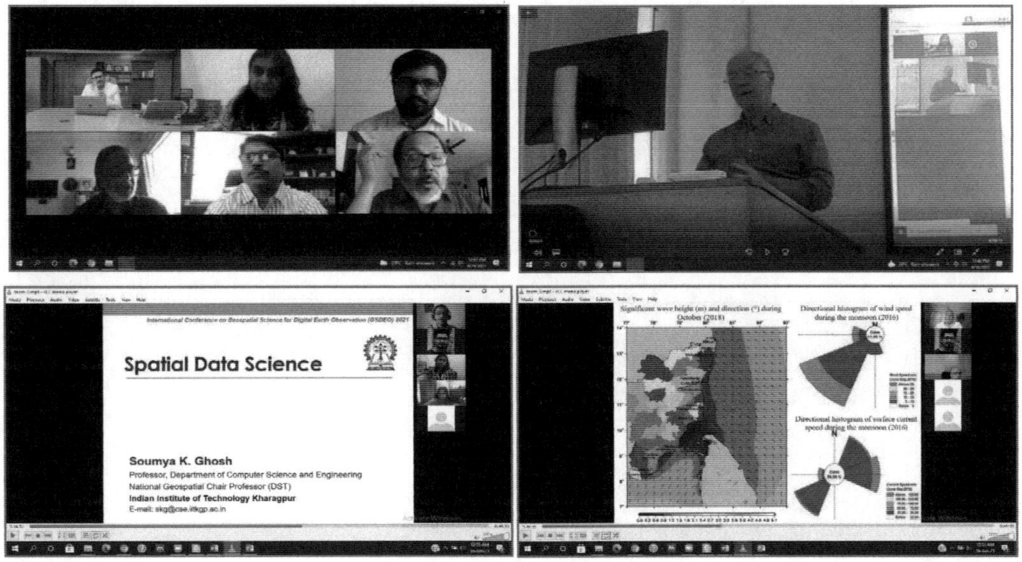

IOP Conf. Series: Earth and Environmental Science 1164 (2023) 011002 doi:10.1088/1755-1315/1164/1/011002

# Peer Review Declaration Form Guidance

On the following page you will find the declaration form.

- Please answer each question.
- You should submit the form along with the rest of your submission files.
- The deadline is the submission date written in your publishing agreement.

**All conference organisers/editors are required to declare details about their peer review.**

**We will publish the information you provide as part of your proceedings.**

IOP Conf. Series: Earth and Environmental Science 1164 (2023) 011002 doi:10.1088/1755-1315/1164/1/011002

# Peer review declaration

All papers published in this volume of IOP Conference Series: Earth and Environmental Science have been peer reviewed through processes administered by the Editors. Reviews were conducted by expert referees to the professional and scientific standards expected of a proceedings journal published by IOP Publishing.

- **Type of peer review: Single-blind / Double-blind / Triple-blind / Open / Other (please describe)**
  Single-blind: authors' identities are known to the reviewers, but the reviewers' identities are hidden from the authors.

- **Conference submission management system:** Full paper submission was fully managed by the editorial board of the Conference. The steps of the editorial process are as follows:

  1. The abstracts were submitted by the participants through the online system which was available on the conference web page. https://science.adamasuniversity.ac.in/gsdeo2021/.
  2. The submitted abstracts were then reviewed by the editorial team of the conference.
  3. The authors of the selected abstracts were asked to modify and resubmit their abstracts following the reviewer's comments.
  4. Updated abstracts were received from the authors.
  5. The authors of the selected abstracts presented their research at the conference.
  6. After the conference, the notification of full paper submission was sent to the authors who attended and presented their research activities during the conference.
  7. The full papers were submitted by the authors through the conference email gsdeo2021@adamasuniversity.ac.in
  8. The submitted full papers were then collected by the editorial team for review regarding the quality, format, language, plagiarism etc.
  9. Based on the quality of the papers, the editorial board selected some papers for the review process.
  10. Each paper (selected by the editors in the previous step) was sent to two reviewers (1st and 2nd) based on the topic of the paper and the reviewer's competency.

IOP Conf. Series: Earth and Environmental Science 1164 (2023) 011002 doi:10.1088/1755-1315/1164/1/011002

11. After the first review process, a full paper acceptance email was sent to the authors of those submissions only, which passed the first review and were accepted by the reviewers and editors.

12. The reviewer's comments were sent to the authors and the authors were asked to modify and resubmit their papers following the reviewer's comments.

13. After receiving the modified papers, each paper was then assigned to a 3rd reviewer to check the modification done by the authors and for further evaluation.

14. The comments of the third reviewer were then sent to the authors and the authors were again asked to modify and resubmit their papers following the reviewer's comments.

15. After receiving the updated version of the papers, each paper was then assigned to one editor for final format checking.

16. After receiving the final formatted papers from the editors, the chief editor then again checked the formats of the papers and did necessary modifications to make the papers camera ready.

17. The camera-ready papers along with the supporting documents were then submitted to IOP for publication.

- **Number of submissions received: 44**

- **Number of submissions sent for review: 25**

- **Number of submissions accepted: 17**

- **Acceptance Rate (Number of Submissions Accepted / Number of Submissions Received X 100): 38.64%**

- **Average number of reviews per paper: 3**

- **Total number of reviewers involved: 9**

- **Any additional info on review process: NA**

- **Contact person for queries:**
Name : Dr. Tuhin Bhadra
Affiliation: Department of Geography, School of Basic and Applied Sciences, Adamas University, Kolkata
Email: tuhin.bhadra@adamasuniversity.ac.in, tuhinbhadra.au@gmail.com

# Assessing the Urban Environmental Quality: A Case Study of Kolkata Metropolitan Area, India

**D Paul[1], K Mukherjee[1*], J K Pandey[2], A Dutta Roy[3]**

[1]Department of Geography, School of Basic and Applied Science, Adamas University, Kolkata-700126, India
[2] School of Basic and Applied Science, Adamas University, Kolkata-700126, India
[3]Department of Geography, Bankura Zilla Saradamani Mahila Mahavidyapith, Bankura

*Corresponding Author Email: kasturi.mukherjee@adamasuniversity.ac.in

**Abstract.** Rapid unplanned urbanization is always associated with environmental degradation which plays a direct role in the quality of life. Poor environmental quality is linked to premature death, cancer, long-term damage to respiratory and cardiovascular systems, etc. Using a case study of the Kolkata Metropolitan Area (KMA), the Urban Environmental Quality (UEQ) index has been constructed using a geoinformatics platform in the dimensions of the physical environment and built environment encompassing urban greenery, $NO_2$, SO2, PM 2.5, PM 10, road induced pollution, land surface temperature, built-up area and noise followed by an exploration of its spatial pattern. KMA is the largest urban agglomeration in eastern India and the second-largest in India with a population of 13.2 million as per the 2001 census. Using the Urban Environmental Quality (UEQ) index, this study tried to find out the poor Environmental Quality area through multi-criteria analysis in KMA. Based on the relative importance of controlling the outdoor environment, criteria weights for nine parameters have been derived through the analytical hierarchy process (AHP). The UEQ index was prepared by combining parameters rating and corresponding weight. Polycentric growth of urban centers leads to appropriate spatial matches to low environmental quality. Although the UEQ in the peripheries is being affected by suburbanization in a few instances, the majority of the areas have better environmental quality.

**Keywords:** Urban Environmental Quality Index, Air Pollution, Urbanization, Land Surface Temperature, Built Environment, Noise Pollution, Urban Greenery

## 1. Introduction

Predominantly the expression "environmental quality" refers to an assemblage of factors, including noise [1, 2], access to open space [3, 4], the aesthetic impression of buildings [5, 6], and the possible influence which these factors may have on physical and psychological health [7]. The quality of life, the number of years spent in good health, and health inequities are all significantly influenced by the environment. Early death, a cancerous growth, and long-term impairment of the cardiovascular and pulmonary systems are associated with poor environmental quality [8].

Since the urban environment is getting worse across India, the sustainability of cities is seriously threatened [9–12]. Regardless of the environmental carrying capacity, urban expansion in terms of people and economy will negatively impact the environment's key functions [13]. As a result, the urban poor is constantly experiencing a decline in the quality of the environment, human health, and quality of life (QOL) [14]. With a focus on Gorakhpur, researchers [9] examined the seasonal levels of benzene,

IOP Conf. Series: Earth and Environmental Science 1164 (2023) 012001 doi:10.1088/1755-1315/1164/1/012001

toluene, and xylene in the Indian Terai. The fluctuation of metals/metalloids and polycyclic aromatic hydrocarbons (PAH) in the air of Delhi, one of the most polluted cities in India and the globe, has been demonstrated by authors [12]. Vehicle emissions and biogenic emissions are the sources of PAHs. The research indicated that it had a cancerous effect on people's health. Ahmedabad, located in the western region of India, experienced a similar destiny, as evidenced by the works of [15], where secondary organic carbon (SOC) concentrations were examined. Black carbon dynamics and trajectories in the peninsular cities of Anantpur and Andhra Pradesh were investigated by researchers [16]. Aside from air pollution, India's metropolitan areas also have significant noise pollution [11], soil pollution [17], and water contamination [18–20]. Thus, assessing environmental quality becomes the most crucial component of effective urban environment planning and management. It can be viewed as an abstract and complex condition that continuously changes across the urban environment [21]. There is currently no standardized method for determining environmental quality, although numerous research have pushed for an integrated strategy that includes both objective and subjective measurements [22].

Kolkata is the 55th most polluted city in the world and the 7th largest city in India, according to the 2011 Census [23]. Over the past 20 years, thousands of research projects on Kolkata's environmental quality have been conducted in an effort to comprehend the spatiotemporal dynamics of the quality of air [11,24-26]. As can be observed in the works of [27-29], studies on water pollution in Kolkata have largely focused on the EKW (East Kolkata wetlands), which is a Ramsar site, the Hooghly River, and a few landfill sites. Utilizing multidimensional data sources, extensive works of literature on Kolkata's urban sprawl and accompanying heat island influence have been uncovered [30–32]. However, there hasn't been a study that combines all the various environmental factors to address the issue of environmental deterioration holistically so that timely mitigation measures can be taken.

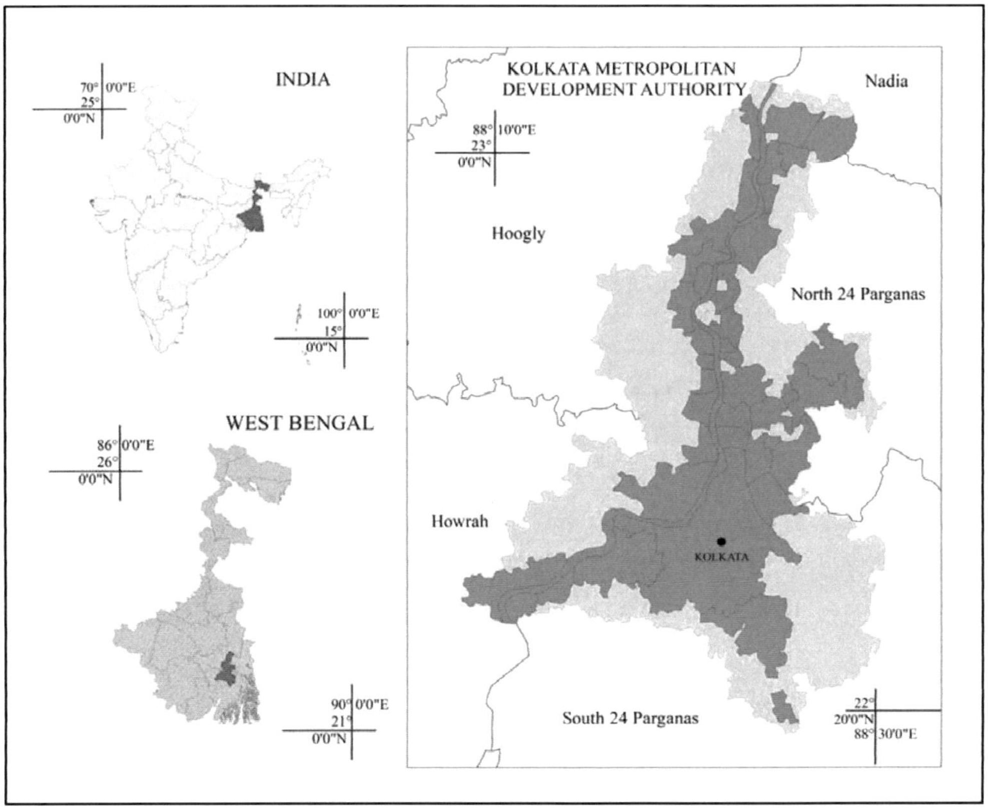


**Figure 1.** Location of the study area on a map

## 2. Study area

With remarkably high population growth from 2001 (14.77 million) to 2021 (20 million) and 7950 people per square kilometer, the KMA has been recognized as the third-biggest megacity in India and the biggest urban agglomeration of West Bengal (figure 1). The Kolkata Metropolitan Development Authority is responsible for overseeing the area (KMDA). KMA covers 1851.40 km$^2$ and is located between latitudes 22°0′19″ and 23°0′01″ north and latitudes 88°0′04″ and 88°0′33″ east. The region includes 39 municipalities, 1 cantonment, 4 municipal corporations, and portions of 24 panchayat samite. Previous studies [33, 34] have demonstrated that Kolkata has been undergoing significant unpremeditated urban growth as a result of the built-up conversion of vegetative cover, and open wetland, which has led to pollution, traffic congestion, poverty, and related socio-economic issues.

## 3. Data

The entire Landsat archive is now available free of cost for scientific research, which is a treasure of information used to calculate the LST, NDVI, and NDBI [31,35]. The specifics of the Landsat images used in this investigation are shown in Table 1. All of the available summer and winter scenes of 2018 were collected and the averages of LST, NDVI, and NDBI were estimated.

**Table 1**. Description of satellite datasets used in the study

| Data | Spatial Resolution | Spectral Resolution | Radiometric Resolution | Period | Purpose |
|---|---|---|---|---|---|
| Landsat 8 OLI &TIRS | OLI Band 1- 7: 30 m<br>OLI panchromatic band: 15 m<br>TIRS Band 10 and 11: 100 m | 11 | 8 | Summer and winter 2018 | LST, NDVI, NDBI |

Data on outdoor air quality characteristics have been initially gathered from the WBPCB's (West Bengal Pollution Control Board) website to generate the UEQ index for the Kolkata Metropolitan Area. For the 34 sites in the KMA, the WBPCB primarily monitors the three key air quality parameters $NO_2$, $SO_2$, and PM10. PM 2.5 is occasionally also monitored, although not for all the stations. For a small number of locations, noise data were available. The stations for 2018 for which the data were gathered are listed in table 2. Data were accessible at roughly two- to three-day intervals. For further usage, these data were collated, and annual averages were calculated.

**Table 2**. Location of WBPCB monitoring sites and data availability

| Sl No | Monitoring Station | Latitude | Longitude | Data available for 2018 |
|---|---|---|---|---|
| 1. | Uluberia | 22.478803 | 88.092591 | $NO_2$, $SO_2$, PM10 |
| 2. | Ultadanga | 22.591334 | 88.393243 | |
| 3. | Tribeni | 22.984414 | 88.386432 | |
| 4. | Topsia | 22.542893 | 88.398358 | |
| 5. | Tollygunge | 22.517052 | 88.345838 | |
| 6. | Shyambazar | 22.601898 | 88.373744 | $NO_2$, $SO_2$, PM10, PM2.5 |
| 7. | Sankrial | 22.560488 | 88.226898 | |
| 8. | Saltlake | 22.583478 | 88.418718 | $NO_2$, $SO_2$, PM10 |
| 9. | Rishra | 22.72119 | 88.354404 | |
| 10. | Rajarhat | 22.619392 | 88.464474 | |
| 11. | Picnic garden | 22.524544 | 88.382394 | |
| 12. | Paribesh bhawan | 22.56193 | 88.408315 | |
| 13. | Moulali | 22.561392 | 88.367121 | |
| 14. | Mominpur | 22.527127 | 88.32596 | |

| | | | | |
|---|---|---|---|---|
| 15. | Mintopark | 22.541189 | 88.354835 | $NO_2$, $SO_2$, PM10, PM2.5 |
| 16. | Khardah | 22.72399 | 88.373565 | $NO_2$, $SO_2$, PM10 |
| 17. | Kalyani | 22.972339 | 88.45877 | |
| 18. | Hyderoad | 22.533876 | 88.308148 | |
| 19. | HMC | 22.582244 | 88.33703 | |
| 20. | Ghusuri | 22.610926 | 88.354062 | $NO_2$, $SO_2$, PM10, PM2.5 |
| 21. | Gariahut | 22.516249 | 88.366304 | $NO_2$, $SO_2$, PM10 |
| 22. | Dunlop bridge | 22.654072 | 88.377148 | |
| 23. | Dumdum | 22.622828 | 88.414265 | |
| 24. | Dhulagarh | 22.573805 | 88.175891 | |
| 25. | Dankuni | 22.680887 | 88.297044 | |
| 26. | Chinsurah | 22.88809 | 88.39938 | |
| 27. | Beleghata | 22.56585 | 88.37725 | |
| 28. | Baruipur | 22.362809 | 88.43077 | |
| 29. | Behala Chowrasta | 22.487082 | 88.313237 | $NO_2$, $SO_2$, PM10, PM2.5 |
| 30. | Bator | 22.575403 | 88.301843 | $NO_2$, $SO_2$, PM10 |
| 31. | Barrackpore | 22.760488 | 88.37032 | |
| 32. | Barasat | 22.723139 | 88.478495 | |
| 33. | Baishnabghata | 22.466192 | 88.392815 | |
| 34. | Badhaghat | 22.601934 | 88.342139 | |
| 35. | Golpark | 22.51714 | 88.36518 | Noise |
| 36. | Newmarket | 22.56008 | 88.35247 | |
| 37. | Patuli | 22.47149 | 88.38443 | |
| 38. | SSKM | 22.53967 | 88.34218 | |
| 39. | WBPCB | 22.56197 | 88.409 | |
| 40. | Birati | 22.6636 | 88.42728 | |
| 41. | Rgkar | 22.60409 | 88.37811 | |
| 42. | Tollygunge | 22.49864 | 88.34539 | |
| 43. | Bag Bazar | 22.60261 | 88.3659 | |
| 44. | Taratala | 22.51687 | 88.29909 | |

## 4. Methodology

*4.1. Estimation of LST*

For estimating LST from Landsat 7, the single-channel algorithm has been used. This process includes the following 4 stages [36]:

Step 1: Conversion of DN to sensor spectral radiance;
Step 2: Conversion of spectral radiance to surface-escaping radiance;
Step 3: Conversion of radiance to ambient radiant temperature; and
Step 4: Conversion of ambient radiant temperature to surface temperature.

*4.2. Estimation of Normalized Difference Vegetation Index (NDVI)*

For estimating urban greenness NDVI is a popular solution to employ. Green leaves strongly reflect the Near-Infrared (NIR) band (wavelength range of 700-1100 nm) radiation. The high reflectance of the NIR band as compared to the red band from healthy vegetation [37] is a crucial property that is used in remote sensing for analyzing vegetation cover by calculating NDVI. The NDVI value varies from -1 to 1. The positive values denote areas covered by vegetation and negative values denote surfaces without vegetation like waterbodies, bare ground, etc. The NDVI value of water is often less than 0, bare soil ranges between 0 and 0.1. High values, such as 0.5 and above, often correspond to dense vegetation, such as rainforests, whereas moderate values, around 0.2 and 0.3, typically correspond to shrubs and grasslands [38]. In this study, NDVI is calculated using equation (1) [39] for estimating the average NDVI of 2018.

$$\text{NDVI} = (\text{NIR} - \text{Red}) / (\text{NIR} + \text{Red}) \tag{1}$$

### *4.3. Estimation of Normalized Difference Built-up Index (NDBI)*

NDBI has been used to classify the built-up areas. The value ranges from -1 to 1. A higher NDBI score denotes better imperviousness, less green cover, and lower moisture content since NDBI characterizes built-up intensity. Built-up lands exhibit more reflectance in the MIR spectral region than in the NIR spectral region when compared to other land surfaces. Equation (2) is used to estimate the annual NDBI for 2018.

$$\text{NDBI} = (\text{SWIR-NIR}) / (\text{SWIR+NIR}) \tag{2}$$

### *4.4. $NO_2$, $SO_2$, $PM_{2.5}$, $PM_{10}$, and Noise*

To best explain the geographical variability of the air quality parameters, like $NO_2$, $SO_2$, $PM_{2.5}$, $PM_{10}$, and noise data were interpolated on the GIS platform using IDW (Inverse Distance Weighted) analysis. Researchers [40] assume that wind direction and speed have insignificant influence on it. A point-based interpolation technique is IDW. The formula stated in equation (3) is used to calculate the pollutant concentration value at point i ($N_0$).

$$N_0 = \frac{\sum_{i=1}^{n} N_i \, P_i}{\sum_{i=1}^{n} N_i} \tag{3}$$

Where $N_i$ is the value of the parameter at location i, $P_i$ is the weight of the pollutant value at the position i, and N is the total number of measurement points.

Pi weights are estimated using Equation (4), which demonstrates that pi weights depend on the distance between the point of reference and the interpolation point. This establishes that closer points have a greater impact than farther ones.

$$P_i = \frac{1}{d_i^k} \tag{4}$$

In the following equation (5), k is the power of the distance, and di is the horizontal distance within the interpolation point at ($x_0$, $y_0$) and the points of reference at ($x_i$,$y_i$).

$$d_i = \sqrt{((x_i - x_i)^2)} + (y_i - y_0)^2 \tag{5}$$

### *4.5. Experts' scoring for weights*

One of the most essential requisites of a multi-criteria analysis is the weighting method. In multi-criteria decision-making, mathematical procedures have to be followed in order to solve problems by either addressing or combining various techniques [41]. These techniques include clustering of weights that could be estimated through different methods. Several methods currently exist for the measurement of weights. Saaty, put forth the Analytic Hierarchy Process (AHP) in 1980 [42] to aid in multi-criteria decision-making. The AHP is a semi-quantitative method that evaluates the relative importance of many interconnected criteria arranged in a hierarchical order in order to analyze complex problems. The approach establishes a hierarchical structure using the problem. Experts evaluate the relative relevance of decision criteria using pairwise comparisons allocated from a predetermined scale of comparative significance based on this framework. The scale uses integers from 1 to 9. Calculating the matrix's eigenvector can be used to estimate the weightage of all criteria at the same level [42]. To

IOP Conf. Series: Earth and Environmental Science 1164 (2023) 012001 doi:10.1088/1755-1315/1164/1/012001

decide the relative weight of all criteria, the criteria are assessed for each level. The subsequent criterion level can be derived by considering the bottom-level criteria and multiplying by each weight. The approach is recurrent up til the top-level overall decision target is achieved [43–48]. The AHP makes the unrealistic assumption that all criteria—both on the outer and inner dimensions—are independent of one another [43]. However, the weight-based index is frequently calculated in numerous research using the Analytical Hierarchy Process (AHP). Five experts on urban-environment geography contributed their opinions to this study, and the experts' opinions were used to create a pairwise comparison table. After each iteration of a comparison matrix, experts are permitted to modify the score, considering the advice of earlier studies [42,49], till the consistency ratio (CR) has reached a value lower than 10%. Each expert's comparison matrix is averaged into a pairwise matrix (table 3), and the findings are afterward normalized to establish the weight of each criterion. Then, the normalized matrix's corresponding row's average was used to calculate each criterion's weight.

**Table 3.** Pairwise comparison weight matrix for AHP

| | NO2 | SO2 | PM 2.5 | PM 10 | NOISE | LST | NDVI | NDBI |
|---|---|---|---|---|---|---|---|---|
| NO2 | 1 | 1 | 0.2 | 0.5 | 2 | 0.2 | 5 | 5 |
| SO2 | 1 | 1 | 1 | 1 | 5 | 1 | 9 | 5 |
| PM 2.5 | 5 | 1 | 1 | 3 | 6 | 2 | 9 | 7 |
| PM 10 | 2 | 1 | 0.33 | 1 | 5 | 1 | 8 | 6 |
| NOISE | 0.5 | 0.2 | 0.17 | 0.2 | 1 | 0.2 | 5 | 5 |
| LST | 5 | 1 | 0.5 | 1 | 5 | 1 | 7 | 4 |
| NDVI | 0.2 | 0.11 | 0.11 | 0.12 | 0.2 | 0.14 | 1 | 0.5 |
| NDBI | 0.2 | 0.2 | 0.14 | 0.17 | 0.2 | 0.25 | 2 | 1 |

**CR = 6.7%**

*4.6. Consistency check*

Equation (6) was used to assess the consistency of the criteria and gauge the resilience of the AHP technique:

$$CR= CI/RI \tag{6}$$

Where RI is the random index given [42] for a certain number of criteria assessed for a study, CR is the consistency ratio, CI is the consistency index. As a result, this study's RI value, which was calculated by comparing five criteria, is 1.12. CI is estimated using equation (7), where max is the highest eigenvalue of the comparison matrix and n is the total number of criteria.

$$CI= (\lambda max-n)/(n-1) \tag{7}$$

Saaty (1990) provides precise mathematical formulas for the consistency check and max for a deeper understanding. The current investigation discovered that $n = 5$, $max = 5.06$, $CI = 0.014$, and $CR = 0.01$.

The acceptance threshold limit of CR value is less than 0.1, otherwise, re-evaluation of criteria weight is necessary, defined in the AHP theory. The CR value, which is 0.01, supports the accuracy of the weights assigned to the various criteria in the pairwise matrix. Table 4 displays the facility criterion weights obtained by AHP for computing the Urban Environmental Quality (UEQ) index.

**Table 4.** Normalized weights of the criteria, and their average weights

| Sl. no. | Parameters | Priority | Rank |
|---|---|---|---|
| 1 | $NO_2$ | 9.00% | 5 |
| 2 | $SO_2$ | 17.10% | 3 |
| 3 | PM 2.5 | 28.50% | 1 |
| 4 | PM 10 | 16.10% | 4 |
| 5 | NOISE | 5.60% | 6 |
| 6 | LST | 18.90% | 2 |
| 7 | NDVI | 1.80% | 8 |
| 8 | NDBI | 2.80% | 7 |

### *4.7. Parameter categorization for rating*

To increase rationale, the value range of all criteria was categorized. The parameters have been clustered by using the natural break grading method. Each of the eight parameters was classified into five categories. Each parameter's pixels in each group were given a value ranging from 2 (for maximum impact) to 10 (for minimum impact) based on how important each parameter is in determining environmental quality (table 5). Pixels with a high LST value, for example, were assigned a value of 2, and pixels with a high NDVI value were assigned a value of 10.

**Table 5.** Criteria classes and weights used in the UEQ Index

| Parameters | Class | Rating | Weight |
|---|---|---|---|
| $NO_2$ | 55.40 - 92.46 | 2 | 0.09 |
| | 47.94 - 55.40 | 4 | |
| | 43.19 - 47.94 | 6 | |
| | 38.52 - 43.19 | 8 | |
| | 31.46 - 38.52 | 10 | |
| $SO_2$ | 48.53 - 90.05 | 2 | 0.17 |
| | 27.73 - 48.53 | 4 | |
| | 15.81 - 27.73 | 6 | |
| | 8.41 - 15.81 | 8 | |
| | 3.61 - 8.41 | 10 | |
| PM 2.5 | 74.29 - 138.18 | 2 | 0.285 |
| | 38.21 - 74.29 | 4 | |
| | 17.87 - 38.21 | 6 | |
| | 6.02 - 17.87 | 8 | |
| | 0.00 - 6.02 | 10 | |
| PM 10 | 155.80 - 209.40 | 2 | 0.161 |
| | 138.78 - 155.80 | 4 | |
| | 124.86 - 138.78 | 6 | |
| | 109.63 - 124.86 | 8 | |
| | 83.69 - 109.63 | 10 | |
| NOISE | 65.98 - 70.33 | 2 | 0.056 |
| | 64.45 - 65.98 | 4 | |
| | 63.58 - 64.45 | 6 | |
| | 62.35 - 63.58 | 8 | |
| | 60.12 - 62.35 | 10 | |

| | | | |
|---|---|---|---|
| LST | 22.67 - 28.33 | 2 | 0.189 |
| | 21.37 - 22.67 | 4 | |
| | 20.37 - 21.37 | 6 | |
| | 19.32 - 20.37 | 8 | |
| | 15.56 - 19.32 | 10 | |
| NDVI | -0.18 - (-0.01) | 2 | 0.018 |
| | -0.01 - 0.08 | 4 | |
| | 0.08 - 0.14 | 6 | |
| | 0.14 - 0.20 | 8 | |
| | 0.20 - 0.40 | 10 | |
| NDBI | -0.01 - 0.45 | 2 | 0.028 |
| | -0.05 - (-0.01) | 4 | |
| | -0.10 - (-0.05) | 6 | |
| | -0.14 - (-0.10) | 8 | |
| | -0.34 -(-0.14) | 10 | |

*4.8. Calculating Urban Environmental Quality (UEQ) Index*
The following equation (8) has been used to construct the Urban Environmental Quality (UEQ) index after determining the weights and ratings of each parameter.

$$UEQ = \sum_{i=1}^{8}\left(\sum_{j=1}^{5} r_{ij}\right) \times w_i \quad (8)$$

Where $r_{ij}$ is the rating of the i^th^ parameter for the j^th^ category, $w_i$ is the weight of the i^th^ parameter attained through AHP, and n is the number of criteria (parameters). Finally, the UEQ index layer was classified into five classes to generate different environmental quality zones.

## 5. Result and Discussion

### *5.1. Spatial variability of parameters*

5.1.1. *LST.* Urban climatology research is increasingly concerned with the size and distribution of Urban Heat Island (UHI) effects, which are most frequently quantified by Land Surface Temperature (LST). The rainfall and wind characteristics of urban areas, apart from the abnormality in temperature, can be changed by the heat waves generated under the UHI effect [50]. High temperature coupled with air pollution can potentially influence the health of urban residents like respiratory problems, discomfort, heat strokes, cramps, exhaustion, and sometimes death [51,52]. By visualizing the spatial variation (figure 2) it can be understood that the entire area around KMC is characterized by high LST. The temperatures have ranged in these areas between 25.00 to 20.33 ºC. The constant urban development and high building density, lack of vegetation, and water bodies are enhancing the LST of this region, leading to the formation of UHI. Temperatures are lowest in the East Kolkata wetlands and the river Hugli. The thermal contrast between the KMC surroundings and suburban is extremely prominent on the maps (figure 2).

5.1.2. *NDVI.* In this study area, the overall NDVI value is low varying from -0.18 to 0.40 (figure 2) which signifies the presence of sparsely vegetative areas and vegetation-free bare surfaces. Low values of NDVI also correspond to the built-up in the center of KMC and the turbid waters of River Hugli.

*5.1.3. NDBI.* The average NDBI value ranges between -0.18 and 0.4 (figure 2). The areas with high NDBI values between 0.1 and 0.3 show the built-up area, whereas NDBI > 0.25 shows the barren lands

[53]. The areas along river Hugli where most of the dense built-up, industrial centers are located, have NDBI values between 0.1 and 0.3, characterizing high-intensity built-up areas.

5.1.4. *$NO_2$, $SO_2$, PM 2.5, and PM 10.* The air quality standard for $NO_2$ and $SO_2$ is 80μg/m$^3$ decided by the CPCB of India. Except for Bandaghat and HMC, all monitoring stations confirmed $NO_2$ levels within the acceptable range of 30-60μg/m$^3$, which is within the Ambient Air Quality standard prescribed by CPCB (figure 2). In the majority of KMA stations, the SO2 concentration levels are within acceptable limits. Except for Behala Chowrasta and Bator, all other monitoring stations have recorded $SO_2$ concentrations within 40μg/m$^3$ (figure 2). The study area's $PM_{10}$ concentration is a major point of concern. The permitted levels of 60 g/m$^3$ for all stations are far exceeded. For $PM_{2.5}$, the maximum permissible limit is also 60μg/m$^3$, which has been exceeded in a few stations. Except for $SO_2$, all the other air pollutants considered for this study have surpassed the WHO permissible limits-NO $_2$-being 10 μg/m$^3$, $SO_2$ - 40 μg/m$^3$, PM10 - 15 μg/m$^3$, PM 2.5 - 5 μg/m$^3$ (figure 2).

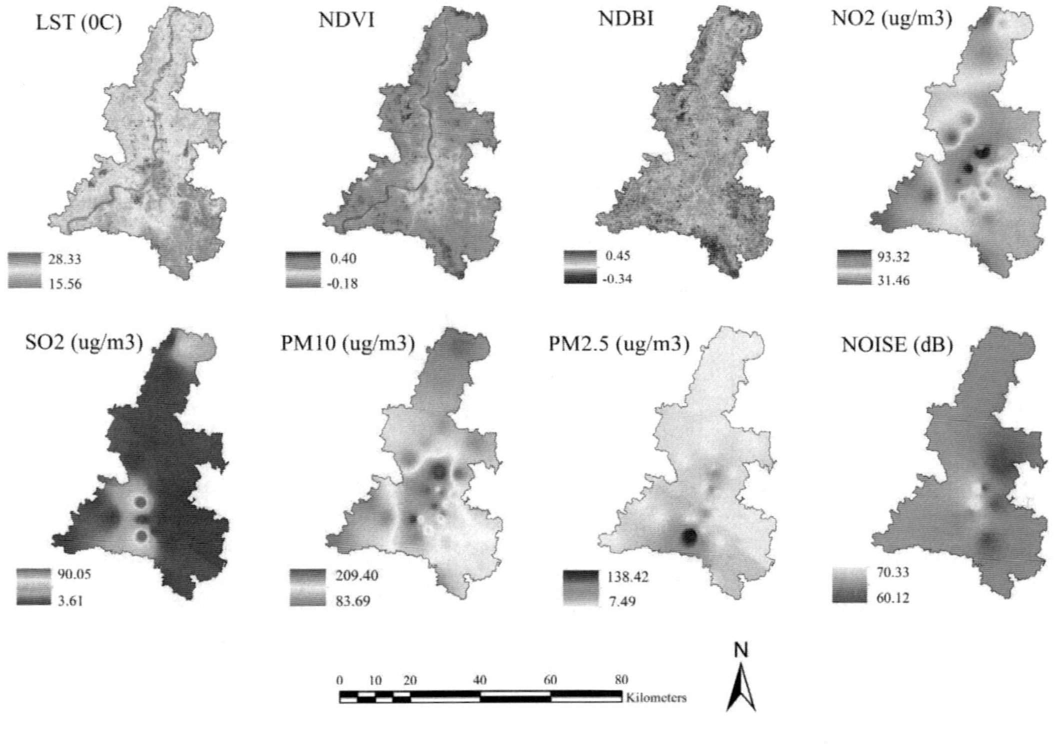


**Figure 2.** Spatial variation of LST, NDVI, NDBI, $NO_2$, $SO_2$, PM10, PM2.5, and Noise in KMA

*5.1.5. Noise.* Guidelines for the regulation and management of noise in India were defined by the Noise Pollution (Regulation and Control) Rules, 2000. The industrial area's noise restriction has been set at 75 dB during the day and 70 dB at night. In contrast, the commercial area's noise limit is 65 dB and 55 dB, and the residential area's noise limit is 55 and 45 dB, respectively, during the day and night. It should be quieter than 50 dB during the day and less than 40 dB at night in the silent zone (100 meters near hospitals and schools). In KMA the annual average of day and night time noise data revealed that the range is between 60-70 dB (figure 2) which is high for the residential and silent zones. The New Market and Bag Bazar areas experience maximum noise pollution.

5.1.6. *Urban Environmental Quality (UEQ).* Table 5 presents the weights of the criteria for MCDA that were generated using AHP. The UEQ index was developed by combining eight characteristics (figure

2) and their related weights. There are five categories for the UEQ index (figure 3). It is satisfactory to note that the very poor UEQ zone only consists of 6.44% of the area, followed by a poor UEQ index zone which covers 11.16% of the area. The majority of the KMA is covered by a good UEQ zone (41.32%). The Vey good and moderate UEQ zones cover 19.71% and 21.36% of the area.

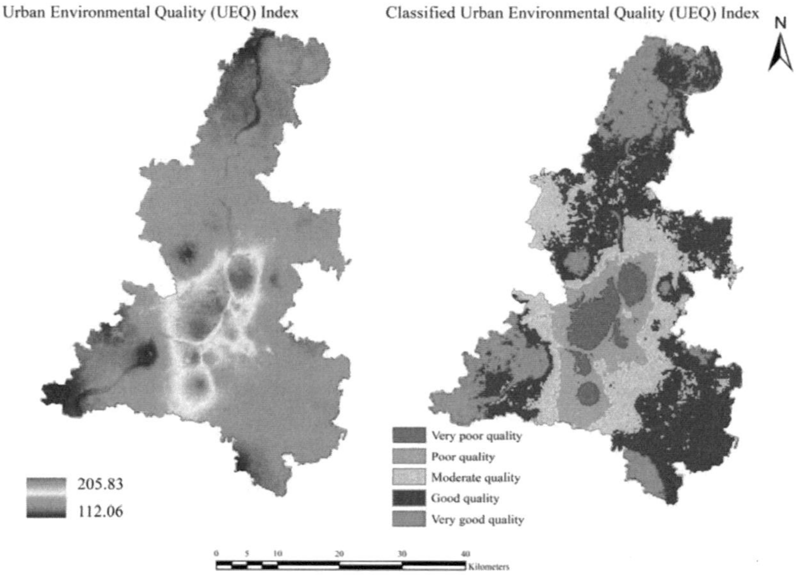


**Figure 3.** Spatial variation of urban environmental quality

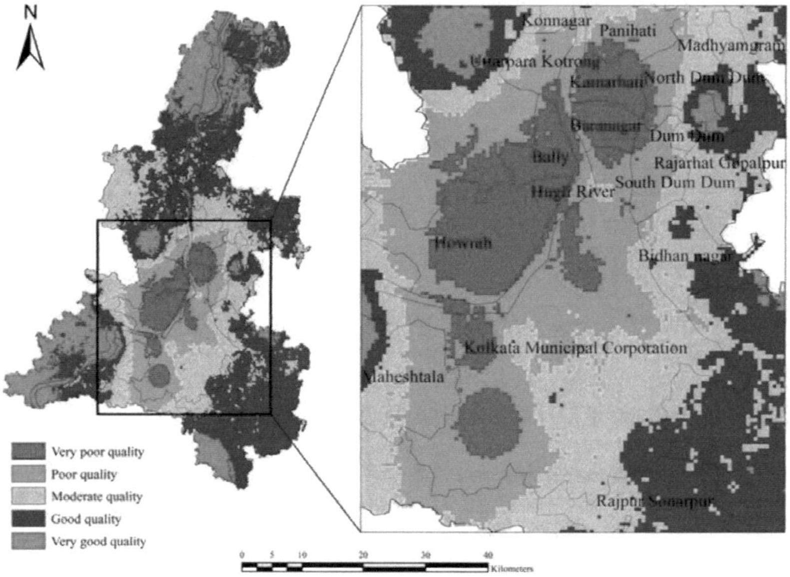


**Figure 4.** Cities with very poor UEQ index value

Four municipalities viz. Kamarhati, Baranagar, Bally, and Howrah are almost fully covered by the very poor UEQ zone whereas KMC is covered mostly by poor and moderate UEQ zones with a little share of the very poor zone. The surrounding municipalities like North and South Dumdum, Panihati, and Uttarpara, are partially covered by the poor UEQ zones highlighted the figure 4.

## 6. Conclusion

Most of the works of literature have taken into account individual environmental parameters to assess their concentrations, distributions, and densities spatially in urban centers. Rarely any literature has considered delving into the overall holistic urban environmental quality at a micro-level having an amalgamation of all the environmental variables such as air, water quality, noise pollution, building density, and LST. The combined impact of all these parameters highlights the areas which require immediate and maximum interventions toward mitigating the degradation. Remote sensing has played an indispensable role in intricately capturing spatial variations. The AHP technique has helped in identifying and rather ranking the environmental parameters in an order to its influence on the quality of the environment and the UEQI has helped in the assessment of the regions which sought policy planning. The heartland of KMA which incorporates mainly Kamarhati, Baranagar, Howrah, and Bally municipal corporations has proved to be the one with the worst environmental conditions. The PM2.5, PM10 concentrations, and LST have been majorly responsible for deteriorating the scenario. These areas have old colonial buildings and streets, within which new apartments and high rises have cropped up in a very unfriendly manner. Moreover, these regions have high traffic and road density adding up to overcrowded and populated streets. All these factors have aggravated the situation. Green buildings, rooftop greeneries, and proper traffic regulations could be the possible alternatives for mitigating the problem as demolishing buildings, widening roads, or introducing green patches, or water bodies where no open space is left has a meager chance of sustainable development.

## References

[1] Sun Y, Li Y, Yu T, Zhang X, Liu L and Zhang P 2021 Resource extraction, environmental pollution and economic development: Evidence from prefecture-level cities in China *Resources Policy* **74** 102330

[2] Chiarini B, D'Agostino A, Marzano E and Regoli A 2020 The perception of air pollution and noise in urban environments: A subjective indicator across European countries *Journal of Environmental Management* **263** 110272

[3] Rodiek J 2008 Protecting ecosystems and open spaces in urbanizing environments *Landscape and Urban Planning* **84** 3–6

[4] Huang J, Zhou C, Zhuo Y, Xu L and Jiang Y 2016 Outdoor thermal environments and activities in open space: an experiment study in humid subtropical climates *Build Environ* **103** 238–49

[5] Wang S, Chen B, Suo J and Zhao J R 2021 Impact of building morphology and outdoor environment on light and thermal environment in campus buildings in cold region during winter *Building and Environment* **204** 108074

[6] Jo H I and Jeon J Y 2022 Perception of urban soundscape and landscape using different visual environment reproduction methods in virtual reality *Applied Acoustics* **186** 108498

[7] Guan Y, Xiao Y, Rong B, Zhang N and Chu C 2021 Long-term health impacts attributable to PM2. 5 and ozone pollution in China's most polluted region during 2015–2020 *Journal of Cleaner Production* **321** 128970

[8] Wolf K, Hoffmann B, Andersen Z J, Atkinson R W, Bauwelinck M, Bellander T, Brandt J, Brunekreef B, Cesaroni G and Chen J 2021 Long-term exposure to low-level ambient air pollution and incidence of stroke and coronary heart disease: a pooled analysis of six European cohorts within the ELAPSE project *The Lancet Planetary Health* **5** e620–32

[9] Masih A, Lall A S, Taneja A and Singhvi R 2018 Exposure levels and health risk assessment of ambient BTX at urban and rural environments of a terai region of northern India *Environmental pollution* **242** 1678–83

[10] Guleria A, Gupta S K, Gupta I, Swami D and Shukla D P 2019 Understanding the spatial and temporal dependence of the migration of conservative contaminant plume in urban groundwater environment in Panchkula region, Haryana, India *Groundwater for Sustainable Development* **8** 93–103

[11] Das P, Talukdar S, Ziaul S, Das S and Pal S 2019 Noise mapping and assessing vulnerability in meso level urban environment of Eastern India *Sustainable Cities and Society* **46** 101416

[12] Hazarika N, Srivastava A and Das A 2017 Quantification of particle bound metallic load and PAHs in urban environment of Delhi, India: Source and toxicity assessment *Sustainable Cities and Society* **29** 58–67

[13] Gurram S, Stuart A L and Pinjari A R 2015 Impacts of travel activity and urbanicity on exposures to ambient oxides of nitrogen and on exposure disparities *Air Quality, Atmosphere & Health* **8** 97–114

[14] Leitmann J, Bartone C and Bernstein J 1992 Environmental management and urban development issues and options for Third World cities *Environment and Urbanization* **4** 131–40

[15] Rengarajan R, Sudheer A K and Sarin M M 2011 Aerosol acidity and secondary organic aerosol formation during wintertime over urban environment in western India *Atmospheric Environment* **45** 1940–5

[16] Hussain S N, Rao T C, Balakrishnaiah G, Gopal K R, Reddy K R O, Reddy N S K, Reddy T L, Kumari S P, Reddy P R and Reddy R R 2018 Investigation of black carbon aerosols and their characteristics over tropical urban and semi-arid rural environments in peninsular India *Journal of Atmospheric and Solar-Terrestrial Physics* **167** 48–57

[17] Sarkar A, Deb S, Ghosh S, Mandal S, Quazi S A, Kushwaha A, Hoque A and Choudhury A 2022 Impact of anthropogenic pollution on soil properties in and around a town in Eastern India *Geoderma Regional* **28** e00462

[18] Victoria N S, Kumari T S D and Lazarus B 2022 Assessment on impact of sewage in coastal pollution and distribution of fecal pathogenic bacteria with reference to antibiotic resistance in the coastal area of Cape Comorin, India *Mar Pollut Bull* **175** 113123

[19] Rahman K, Barua S and Imran H M 2021 Assessment of water quality and apportionment of pollution sources of an urban lake using multivariate statistical analysis *Cleaner Engineering and Technology* **5** 100309

[20] Aithani D, Jyethi D S, Siddiqui Z, Yadav A K and Khillare P S 2020 Source apportionment, pollution assessment, and ecological and human health risk assessment due to trace metals contaminated groundwater along urban river floodplain. *Groundwater for Sustainable Development* **11** 100445

[21] Nichol J E, Bilal M, Nazeer M and Wong M S 2021 Urban Pollution pp 243–58

[22] Van Kamp I, Leidelmeijer K, Marsman G and de Hollander A 2003 Urban environmental quality and human well-being *Landscape and Urban Planning* **65** 5–18

[23] WHO 2018 *Ambient (Outdoor) Air quality database (update 2018)* (Geneva, Switzerland)

[24] Karar K and Gupta A K 2007 Source apportionment of PM10 at residential and industrial sites of an urban region of Kolkata, India *Atmospheric Research* **84** 30–41

[25] Dutta C, Chatterjee A, Jana T K, Mukherjee A K and Sen S 2010 Contribution from the primary and secondary sources to the atmospheric formaldehyde in Kolkata, India *Science of the total environment* **408** 4744–8

[26] Rao K R M, Kant Y, Gahlaut N and Roy P S 2012 Assessment of quality of life in Uttarakhand, India using geospatial techniques *Geocarto International* **27** 315–28

[27] Dutta J, Zaman S, Thakur T K, Kaushik S, Mitra A, Singh P, Kumar R, Zuan A T K, Samdani M S and Alharbi S A 2022 Assessment of the bioaccumulation pattern of Pb, Cd, Cr and Hg in edible fishes of East kolkata Wetlands, India *Saudi Journal of Biological Sciences* **29** 758–66

[28] Maiti M, Chaudhuri S and Biswas A 2021 Activities of idol immersion leads to heavy metal contamination in river Hooghly in and around the city of Kolkata *Journal of the Indian Chemical Society* **98** 100223

[29] Biswas A K, Kumar S, Babu S S, Bhattacharyya J K and Chakrabarti T 2010 Studies on environmental quality in and around municipal solid waste dumpsite *Resources, Conservation and Recycling* **55** 129–34

[30] Bera B, Shit P K, Saha S and Bhattacharjee S 2021 Exploratory analysis of cooling effect of urban wetlands on Kolkata metropolitan city region, eastern India *Current Research in Environmental Sustainability* **3** 100066

[31] Halder B, Bandyopadhyay J and Banik P 2021 Monitoring the effect of urban development on urban heat island based on remote sensing and geo-spatial approach in Kolkata and adjacent areas, India *Sustainable Cities and Society* **74** 103186
[32] Mandal J, Patel P P and Samanta S 2022 Examining the expansion of Urban Heat Island effect in the Kolkata Metropolitan Area and its vicinity using multi-temporal MODIS satellite data *Advances in Space Research* **69** 1960–77
[33] Dey F 2021 Dynamics Qf Urban Growth And Environmental Challenges: A Case Qf Kolkata, India *Geography, Environment, Sustainability* **14** 24–31
[34] Dinda S, das Chatterjee N and Ghosh S 2021 An integrated simulation approach to the assessment of urban growth pattern and loss in urban green space in Kolkata, India: A GIS-based analysis *Ecological Indicators* **121** 107178
[35] Kamali Maskooni E, Hashemi H, Berndtsson R, Daneshkar Arasteh P and Kazemi M 2021 Impact of spatiotemporal land-use and land-cover changes on surface urban heat islands in a semiarid region using Landsat data *International Journal of Digital Earth* **14** 250–70
[36] Mukherjee K and Das P 2018 Modelling the Relationship between Urban Growth Modes and the Thermal Environment-A Case Study of the Barasat Municipality, West Bengal *Journal of Geography, Environment and Earth Science International* 1–19
[37] Gates D M 1980 *Biophysical ecology* (New York: Springer-Verlag)
[38] USGS 2018 Remote Sensing Phenology NDVI the Foundation *Available at: https://phenology.cr.usgs.gov/ndvi_foundation.php [Accessed 17 May 2020]*
[39] Cracknell A P 1997 *Advanced very high resolution radiometer AVHRR* (CRC Press)
[40] Zhou X, Strezov V, Jiang Y, Kan T and Evans T 2022 Temporal and spatial variations of air pollution across China from 2015 to 2018 *Journal of Environmental Sciences* **112** 161–9
[41] Dash P and Sar J 2020 Identification and validation of potential flood hazard area using GIS-based multi-criteria analysis and satellite data-derived water index *Journal of Flood Risk Management* **13**
[42] Satty T L 1980 *The analytic hierarchy process: Planning, priority setting, resource allocation* (New York: McGraw-Hill)
[43] Wang Y, Li Z, Tang Z and Zeng G 2011 A GIS-based spatial multi-criteria approach for flood risk assessment in the Dongting Lake Region, Hunan, Central China *Water resources management* **25** 3465–84
[44] Wang Y-M, Liu J and Elhag T M S 2008 An integrated AHP–DEA methodology for bridge risk assessment *Computers & Industrial Engineering* **54** 513–25
[45] Gervásio H and da Silva L S 2012 A probabilistic decision-making approach for the sustainable assessment of infrastructures *Expert Systems with Applications* **39** 7121–31
[46] Farkas A 2011 Multi-criteria comparison of bridge designs *Acta Polytechnica Hungarica* **8** 173–91
[47] Abu Dabous S and Alkass S 2008 Decision support method for multi-criteria selection of bridge rehabilitation strategy *Construction Management and Economics* **26** 883–93
[48] Huang Q, Ren Y and Lin Y-Z 2007 Application of uncertain type of AHP to condition assessment of cable-stayed bridges *Journal of Southeast University (English Edition)* **23** 599–603
[49] Papaioannou G, Vasiliades L and Loukas A 2015 Multi-criteria analysis framework for potential flood prone areas mapping *Water resources management* **29** 399–418
[50] Baik J-J, Kim Y-H and Chun H-Y 2001 Dry and Moist Convection Forced by an Urban Heat Island *Journal of Applied Meteorology* **40** 1462–75
[51] Devanathan P and Devanathan K 2011 Heat island effects *Green building with concrete: sustainable design and construction* 175–226
[52] US Environmental Protection Agency 2012 Heat Island Impacts *Available at: http://www.epa.gov/hiri/impacts/index.htm [Accessed 20 Oct. 2020]*
[53] Zha Y, Gao J and Ni S 2003 Use of normalized difference built-up index in automatically mapping urban areas from TM imagery *International Journal of Remote Sensing* **24** 583–94

# Assessing the Tidal Influences on the Coastal Erosion-Accretion Processes in the Indian Sundarban Delta using Hydrodynamic Models and Geoinformatics

**K Bhui[1] *, T Bhadra[2*], S Samanta[1], S Hazra[1]**

[1]School of Oceanographic Studies, Jadavpur University, Kolkata, India
[2]Department of Geography, Adamas University, Kolkata, India

*Corresponding Author Email:
iamkoushikbhui@gmail.com, tuhinbhadra.au@gmail.com

**Abstract.** The Sundarbans is currently a subject of serious concern due to the massive coastline erosion that is responsible for reshaping the geomorphological content of the islands. The coastal islands of the Sundarbans have become more susceptible to sea-level rise, cyclonic storms, and coastal flooding, which further accelerate the process of coastal erosion. Identification of erosion-prone regions is highly essential for integrated coastal zone management. Till date, no scientific study has been done to relate the coastal erosion accretion process with tidal velocity and current direction in the Indian Sundarbans. In this perspective, the present study identifies the erosion-accretion zones in the Indian Sundarbans using Landsat imageries of 2000 and 2020 using geoinformatics. It investigates the impact of tidal velocity and direction on the erosion accretion processes. MIKE 21 two-dimensional hydrodynamic model was simulated to predict tidal velocities and directions. The study finds that the sea-facing islands of the Indian Sundarbans experience the maximum amount of erosion. A significant shrinking of the land area takes place in the Bulchery, Bhangaduiani, and Dalhousie islands. The rates of erosion of these islands are 485.99 m/year, 487 m/year, and 480.65 m/year, respectively. Places with high velocity are found to be erosion-prone, while the rate of erosion significantly varied with the flow directions. Two islands, Bulchery and Dalhousie, on the south-eastern margin of the delta, suffered maximum erosion over the past decades due to the impact of high-velocity currents from the southeast and southwest, respectively. Alternatively, minor accretions were observed along with the sheltered island precincts where velocities were comparatively low. Such studies on the current drive erosion–accretion processes along the delta margin during sea-level rise appear to be of utmost importance for sustainable coastal protection and management.

**Key Words:** Coastal Accretion, Coastal Erosion, Coastal Accretion, Geoinformatics, Hydrodynamic Model, Sundarban, Tidal Current, Tidal Velocity.

## 1. Introduction

There are two regular phenomena: Erosion and accretion processes that make the fluvio-marine coastal environment a dynamic system [1, 2]. The erosion process is an important agent that erodes the coastal land and causes the coastline to retreat [3]. On the other hand, the process of land accretion happens due to the deposition of erosional material in another place [4]. It is imperative to estimate erosion and accretion patterns accurately with their geospatial extents for monitoring erosion-accretion and understanding the morphological processes [5].

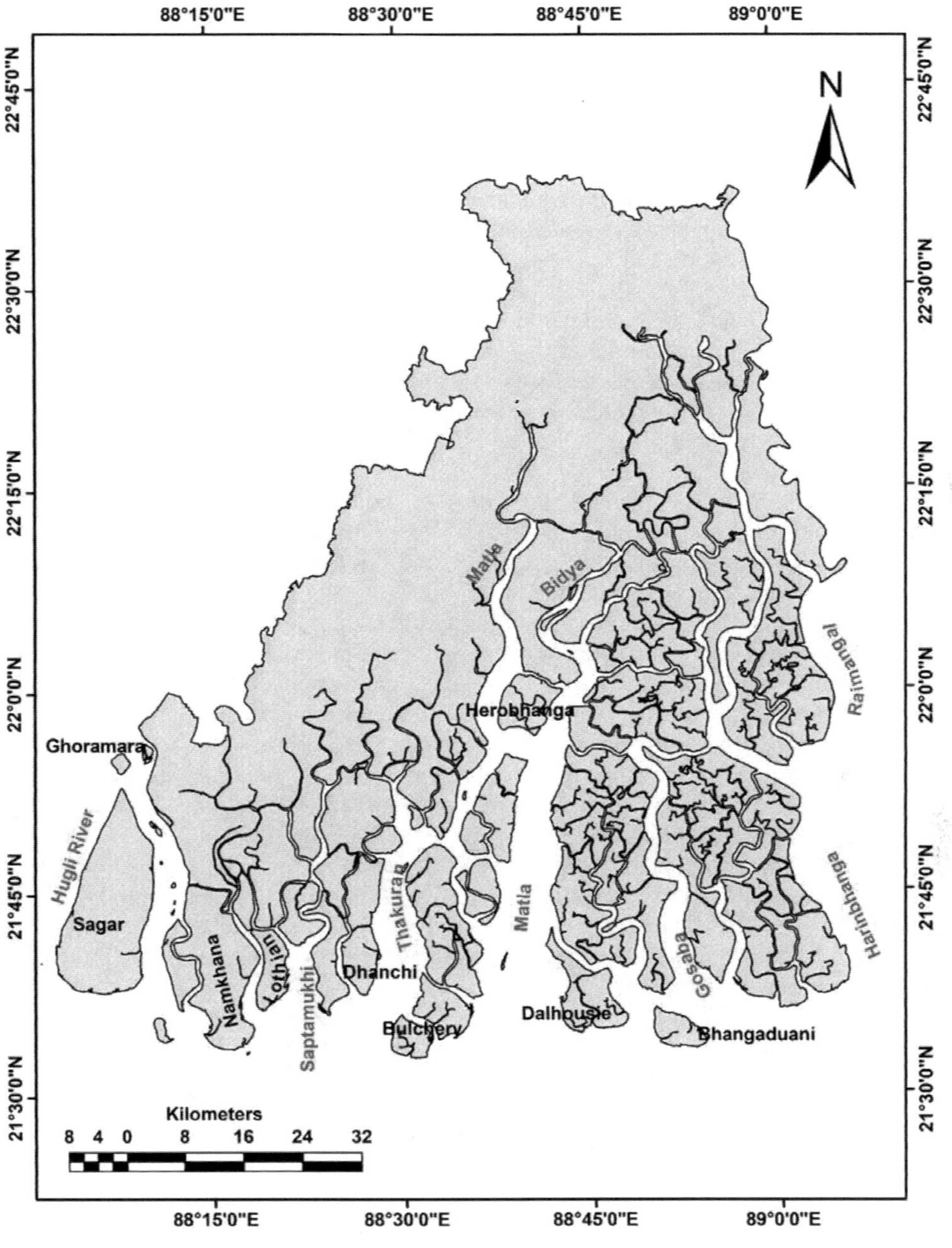


**Figure 1.** Islands and Coastlines of the Indian Sundarbans.

Sundarbans, the largest mangrove forest in the world, also a tide-dominated island, is currently under serious threat due to severe coastline erosion along with other coastal hazards like tropical cyclones, storm surges, coastal flooding, etc., which are constantly reshaping the islands of the region [6, 7]. The dynamic delta, Sundarbans, shown in figure 1, is characterized by numerous tidal channels

IOP Conf. Series: Earth and Environmental Science 1164 (2023) 012002 doi:10.1088/1755-1315/1164/1/012002

and creeks with rapid morphological changes that are significantly controlled by sediment supply and changes in sea level [8-10]. Similarly, wave energy, wave height, tidal range, and littoral drift are important physical factors that also significantly contribute to the alteration of the coastal landform [11]. Numerous studies [5, 10, 12-15] have been conducted globally for different coastal regions to identify erosion and accretion processes using different geoinformatics methods. Many researchers have tried to explore the geomorphological changes of the Indian Sundarbans. Using the tools of geoinformatics Kundu and Halder [16] attempted to track changes in the shorelines of the Sundarbans for the duration of 1975-2018. Hazra et al. [17] tried to understand the geomorphological changes in the Sundarbans caused by the sea level over 10-20 years. The erosion-accretion pattern and allied changes in the islands of the Indian Sundarbans were studied by many researchers [7, 18-20] using RS and GIS techniques. Most of the previous studies have focused on the identification of erosional and accretional zones using multi-temporal satellite images. No scientific study has been carried out till now to relate the coastal erosion and accretion processes with tidal velocities and current directions in the Indian Sundarbans. Therefore, the key objectives of the present study comprise identifying the erosion accretion zones over 20 years in the Indian Sundarbans delta and analyzing the effects of tidal velocities and current directions on the erosion and accretion processes. Such studies on the current drive erosion–accretion process along the delta margin during sea-level rise appears to be of utmost importance for sustainable coastal protection and management.

## 2. Data and Methodology

To estimate the morphological changes, RS and GIS are the most useful techniques in modern research [20, 21]. In the present study, Landsat 7 ETM+ (30m) for 2000 (March) and Landsat 8 OLI (30m) for 2020 (December) satellite images of the Indian Sundarbans were retrieved from www.earthexplorer.usgs.gov. Both images were acquired during the low tidal condition and considered for downloading to maintain the similarity of water level conditions and prepare an error-free output. The Landsat data were good to consider for the present study area because it has multi-spectral and multi-temporal capacities that can track the various morphological changes of the estuarine and coastal regions [20]. Erosion and accretion were estimated by using ArcGIS 10.3 software, manually digitizing the islands of the Indian Sundarbans. All southern islands and the Herobhanga island in the north of the region were considered to answer the question of the present research. Furthermore, the 'Erase' tool of ArcGIS software was applied to extract the erosional and accretional areas of the islands from the Landsat images of 2000 and 2020.

In order to evaluate the tidal influences on the erosion and accretion processes, speeds and directions of tidal current were predicted using the hydrodynamic (HD) module of MIKE 21, a two-dimensional (2D) software. To illustrate the changes in current flow and water level movement, the model uses the following equations (1), (2), and (3), derived based on the principles of the conservation of mass and momentum [22].

Equation of the continuity:

$$\frac{\partial h}{\partial t}+\frac{\partial h\bar{u}}{\partial x}+\frac{\partial h\bar{v}}{\partial y}=hS \tag{1}$$

Equation of the momentum in the direction of x:

$$\begin{aligned}\frac{\partial h\bar{u}}{\partial t}+\frac{\partial h\bar{u}^{2}}{\partial x}+\frac{\partial h\bar{v}\bar{u}}{\partial y}=f\bar{v}h-gh\frac{\partial \eta}{\partial x}-\frac{h}{\rho_0}\frac{\partial p_a}{\partial x}-\frac{gh^2}{2\rho_0}\frac{\partial \rho}{\partial x}+\frac{\tau_{sx}}{\rho_0}-\frac{\tau_{bx}}{\rho_0}\\-\frac{1}{\rho_0}\left(\frac{\partial s_{xx}}{\partial x}+\frac{\partial s_{xy}}{\partial y}\right)+\frac{\partial}{\partial x}\left(hT_{xx}\right)+\frac{\partial}{\partial y}\left(hT_{xy}\right)+hu_sS\end{aligned} \tag{2}$$

Equation of the momentum in the direction of y:

$$\frac{\partial h\bar{v}}{\partial t}+\frac{\partial h\bar{v}^{2}}{\partial y}+\frac{\partial h\overline{uv}}{\partial x}=-f\bar{u}h-gh\frac{\partial \eta}{\partial y}-\frac{h}{\rho_0}\frac{\partial p_a}{\partial y}-\frac{gh^2}{2\rho_0}\frac{\partial \rho}{\partial y}+\frac{\tau_{sy}}{\rho_0}-\frac{\tau_{by}}{\rho_0}$$
$$-\frac{1}{\rho_0}\left(\frac{\partial s_{yx}}{\partial x}+\frac{\partial s_{yy}}{\partial y}\right)+\frac{\partial}{\partial x}\left(hT_{xy}\right)+\frac{\partial}{\partial y}\left(hT_{yy}\right)+hv_sS \quad (3)$$

Where,

| | | | |
|---|---|---|---|
| $t$ | Time | $p_a$ | Pressure of atmosphere |
| $x$ and $y$ | Cartesian coordinates | $\rho$ | Water density |
| $h=\eta+d$ | Total depth, ($\eta$ stands for the elevation of the water surface and $d$ represents the depth of still water) | $\rho_0$ | The ratio of the water density and air density |
| $\bar{u}$ and $\bar{v}$ | Elements of the velocity in the direction of the cartesian coordinates | ($\tau_{sx}$, $\tau_{sy}$) and ($\tau_{bx}$, $\tau_{by}$) | Elements of the Surface wind and bottom stress in the x and y direction |
| $S$ | Size of the discharge from point sources | $s_{xx}$, $s_{xy}$, $s_{yx}$ and $s_{yy}$ | Elements of the radiation stress tensor |
| $f$ | Parameter of coriolis | $T_{xx}$, $T_{xy}$, $T_{yx}$ and $T_{yy}$ | Elements of the effective shear stress |
| $g$ | Gravity induced acceleration | ($u_s$, $v_s$) | Elements of velocity in the cartesian coordinates |

In the present research, an HD model, verified (calibrated and validated) by Bhui, K. [23] was used to collect the velocity and current direction data for the study. The model was set up using the surveyed bathymetry and General Bathymetry Chart of the Ocean (GEBCO) data and other default input parameters. The predicted data were matched with the observed data recorded using an Acoustic Doppler Current Profiler (ADCP) for the calibration and validation of the model. The graphical (Time series and Scatter graph) and statistical (NSE, PBIAS, RMSE, and $R^2$) techniques were utilized to verify the model. The values of $R^2$ found by Bhui et al. [23] during the calibration and validation, were more than 0.75 indicating good model performance. In addition, Bhui et al. [23] estimated acceptable NSE, PBIAS, and RMSE values when the calibration and validation were being conducted. The calculated values were within 0.65 to 0.75 for NSE, 0.50 to 0.60 for RSR, and less than 10 for PBIAS indicating a good to a very good level of accuracy for the HD model [23].

## 3. Result and Discussion

In this part, the morphological changes of the tide-dominated islands in the river systems of the Indian Sundarbans over 20 years (2000-2020) have been investigated and discussed in detail. Understanding the impact of the tide on the mechanism of erosion and accretion, the profiles of tidal velocities and current directions along the erosional zones and accretional zones have been reviewed and discussed.

*3.1. Pattern of Erosion and Accretion in the Sundarbans Islands of India*

Previous studies have shown that the Indian Sundarbans' southernmost islands, which are exposed to the ocean's daily surges, are more prone to erosion than the enclosed northern islands [17, 24, 25]. In the current research, 11 southern islands along with one northern island of the Indian Sundarbans have been investigated. The investigation of the morphological changes from 2000 to 2020, shown in figure 2, reveals that the seaward islands are subject to a substantial erosion process. The comparison between the Landsat images 0f 2000 and 2020 indicates that the Bulchery, Bhangaduani, and Dalhousie islands, in spite of having dense mangroves, are facing severe erosion, particularly in the southern parts of the islands over these 20 years, shown in the figure 2. The rates of erosion of these islands are 485.99 m/year, 487 m/year, and 480.65 m/year, respectively. The Sagar island and Mousuni island are also under threat of erosion with 375.35 m/year and 127.60 m/year erosion rates, respectively. It is observed that Sagar island is eroding from its south, whereas Mousuni island is facing erosion in its northern region. Table 1 provides an overview of the overall evaluation of the erosion and accretion of the islands. Moreover, the accretion process is also observed in different parts of the islands, shown in figure 3. Significant accretion rates are found over the Sagar, Namkhana, and Lothian islands.

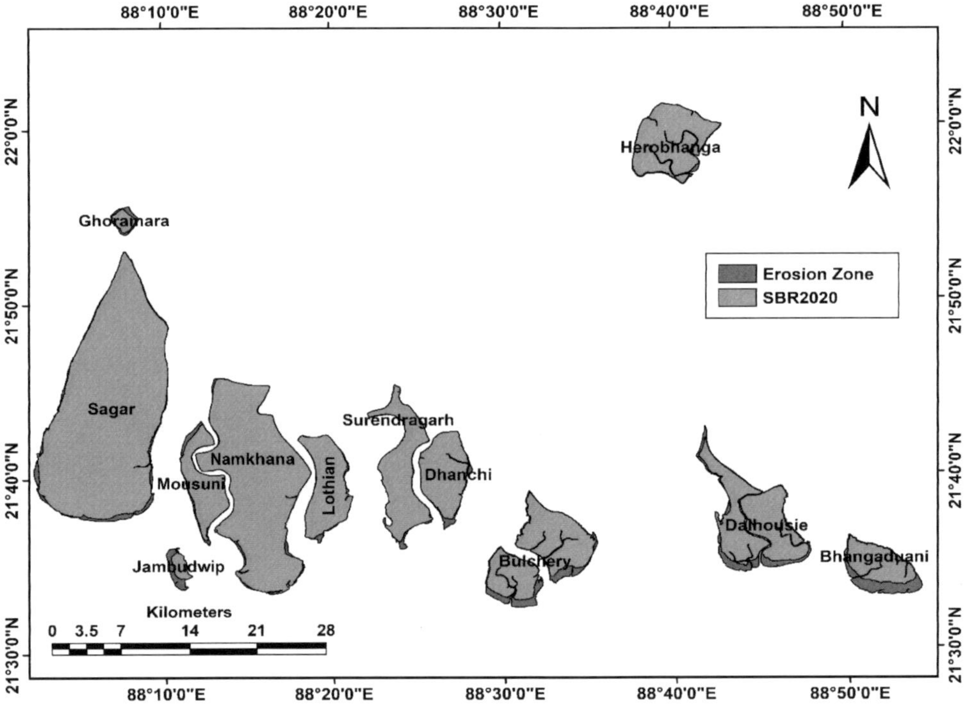


**Figure 2.** Identification of Erosional Zones (2000 - 2020) in Islands of Indian Sundarbans.

However, the rates of net change of these islands are -4.31 sq. km., 0.45 sq. km, and 1.36 sq. km., respectively, where the negative value denotes loss of the land and the positive value indicates gain of the land. In the current research, the net changes of the Bulchery, Dalhousie, Bhangaduani, and Mousuni islands are also found to be -9.34 sq. km., -9.67 sq. km., -8.56 sq. km., and -2.40 sq. km., respectively. The southern parts of the islands are dominated by erosion and submergence that are significantly influenced by the relative sea-level rise [17]. The Ghoramara island is continuously eroding from all sides at a pace of 80.99 m/year. The island structure of the Indian Sundarbans is highly unstable because of the shortage of freshwater flow and undersupply of sediment from the Hooghly River and Ganga-Brahmaputra River [20, 26].

**Table 1**. Estimation of Erosion-Accretion and Net Changes in the Southern Islands of the Indian Sundarbans during 2000 - 2020.

| Name | Erosional region (sq km) | Rate of erosion (m/year) | Accretional region (sq km) | Rate of Accretion (m/year) | Net Change (sq km) |
|---|---|---|---|---|---|
| Jambudwip | 3.09 | 154.28 | 1.21 | 60.51 | -1.88 |
| Bulchery | 9.72 | 485.99 | 0.38 | 19.04 | -9.34 |
| Bhangaduani | 9.74 | 487.00 | 0.07 | 3.44 | -9.67 |
| Dalhousie | 9.61 | 480.65 | 1.05 | 52.62 | -8.56 |
| Lothian | 1.34 | 67.09 | 2.70 | 134.99 | 1.36 |
| Dhanchi | 1.91 | 95.57 | 0.79 | 39.40 | -1.12 |
| Mousuni | 2.55 | 127.60 | 0.15 | 7.46 | -2.40 |
| Surendragarh | 1.38 | 68.85 | 1.36 | 68.25 | -0.01 |
| Namkhana | 2.34 | 116.93 | 2.79 | 139.63 | 0.45 |
| Sagar | 7.51 | 375.35 | 3.20 | 159.94 | -4.31 |
| Ghoramara | 1.62 | 80.99 | 0.00 | 0.00 | -1.62 |
| Herobhanga | 1.62 | 81.03 | 1.17 | 58.47 | -0.45 |

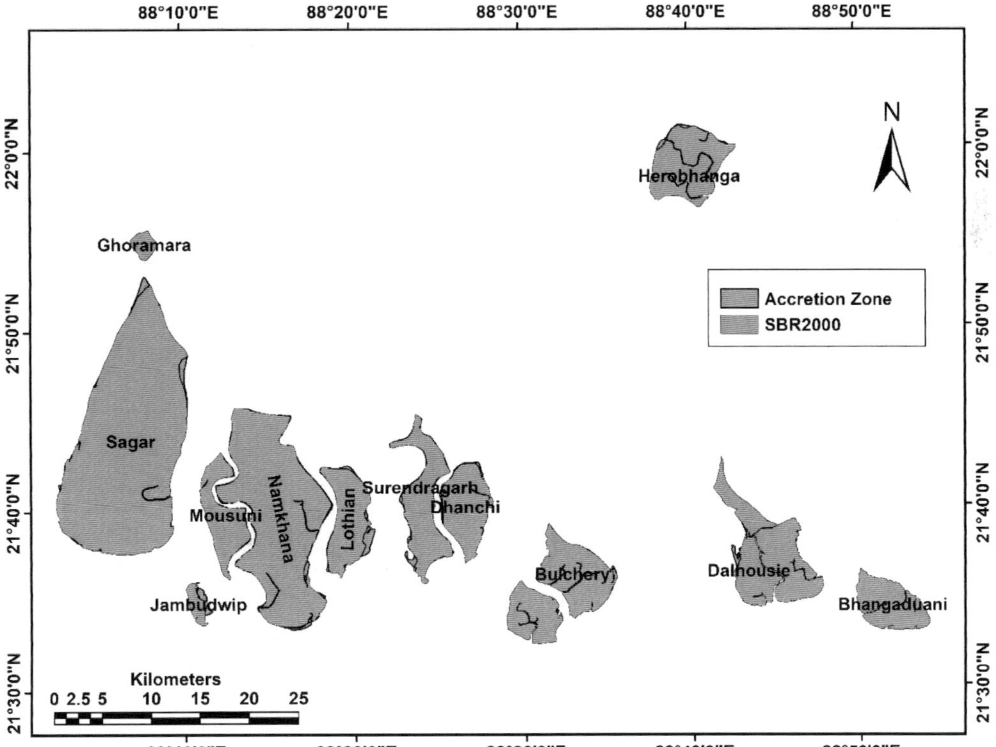


**Figure 3.** Identification of Accretion Zones (2000 - 2020) in Islands of Indian Sundarbans.

### *3.2. Impact of Tidal Phenomena on Erosion and Accretion Processes*

The main emphasis of the current study has been on how tidal events affect erosion and accretion activities. The entire investigation is carried out based on the predicted outcomes of the verified HD model. Semidiurnal tides are observed in the Sundarbans estuary, where the water level fluctuates between +2.5 m to -2.5 m (5.32 m to 0.32 m as per the local datum, which is 2.82 m below the MSL at Sagar) from the highest high tide to lowest low tide (figure 4). The tides amplify towards the north along each estuary. However, the degree amplification is not similar. Flow velocity changes significantly from high tide to low tide. Mike 21 model estimates flow velocity based on water level, bathymetry data, and other parameters (i.e. Manning's M value). The Spatio-temporal variation of flow velocity is assessed from the model output data. The velocity in the system varies from 0.1 m/s to 2.46 m/s [23]. High-velocity regions are identified in the Hugli river system, Muriganga river system near Mousuni Island, Lower Saptamukhi - Curzon Creek and Chaltabani Khal system, Upper Thakuran river system, Matla river system near Canning and Ajmalmari, and Bidyadhari-Raimangal river system near Kumirmari [23]. It is observed that the places with high velocity are erosion-prone, and the rate of erosion significantly varies with the flow direction.

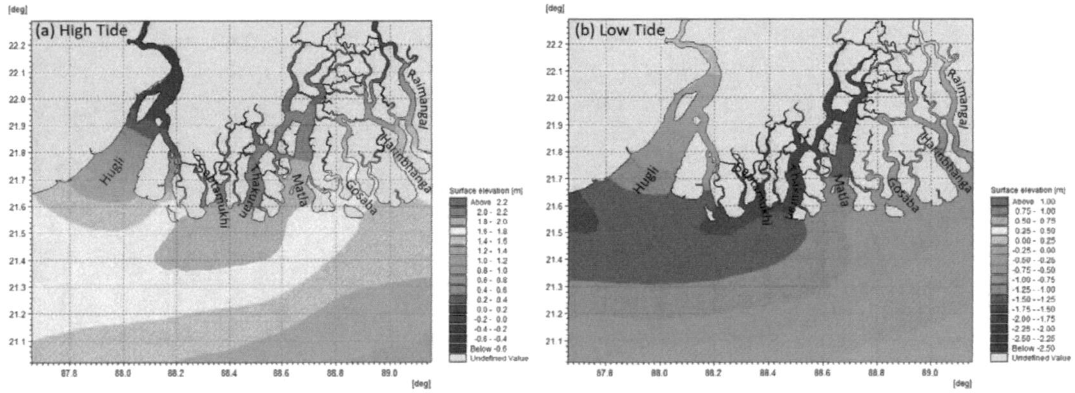


**Figure 4.** Tidal Fluctuation in Indian Sundarbans during (a) High Tide and (b) Low Tide.

The erosion accretion pattern and the surrounding hydrodynamic phenomenon of three southern islands (i.e. Bulchery, Dalhousie, and Bhangaduani) and one inner island (i.e. Herobhanga) of Indian Sundarbans have been studied. The Bulchery island (figure 5 ) is eroding from the south, south-eastern direction. The velocity map (figure 5) shows a high-velocity region in the south and north of this island. During high tide, the current directly hit the island from the south-east (figure 5), which significantly justifies the erosional pattern of the island. The Dalhousie island (figure 6) is also eroding from the south, and northwest as the current directly hits the island from the southern direction during high tide and the northwest direction during low tide (figure 6). The high-velocity zones are noticed in the south and north-eastern portions of this island (figure 6). The Herobhanga is an inner island of the Indian Sundarbans. The rate of erosion on this island is comparatively less than on the southern islands. The island is eroding from the southeast as the high-velocity region exists in the southeastern part of this island (figure 7) and the current directly hits this island from the south-eastern direction during high tide. Erosion takes place at the location of the coast from where the littoral sediment is transported due to the wave action [6]. Marginal accretion along these islands is found where the magnitude of tidal velocity is low and the current direction is parallel to the coast. Therefore, this investigation proves that the process of coastal erosion is considerably caused by high velocity and current direction, perpendicular to the coast. The inconsistency of the degree of erosion and accretion in the coastal regions is also controlled by tidal bores, variability of water outflow amount from the upstream rivers, vegetation cover, compaction of soil, the gradient of beach slope, and human interactions [3, 27].

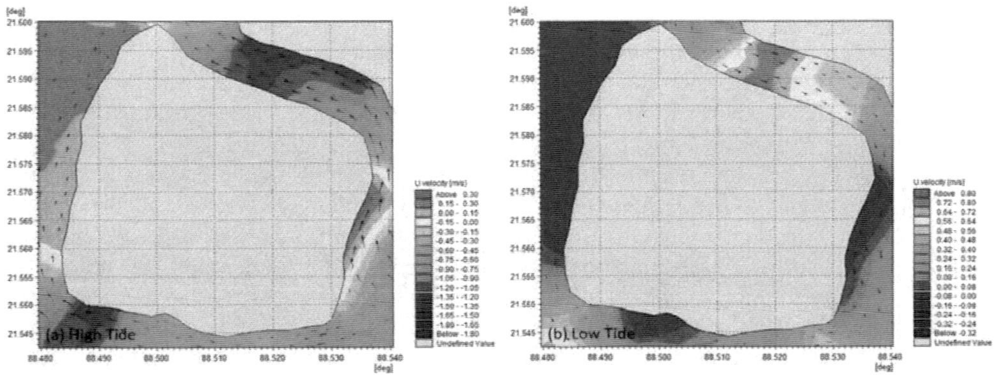


**Figure 5.** Hydrodynamic phenomenon around Bulchery Island during (a) High Tide and (b) Low Tide.

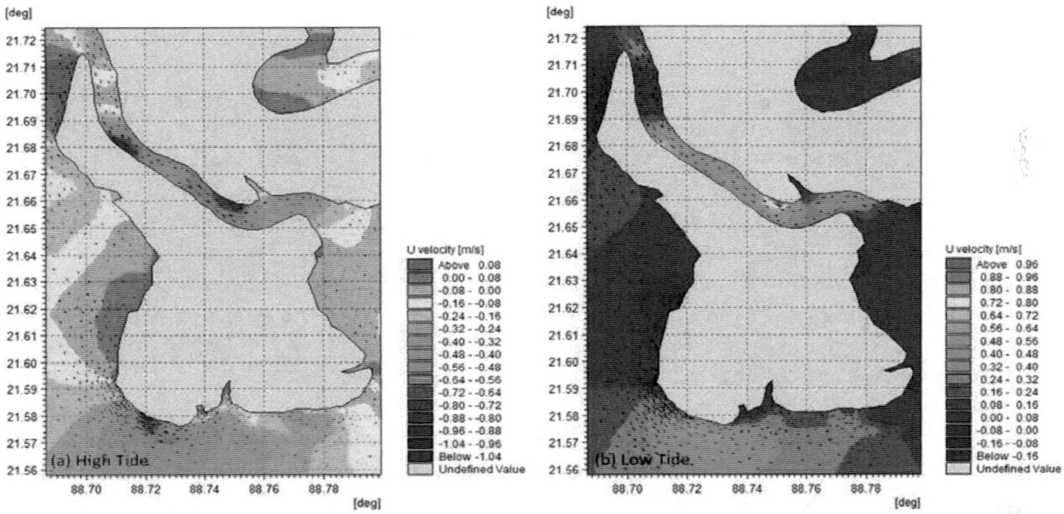


**Figure 6.**Hydrodynamic phenomenon around Dalhousie Island during (a)High Tide and (b) Low Tide.

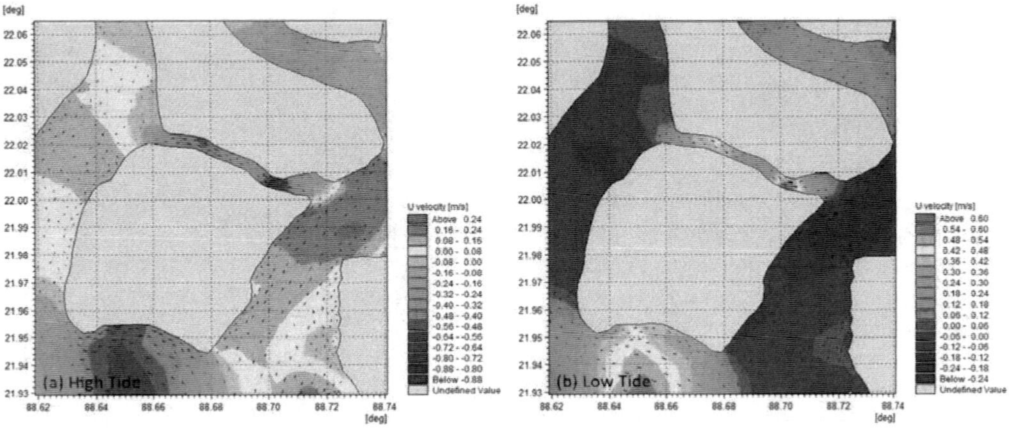


**Figure 7.** Hydrodynamic phenomenon around the Herobhanga Island during (a) High Tide and (b) Low Tide.

## 4. Conclusion

This study has tried to establish the role of tidal phenomena, particularly tidal velocity and current direction, in the Sundarbans Delta that experiences erosion and accretion simultaneously. Geoinformatics and Hydrodynamic Modelling were used in order to complete the current research. The analysis of the Landsat imageries in comparison for 2000 and 2020 using RS and GIS techniques reveals that the southern islands Bulchery, Bhangaduani, Dalhousie, and Sagar are facing severe erosion from the southern direction, whereas marginal accretion is observed in the Lothian and Namkhana Islands. Maximum erosion is identified in the sea-facing southern islands compared to the sheltered northern islands. The model-derived data are used to determine the impact of tidal phenomena, particularly tidal velocity and current direction, on the erosion and accretion processes. This analysis observes that the region with high velocity, where the tidal current hits the island perpendicularly, is prone to severe erosion, whereas the regions where the magnitude of tidal velocity is low and the current directions are parallel to the coast are less erodible. The present study is accomplished using minimal data. This study may get solid ground if more real-time data i.e. surveyed bathymetry, velocity, water level, discharge, and sediment transportation data are used in the future. Such studies will be helpful for sustainable coastal zone management in delta regions.

## References

[1] National Research Council 1990 *Managing coastal erosion* (National Academy Press)
[2] Balica S F, Wright N G and Van der Meulen F 2012 A flood vulnerability index for coastal cities and its use in assessing climate change impacts *Natural hazards* **64(1)** 73-105
[3] Ahmed A, Drake F, Nawaz R and Woulds C 2018 Where is the coast? Monitoring coastal land dynamics in Bangladesh: An integrated management approach using GIS and remote sensing techniques *Ocean & Coastal Management* **151** 10-24
[4] Gibb J G 1978 Rates of coastal erosion and accretion in New Zealand *New Zealand journal of marine and freshwater research* **12(4)** 429-456
[5] Hassan S T, Syed M A and Mamnun N 2017 Estimating erosion and accretion in the coast of Ganges-Brahmaputra-Meghna Delta in Bangladesh *6th Int. Conference on Water & Flood Management* 115-124
[6] Kaliraj S, Chandrasekar N and Magesh N S 2014 Impacts of wave energy and littoral currents on shoreline erosion/accretion along the south-west coast of Kanyakumari, Tamil Nadu using DSAS and geospatial technology *Environmental Earth Sciences* **71(10)** 4523-42
[7] Gopinath G and Seralathan P 2005 Rapid erosion of the coast of Sagar island, West Bengal-India *Environmental Geology* **48(8)** 1058-67
[8] Brammer H, Asaduzzaman M and Sultana P 1996 Effects of climate and sea-level changes on the natural resources of Bangladesh *The Implications of Climate and Sea–Level Change for Bangladesh* 143-203 (Springer, Dordrecht)
[9] Ghosh T, Bhandari G and Hazra S 2003 Application of a 'bio-engineering'technique to protect Ghoramara Island (Bay of Bengal) from severe erosion *Journal of Coastal Conservation* **9(2)** 171-178
[10] Hazra S, Mukhopadhyay A, Chanda A, Mondal P, Ghosh T, Mukherjee S and Salehin M 2016 Characterizing the 2D shape complexity dynamics of the islands of Sundarbans, Bangladesh: a fractal dimension approach *Environmental Earth Sciences* **75(20)** 1-12
[11] Kaliraj S and Chandrasekar N 2012 Geo-processing model on coastal vulnerability index to explore risk zone along the south west coast of Tamilnadu, India *Int J Earth Sci Eng* **5(5)** 1138-47
[12] Sarwar M, Mahabub G and Woodroffe C D 2013 Rates of shoreline change along the coast of Bangladesh *Journal of coastal conservation* **17(3)** 515-526
[13] Adegoke J O, Fageja M, James G, Agbaje G and Ologunorisa T E 2010 An assessment of recent changes in the Niger Delta coastline using satellite imagery *Journal of Sustainable Development* **3(4)** 277

[14] Zoran M and Anderson E 2006 The use of multi-temporal and multispectral satellite data for change detection analysis of the Romanian Black Sea coastal zone *Journal of optoelectronics and advanced materials* **8(1)** 252
[15] Wang L T 2003 Delaware inland bays shoreline extraction using Landsat 7 satellite imagery *Digital Mapping Techniques' 03, Workshop Proceedings: Millersville, Pennsylvania* Vol 3 No 471 (US Department of the Interior, US Geological Survey) p 231
[16] Kundu K and Halder P 2018 Geomorphological Changes in Sundarbans Delta Using GIS and Remote Sensing Data *International Conference on Computational Intelligence, Communications, and Business Analytics* (Springer, Singapore) pp 40-53
[17] Hazra S, Ghosh T, DasGupta R and Sen G 2002 Sea level and associated changes in the Sundarbans *Science and Culture* **68(9/12)** 309-321
[18] Chatterjee N, Mukhopadhyay R and Mitra D 2015 Decadal changes in shoreline patterns in Sundarbans, India
[19] Bera R and Maiti R 2019 Quantitative analysis of erosion and accretion (1975–2017) using DSAS—A study on Indian Sundarbans *Regional Studies in Marine Science* **28** 100583
[20] Adarsa J, Shamina S and Arkoprovo B 2012 Morphological change study of Ghoramara Island, Eastern India using multi temporal satellite data (International Science Congress Association)
[21] Manimaran D 2012 Groundwater Geochmistry Study Using GIS in and Around Vallanadu Hills, Tamilnadu, India *Research Journal of Recent Sciences* 2502
[22] DHI 2014 MIKE21 Flow Model FM User Guide
[23] Bhui K, Hazra S, Bhadra T and Venugopal V 2022 Spatio-Temporal Variability of Tidal Velocities in the Rivers of the Indian Sundarban Delta: A Hydrodynamic Modelling Approach *Journal of The Institution of Engineers (India): Series C* 1-15
[24] Ganguly D, Mukhopadhyay A, Pandey R K and Mitra D 2006 Geomorphological study of Sundarban deltaic estuary *Journal of the Indian Society of Remote Sensing* **34(4)** 431-435
[25] Hazra S and Samanta K 2016 Temporal Change Detection (2001-2008): Study of Sundarban (No. id: 10526)
[26] Allison M and Kepple E 2001 Modern sediment supply to the lower delta plain of the Ganges-Brahmaputra River in Bangladesh *Geo-Marine Letters* **21(2)** 66-74
[27] Krantz M 1999 Coastal erosion on the island of Bhola, Bangladesh (SWEDMAR)

# Spatiotemporal Change Detection Analysis of Land Cover and Land Use using Multi-Temporal Remote Sensed Imagery and Landscape Metrics: A Sub-Catchment Level Case Study of the Bhagirathi-Hugli River, West Bengal, India

**S Das[1,*], R Sarkar[1,*]**

[1]Department of Geography, Adamas University, Kolkata, West Bengal, India.

*Corresponding Author Email: sanatdas9@gmail.com, rajibsarkar07@gmail.com

**Abstract.** To analyse and characterise land use-land cover (LULC) change, remote sensing (RS) data and geographic information system (GIS) approaches are required. The landscape metrics are also extensively used to explain the structure and pattern of the spatial landscape. This case study has been conducted in the area of the Bhagirathi-Hugli River banks in West Bengal. The study area's satellite image data from various periods (1989-2017) were used to analyse the spatiotemporal variation of LULC change. The LULC fragmentation was analysed using a total of six landscape measures. Landscape metrics such as Number of Patches (NP), Largest Patch Index (LPI), Landscape Shape Index (LSI), Mean Shape Index (MSI), Contagion (CONTAG), and Shannon's Diversity Index were used to determine the changes over four different periods. The main features of land use change were the continuous expansion of built-up area (9.30 $km^2$), vegetation (189.14 $km^2$), and the decrease in agricultural fallow land and water bodies from 1989 to 2017. The landscape pattern in the study area has also changed significantly. The landscape metrics indicate that the studied region had a highly fragmented landscape during the study period. The proposed technique is easy to use and can address more than one metric while selecting the appropriate setting can prove problematic.

**Keywords:** Image classification, Land use -land cover, Change detection, Landscape matrices, RS and GIS.

## 1. Introduction

Accumulating data on land use and land cover (LULC) change is important for establishing more agreeable links and influences between humans and the shared environment. It is a crucial source of information for analysing spatiotemporal LULC change. Indeed, well-prepared and explained multi-temporal RS datasets enable the mapping and recognition of landscape changes, resulting in a successful effort toward reasonable landscape planning and management [1].

Land use and land cover are two separate but related aspects. In addition to soil, ice, and water categories, covers include crops, wild plants, grasslands, highways, and structures. Land uses include farming, logging, grazing, and urban growth. Modification and conversion are the two categories used to represent changes in land cover brought about by land use. The modification is a change in a surface cover's state. Alteration in cover type is referred to as a conversion [2]. The river system is the most vulnerable among the landforms that humans have significantly altered. Within the limits of his circumstances, a man uses the land as he sees fit, whether for pleasure or necessity. The desire to live near rivers has existed since the dawn of civilization and man has shaped the land according to its natural features such as geological formations, type of landscape, drainage system, climate, soil types, natural vegetation, groundwater, etc. Human civilization is also influenced by existing LULC patterns. Humans began to modify land cover and land use over time based on social factors (dietary habits, demand, religious motivations, etc.), economic conditions (benefits of irrigation, electricity, transportation, labour supply, capital supply, etc.), and natural phenomena. The Earth's surface is subject to rapid LULC changes due to myraid socioeconomic and natural phenomena [3]. The health of the river is damaged by these combined changes, which also cause bank erosion, flooding, etc. [4]. Growing farms also have a negative impact on bank ecology, water quality, and natural vegetation [5]. Changes in LULC can affect the runoff patterns and sediment supply patterns of rivers [6, 7] and make the catchment more vulnerable to climatic disturbances. However, several societal factors, including several changing land-use patterns, are driving the changes in land cover that affect biodiversity, hydrological and radiative regimes, emissions of trace gases, and other processes that alter the climate and biosphere [2].

Landscape change models are useful tools to understand the factors that shape landscapes. Investigating the effects of projecting short-term landscape changes over a longer period serves as a source of inspiration for the modelling with the fast, reliable, and with most efficient techniques to monitor the change of LU/LC [8, 9, 10]. Dynamic land-use index, transfer matrix, driving force analysis, and integrated dynamic land-use degree were some of the research techniques used in previous publications for dynamic land-use change [11,12]. Land use fragmentation studies often make use of landscape measures. Patch, class, and landscape indices are used to classify the measurements. The study of land-use fragmentation often uses landscape metrics. The measurements are divided into patch, class, and landscape indices. The metrics are rationally ordered within each group based on the feature of the landscape structure to be measured [13, 14, 15].

The Bhagirathi-Hugli River is no different from other major rivers around the world as it is seriously threatened by the rapid changes in land use and cover. The Bhagirathi, Mathabhanga, and Jalangi rivers in the dormant Delta of Bengal inspired the naming of the Nadia rivers [16, 17, 18]. The main river in Nadia district dividing Nadia and Purba Bardhaman districts is the Bhagirathi-Hugli river. It is considered the lifeline of West Bengal [19] and marks the boundary of the district between the two. One of the main tributaries of the Ganga River is the Bhagirathi-Hugli River. River Ganga originates in Murshidabad near Mithipur and empties into the Bay of Bengal after passing through Murshidabad, Nadia, Purba Bardhaman, Hugli, and Medinipur districts.

The study aims to evaluate a section of the Bhagirathi-Hugli river basin in the Nadia and Purba Bardhaman districts of West Bengal. Both rural and urban forms of land cover and land use were discovered along the selected study region. With a wide variety of land uses and land covers including agriculture, vegetation, industry, housing, transportation, fisheries, etc., this area exhibits both rural and urban characteristics. Most of the changes happened randomly in this case. But as the population grows, people are compounding several that are damaging the river's ecosystem, and irrational and unscientific land-use practices are causing erosion and sedimentation, reducing the river's depth. In every monsoon season, this research region is inundated by flood waters, which also leads to bank erosion and river bed changes. To understand the land use fragmentation of the research region, this study used conversion matrices and landscape metrics to analyse the spatiotemporal changes in LULC.

## 2. Study Area

Between Nadia and Purba Bardhaman districts, a selected stretch of the Bhagirathi-Hugli river serves as the district line. This watershed lies within the Bengal Delta [20]. The latitude and longitude of the study area are between 23°12'08.20" N to 23°46'11.88" N and 88°05'05.44" E to 88°29'43.50" E. The study area is 955.45 $km^2$ (Figure 1). This region is an alluvial formation of the Ganga-Hugli river system's rivers. According to the SRTM DEM, the selected portion of Bhagirathi-Hugli river basin has a maximum elevation of around 33 m, which is found in the north and north-western regions. Minimum height of 3 m is noted along the river course, as well as in the south and south-eastern regions (figure 2). Nadia and Purba Bardhaman district lies on the east and the west bank of the river respectively. The

IOP Conf. Series: Earth and Environmental Science 1164 (2023) 012003 doi:10.1088/1755-1315/1164/1/012003

Bhagirathi is commonly known as Hugli, from the confluence of the Jalangi River downstream. The length of the study area within the Bhagirathi- Hugli River basin is 122.4 km approximately.

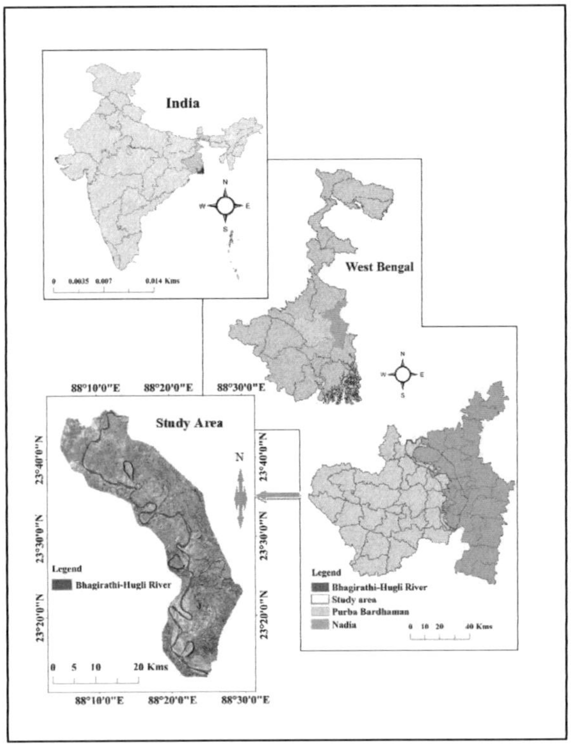


**Figure 1.** Map of the study area

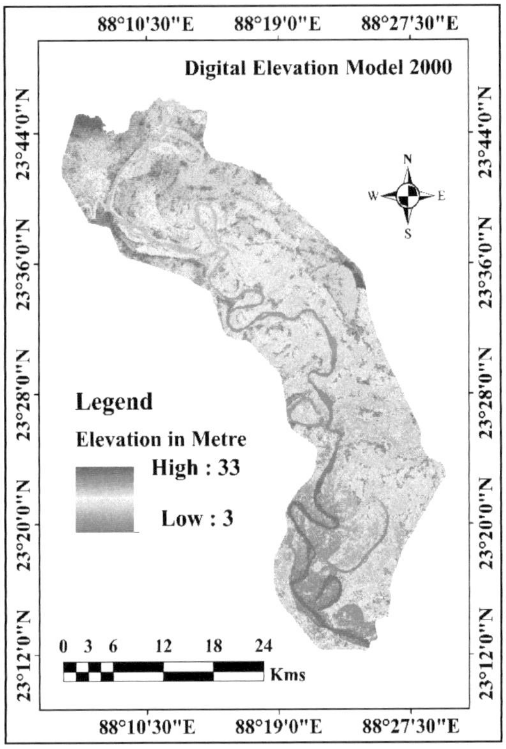


**Figure 2.** DEM of the study region

## 3. Materials and Methodology

### *3.1. Materials*

This study uses Landsat TM and ETM satellite data downloaded from the USGS Erath Explorer website (Table 1). 1972-1973 SOI topographic maps at 1:50000 scale and Google Maps are also used for the boundary map of the study area.

**Table 1.** Data for the investigation and analysis

| Reference system | Sensor | Data acquired | Resolution/ Scale | Map No./Used of Band |
|---|---|---|---|---|
| | Topographical map | 1972-73 | 1:50000 | 79A/1, 79A/2,79A/6, 79A/7,79A/8, 79B/5 |
| WRS_138 & 44 | Landsat_5/TM | 01$^{st}$ April 1989 | 30 m | 1, 2, 3, 4 |
| | Landsat_7/ETM+ | 26$^{th}$ April 2001 | | 1, 2, 3, 5 |
| | Landsat_5/TM | 11$^{th}$ April 2010 | | |
| | Landsat_8/OLI_TIRS | 14$^{th}$ April 2017 | | 2, 3, 4, 5 |
| | SRTM DEM | 11.02.2000 | 90 m | |

Spatial data source: *USGS & SOI*

### *3.2. Methodology*

*3.2.1. Pre-processing of Satellite Image.* Before completing the best image classification analysis, satellite imagery pre-processing is a critical step. Using the Semi-Automatic Classification Plugin (SCP) in QGIS 2.18.14, atmospheric correction is performed across all images using the Dark Object Subtraction (DOS) technique. DOS is preferred over other methods because of its effectiveness and simplicity. After atmospheric correction, each image is retained by the delineated research area boundary established using Survey of India toposheets from 1972-1973.

*3.2.2. Image Classification.* In Erdas Imagine 2014, the categorization of LULC was done using a Maximum Likelihood Classifier (MLS) method with a decision rule. To identify different land cover classes, maximum likelihood classification is frequently utilised. Training regions and test data are used in the supervised classification approach to estimate accuracy [3, 21]. The appropriate adjustments (Recode) were done based on the field verification. Maps of land use and land cover (LULC) were created using classified satellite photos. Waterbodies, agricultural land, agricultural fallow land, vegetation, and built-up areas were the five categories of LULC identified in this study.

*3.2.3. Accuracy Assessment.* Accuracy assessment is an important part of the image classification study. It is used to correctly identify, understand and assess changes [3, 21, 22]. It was done by Erdas Imagine 2014. It mentioned the use of field verification, Google Earth, and processed satellite imagery (FCC imagery). The error matrices reflecting the authorization obtained after eliminating the randomly predictable symmetry of the match were used to determine the overall accuracy, the kappa statistics, the user's accuracy, and the producer's accuracy.

*3.2.4 Conversion Matrix*
The land use structure of the region at a given point in time is represented by the land use/land cover conversion matrix. This matrix quantitatively shows the process of conversion between different land uses and covers as well as the transitions between different land types at the beginning and end of the study [23]. Based on the LULC modifications from 1989 to 2001, 2001 to 2010 and 2010 to 2017, the transition matrix was created. Using MOLUSCE in the QGIS program, the transition matrix created a simple model of the landscape change. The MMULT function in EXCEL can also be used to calculate the matrix product of two arrays [24].

### *3.3. Landscape Metrics*
Landscapes are patchwork quilts of various natural and anthropogenic elements. Ecological processes are influenced by patches, shape, size, variability, and edge features across the landscape [25]. The most comprehensive and related category of landscape elements is metrics. It is fundamental to the functioning of the landscape [26]. Many software programs provide tools for analyzing patterns found in raster grids and remote-sensing image files. Part of this attention is given to the classification of pixels or grid cells to determine different land use classifications. Landscapes are patchwork quilts made up of many artificial and natural components. Patch, shape, size, diversity, and edge features affect ecological processes in the landscape [25]. Metrics are the most comprehensive and relevant type of landscape aspect. It is essential for the functioning of the landscape [26]. Tools for analyzing patterns in raster grids and remote sensing image files are available in many software programs. This focus includes some considerations on classifying pixels or grid cells to define different land use categories. The software program FRAGSTATS Raster Version 4.2.1 was used to create landscape measurements. At both the landscape and class level, it rates terrain using the patch neighbour 8-cell rule option. The spatial resolution entered determines the results of the metric analysis; in this study, a pixel size of 30 m was used. After calculating the landscape metrics, the landscape metrics were selected, including the Number of Patches (NP), Largest Patch Index (LPI), Landscape Shape Index (LSI), Mean Shape Index (MSI), Contagion (CONTAG), Shannon's Diversity Index.

*3.3.1. NP (Number of Patches).* NP is a straightforward indicator of how much the patch type is fragmented. The spatial scale of analysis and the size of the terrain are directly connected to NP [27]. It has a scale of 0 to infinity. When NP is 1, there is just one patch in the landscape.

*3.3.2. Largest Patch Index (LPI).* The biggest patch in the overall landscape area is measured using the LPI, an advantage index [27]. It has a range of 0 to 100 LPI. When the LPI is zero, the size of the largest patch in the landscape gets less and smaller, and when the LPI is one hundred, the largest patch fills the landscape. The primary landscape types may be identified with the aid of LPI, and their value can indicate the direction and intensity of human activity as well as the strength and frequency of interference.

*3.3.3. Landscape Shape Index (LSI).* The edge-to-area ratio for the entire terrain is measured by the LSI index [27]. It has an unlimited LSI range of 1. When the landscape is composed of a single square patch of the appropriate kind, LSI = 1. As the geometry of the terrain grows more erratic and/or as the length of the edge inside the landscape rises, LSI climbs without bounds.

*3.3.4. Mean Shape Index (MSI).* For a class level or landscape level and using the circular or square standard, the MSI is equal to the sum of the perimeters of all patches divided by the square root of the patch area divided by the number of patches [28]. It has an unbounded MSI range of 1. If all terrain consists of square or circular (vector) patches, MSI = 1; Otherwise, MSI will rise forever as patch shapes change.

*3.3.5. Contagion Index (CONTAG).* The degree of lumpiness in the distribution of patch types (class) as opposed to their discontinuous distribution in several small pieces is measured by CONTAG [27]. It offers a methodical way to measure how spatial patterns differ between different environments [29]. It ranges from $0 < \text{CONTAG} \leq 100$. When patch categories are highly disaggregated, CONTAG techniques 0 are used, and when they are excellently aggregated, 100 are used [29, 30].

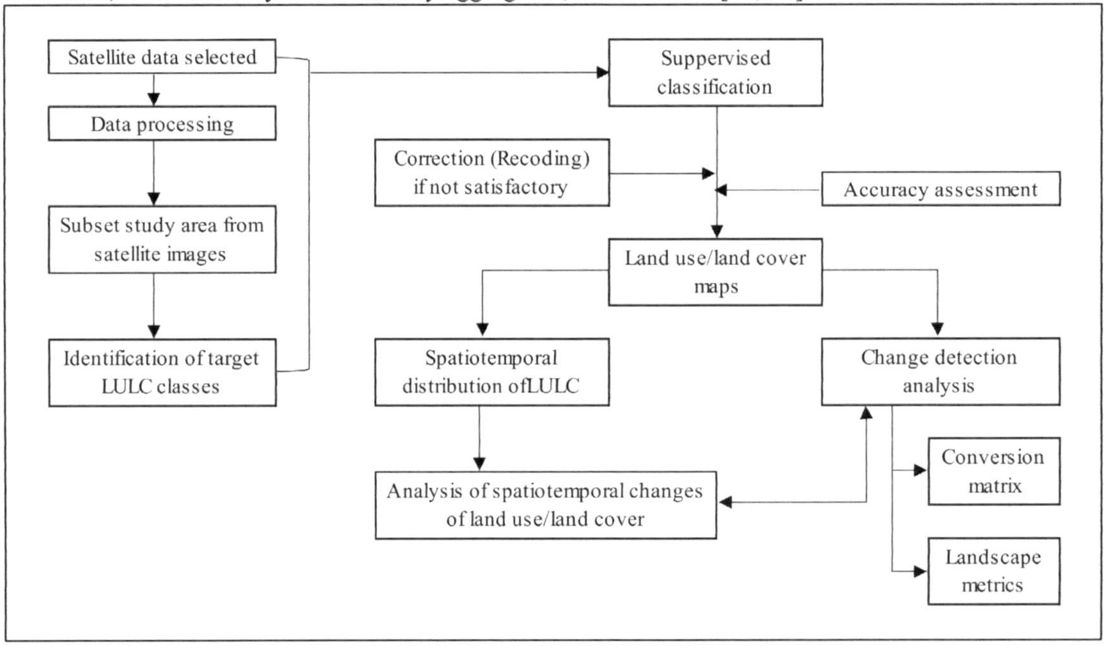


**Figure 3.** The methodology adopted in this study.

*3.3.6. Shannon's Diversity Index (SHDI).* SHDI, a well-liked indicator of diversity in community ecology, is used here to describe landscapes. Compared to Simpson's diversity index, it is a little more sensitive to uncommon patch types [27]. It has an infinite range starting at SHDI 0. When there is only one patch in the landscape, SHDI equals 0. (That is, there is no diversity). When there are more distinct patch types (i.e., patch richness, PR) and/or when the proportionate distribution of area across patch types is more evenly distributed the SHDI rises. Figure 3 depicts the research's methodology.

## 4. Results and Discussion

### *4.1. Analysis of the spatiotemporal land use-land cover distribution*

Between Nadia and Purba Bardhaman districts along the Bhagirathi-Hugli River, land use and land cover change are not consistent. Using the LULC categorization data for the Bhagirathi-Hugli river basin over the period 1989 to 2017, four spatial distribution maps of LULC at four different time points in the study region were constructed (Figure 4). Built-up, vegetation, waterbodies, agricultural, and agricultural fallow land are the five groups of LULC. Table 2 shows the results of these classes. The findings indicate that agricultural fallow land was the predominant LULC along the Bhagirathi-Hugli River, followed by agricultural areas, vegetation, water bodies, and built-up areas. According to Landsat data (LULCC) within the research region, it was demonstrated that vegetation and built-up land increased steadily from 1989 to 2017, while agricultural and agricultural fallow land and water bodies decreased.

Along the left bank of the Bhagirathi-Hugli River, agricultural fallow land and rural communities were intertwined. Between 1989 and 2017, the proportion of agricultural fallow land fell from 46.18 to 41.94 %. Agricultural fallow land decreased by 4.24 %, or 1.41 km²/year, throughout that time. In the research region, agricultural land was the second-largest land use. Rural regions near the riverbanks were where this area spread and concentrated. Agricultural land fell from 39.47 % in 1989 to 25.11 % in 2017. Agriculture land decreased throughout this time by 14.36 %, growing at a pace of 4.91 km²/year. Of all land uses, agricultural land had the greatest reduction. Water bodies also continually shrank from 1989 to 2017.

The third-largest land use in the research region was vegetation. From 4.30 % in 1989 to 24.04 % in 2017, the amount of vegetation increased quickly. Vegetation grew at a rate of 6.67 km²/year, or 19.74 %, throughout this time. Most vegetation was lost during this land use change. The most important land use in this area is the built-up area. The majority of the developed property was located around roadsides, riverbanks, and in and around towns. In the study area's rural villages, there was a noticeable uneven distribution of built-up land. It has steadily increased, rising from 2.55 % in 1989 to 3.52 % in 2017. The built-up land rose by almost 1% over this time at a growth rate of 0.33 km²/year.

**Table 2.** Area transition for land use/land cover classes between the selected years

| LULC Class | Area | 1989 | 2001 | 2010 | 2017 |
|---|---|---|---|---|---|
| Built-up areas | km² | 24.45 | 27.85 | 31.02 | 33.75 |
| | (%) | 2.55 | 2.91 | 3.24 | 3.52 |
| Vegetation | km² | 41.16 | 133.81 | 212.45 | 230.30 |
| | (%) | 4.30 | 13.97 | 22.18 | 24.04 |
| Water bodies | km² | 71.85 | 61.77 | 54.99 | 51.62 |
| | (%) | 7.50 | 6.45 | 5.74 | 5.39 |
| Agricultural land | km² | 378.10 | 360.60 | 255.66 | 240.55 |
| | (%) | 39.47 | 37.64 | 26.69 | 25.11 |
| Agricultural fallow land | km² | 442.41 | 373.95 | 403.84 | 401.76 |
| | (%) | 46.18 | 39.04 | 42.16 | 41.94 |

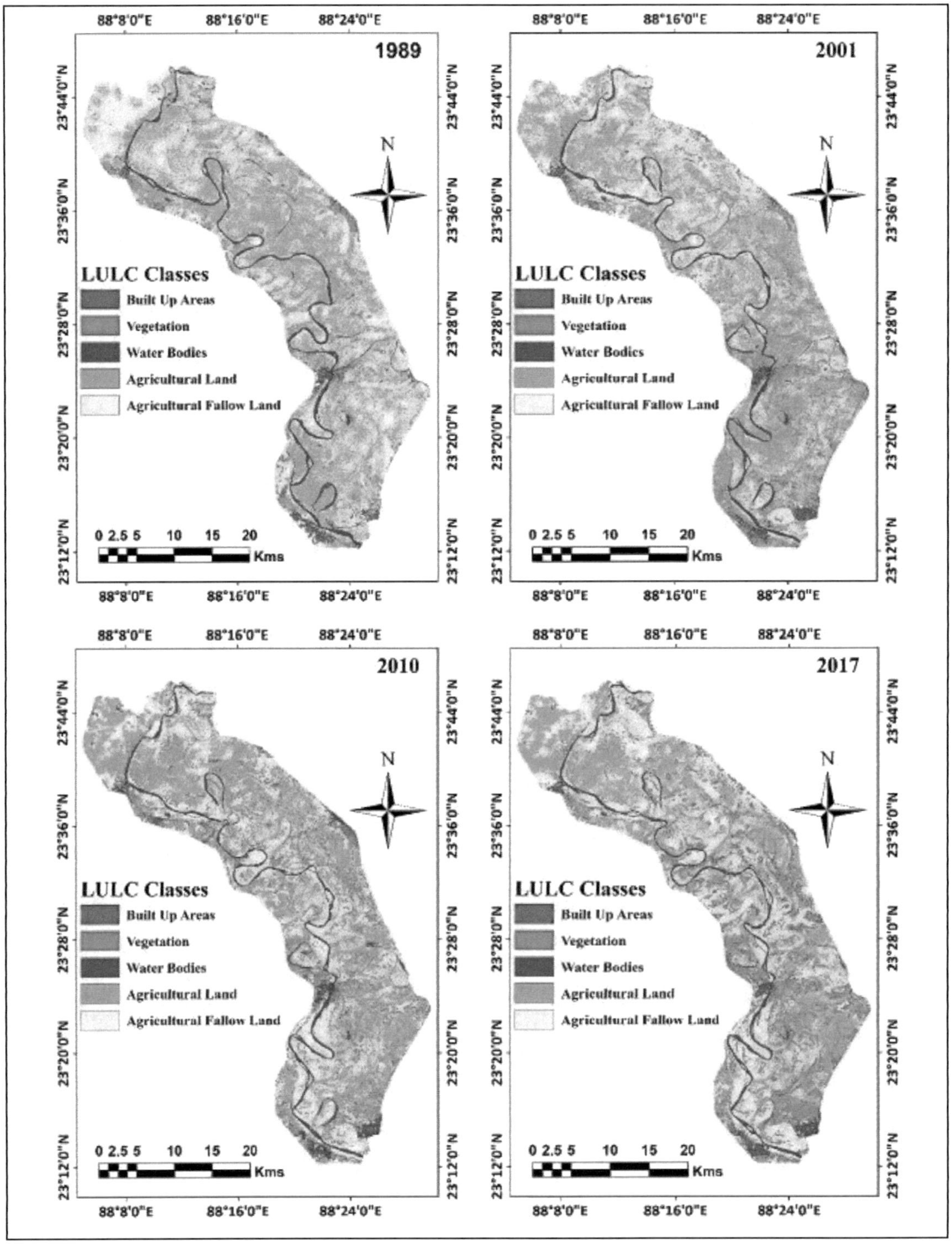


**Figure 4.** Spatiotemporal distribution of land use-land covers from 1989 to 2017.

### *4.2. Change Detection Analysis*

Change detections were performed for the periods 1989–2001, 2001–2010, 2010–2017, and 1989–2017. First, based on the LULC maps, the category changes in the net area between different years were measured (Figure 5). While the agricultural area has declined sharply, the proportion of built-up areas and vegetation has increased noticeably. From 1989 to 2001, the built-up area increased by 3.39 km² at a rate of 0.28 km²/year. In the next two periods, the built-up area increased by 0.35 km²/year and 0.39 km²/year. This indicates a strong upward trend in settlement development from 1989 to 2017. Agricultural areas and bodies of water decreased with fluctuations that varied over time.

Figure 6 shows the actual changes in LULC from 1989 to 2017. Accordingly, from 1989 to 2017, the amount of degraded agricultural land (14.36%), agricultural fallow land (4.24%) and bodies of water (2.11%) increased dramatically, as did vegetation (19.74% or 189.14 km²) and built-up area (0.97% or 9.30 km²).

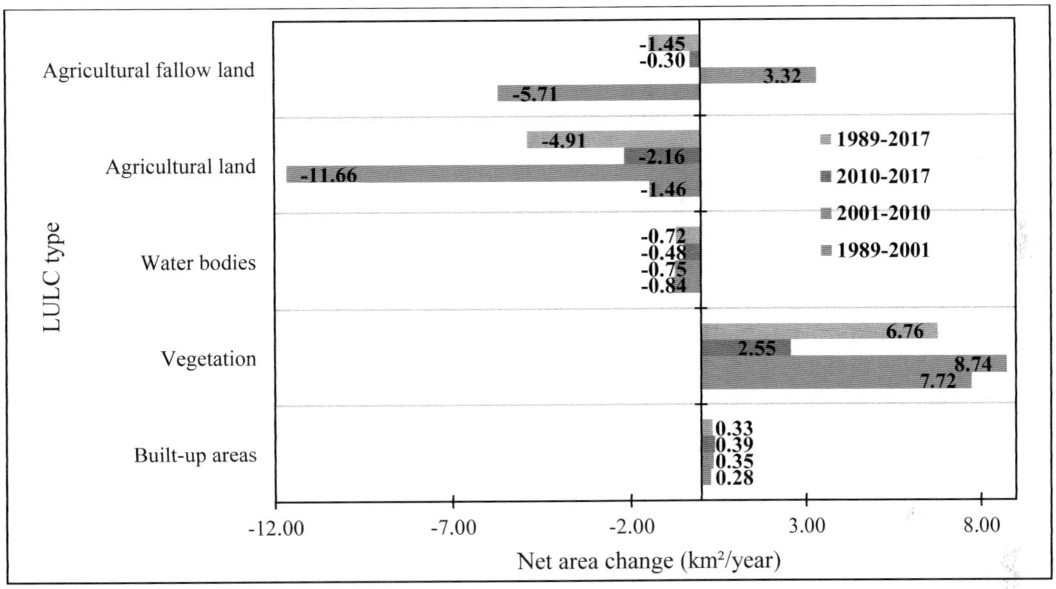


**Figure 5.** Changes in the net area of the LULC class between different years.

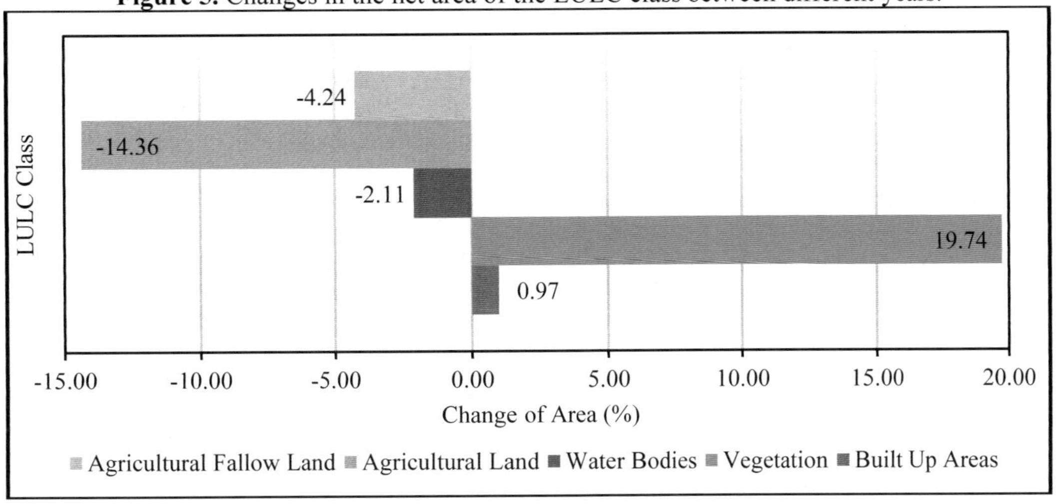


**Figure 6.** Relative change of LULC class during the year 1989-2017.

*4.2.1 Change detection analysis by conversion matrix.* LULC transition matrices for the study region from 1989 to 2001, 2001 to 2010 and 2010 to 2017 are presented in Tables 3-6. This table's columns show how a certain LULC class from 1989 was changed into other classes in 2001, another LULC class from 2001 was changed into other classes in 2010, and a LULC class from 2010 was changed into other classes in 2017. Contrarily, each row indicates how several LULC classes from the years 1989, 2001, and 2010 were changed into a particular class in the same years, as well as in 2001, 2010, and 2017. The unaltered region that belongs to a certain class is indicated by the diagonal values in this matrix.

A large percentage of agricultural land was converted between the years 1989 and 2001 (table 3) across all land use-land cover groups, with the majority of this conversion occurring to built-up, water, and Agricultural fallow land. There was a conversion of 217.91 km², accounting for 22.75% of the agricultural land converted. Agricultural fallow land was mainly converted into areas of agriculture, vegetation, and build-up. This conversion was 187.16 km² or 19.54 % of the area of agricultural fallow in 1989. Significant land conversion occurred to agricultural land from agricultural fallow land. The total area of vegetation converted was 106.02 km² or 11.07% of the conversion. It was mostly turned into agricultural areas, agricultural fallow land, and built-up areas. The built-up area was primarily transformed into vegetation, agricultural fallow, and agricultural land. This conversion had a surface area of 21.72 km², or 2.27 % built-up area in 1989. A tiny portion of water bodies was also turned into areas of agricultural and agricultural fallow. This conversion was 18.71 km², which represented 1.95 % of water bodies in 1989.

During the 2001–2010 period (table 4), agricultural fallow land was most frequently turned into areas of agriculture, vegetation, water bodies, and built-up. This conversion was 172.70 km² or 18.03% of an area of agricultural fallow in 2001. There was a large conversion of agricultural fallow land to agricultural land. The second most changed land use at this time was vegetation, which was mostly turned into agricultural land and agricultural fallow land. A total of 126.98 km² or 13.26% were converted of the total conversion. Most agricultural areas were turned into fallow land and vegetation. This conversion has a surface area of 103.97 km² or 10.85 % of all land used for agriculture in 2001. The built-up areas were primarily transformed into areas of agricultural fallow, agricultural and vegetation. In the shortest length of time, water bodies were also transformed into agricultural land and fallow land. It has been demonstrated that while vegetation and built-up areas rose in comparison to the preceding time frame, agricultural fallow land, agricultural land, and water bodies declined.

**Table 3.** Land use-land cover conversion matrix from 1989 to 2001

| LULC Classes | | Area | 2001 | | | | | Total |
|---|---|---|---|---|---|---|---|---|
| | | | Built-up areas | Vegetation | Water bodies | Agricultural land | Agricultural fallow land | |
| 1989 | Built-up areas | km² | **6.13** | 1.95 | 0.45 | 7.72 | 11.59 | 27.85 |
| | | % | **0.64** | 0.20 | 0.05 | 0.81 | 1.21 | 2.91 |
| | Vegetation | km² | 7.52 | **27.79** | 2.18 | 57.25 | 39.07 | 133.81 |
| | | % | 0.78 | **2.90** | 0.23 | 5.98 | 4.08 | 13.97 |
| | Water bodies | km² | 0.34 | 0.46 | **43.06** | 6.36 | 11.55 | 61.77 |
| | | % | 0.04 | 0.05 | **4.50** | 0.66 | 1.21 | 6.45 |
| | Agricultural land | km² | 5.27 | 7.03 | 12.20 | **142.68** | 193.41 | 360.60 |
| | | % | 0.55 | 0.73 | 1.27 | **14.89** | 20.19 | 37.64 |
| | Agricultural fallow land | km² | 5.19 | 3.93 | 13.96 | 164.09 | **186.79** | 373.95 |
| | | % | 0.54 | 0.41 | 1.46 | 17.13 | **19.50** | 39.04 |
| Total | | km² | 24.45 | 41.16 | 71.85 | 378.10 | 442.41 | **957.97** |
| | | % | 2.55 | 4.30 | 7.50 | 39.47 | 46.18 | **100.00** |

*The value of the bold mark indicates the unchanged area.*

**Table 4.** Land use-land cover conversion matrix from 2001 to 2010

| LULC Classes | | Area | 2010 | | | | | Total |
|---|---|---|---|---|---|---|---|---|
| | | | Built-up areas | Vegetation | Water bodies | Agricultural land | Agricultural fallow land | |
| 2001 | Built-up areas | km² | **8.36** | 5.28 | 0.49 | 6.73 | 10.17 | 31.02 |
| | | % | **0.87** | 0.55 | 0.05 | 0.70 | 1.06 | 3.24 |
| | Vegetation | km² | 5.92 | **85.47** | 1.73 | 73.53 | 45.80 | 212.45 |
| | | % | 0.62 | **8.92** | 0.18 | 7.68 | 4.78 | 22.18 |
| | Water bodies | km² | 0.36 | 0.88 | **42.58** | 3.92 | 7.25 | 54.99 |
| | | % | 0.04 | 0.09 | **4.45** | 0.41 | 0.76 | 5.74 |
| | Agricultural land | km² | 2.56 | 17.99 | 3.83 | **151.70** | 79.59 | 255.66 |
| | | % | 0.27 | 1.88 | 0.40 | **15.84** | 8.31 | 26.69 |
| | Agricultural fallow land | km² | 10.65 | 24.19 | 13.15 | 124.71 | **231.14** | 403.84 |
| | | % | 1.11 | 2.53 | 1.37 | 13.02 | **24.13** | 42.16 |
| Total | | km² | 27.85 | 133.81 | 61.77 | 360.60 | 373.95 | **957.97** |
| | | % | 2.91 | 13.97 | 6.45 | 37.64 | 39.04 | **100.00** |

*The value of the bold mark indicates the unchanged area.*

**Table 5.** Land use-land cover conversion matrix from 2010 to 2017

| LULC Classes | | Area | 2017 | | | | | Total |
|---|---|---|---|---|---|---|---|---|
| | | | Built-up areas | Vegetation | Water bodies | Agricultural land | Agricultural fallow land | |
| 2010 | Built-up areas | km² | **9.02** | 7.76 | 0.49 | 1.74 | 14.75 | 33.75 |
| | | % | **0.94** | 0.81 | 0.05 | 0.18 | 1.54 | 3.52 |
| | Vegetation | km² | 5.78 | **126.23** | 2.57 | 33.49 | 62.24 | 230.30 |
| | | % | 0.60 | **13.18** | 0.27 | 3.50 | 6.50 | 24.04 |
| | Water bodies | km² | 0.17 | 0.68 | **41.24** | 1.84 | 7.69 | 51.62 |
| | | % | 0.02 | 0.07 | **4.31** | 0.19 | 0.80 | 5.39 |
| | Agricultural land | km² | 3.27 | 32.64 | 1.68 | **148.97** | 53.99 | 240.55 |
| | | % | 0.34 | 3.41 | 0.18 | **15.55** | 5.64 | 25.11 |
| | Agricultural fallow land | km² | 12.79 | 45.15 | 9.02 | 69.63 | **265.18** | 401.76 |
| | | % | 1.34 | 4.71 | 0.94 | 7.27 | **27.68** | 41.94 |
| Total | | km² | 31.02 | 212.45 | 54.99 | 255.66 | 403.84 | **957.98** |
| | | % | 3.24 | 22.18 | 5.74 | 26.69 | 42.16 | **100.00** |

*The value of the bold mark indicates the unchanged area*

During the 2010–2017 period (table 5), the matrix of land use-land cover conversion was different from the prior period. In terms of the different land uses, the area that was transformed from a built-up area to agricultural fallow land and vegetation showed consistent growth. In 2010, there was a conversion of 24.74 km² or 2.58 % of the built-up areas. In comparison to the preceding time frame, there was a decline in the conversion of vegetation, water bodies, agricultural land, and agricultural fallow land. The majority of the agricultural fallow land was turned into agricultural land, vegetation, and populated (built-up) regions. This conversion, which had a surface area of 136.58 km², represented 14.26 % of agricultural fallow land in 2010. The vegetation was mostly transformed into built-up regions, agricultural fallow land, and agricultural land. This conversion covered an area of 104.07 $km^2$ or 22.75 % of the conversion's vegetation. The majority of agricultural land was turned into fallow land. There was a conversion of 91.58 $km^2$ of agricultural land into areas of agricultural fallow, vegetation, built-up, and water bodies. Water bodies were also the least converted to other classes.

**Table 6.** Report on the classification evaluation of the confined images from 1989 to 2017

| Class of LULC | 1989 | | 2001 | | 2010 | | 2017 | |
|---|---|---|---|---|---|---|---|---|
| | PA (%) | UA (%) | PA (%) | UA (%) | PA (%) | UA (%) | PA (%) | UA (%) |
| Built-up areas | 100 | 75 | 87.5 | 70 | 88.24 | 75 | 93.75 | 75 |
| Vegetation | 100 | 100 | 76.19 | 80 | 88.89 | 80 | 79.17 | 95 |
| Water bodies | 100 | 100 | 100 | 85 | 90.91 | 100 | 100 | 100 |
| Agricultural land | 69.23 | 90 | 76.19 | 80 | 86.36 | 95 | 82.61 | 95 |
| Agricultural fallow land | 87.5 | 70 | 78.26 | 90 | 90.48 | 95 | 88.24 | 75 |
| *Overall accuracy* | 87.50% | | 81.00% | | 89% | | 88% | |
| *Kappa Statistics* | 0.84 | | 0.76 | | 0.86 | | 0.85 | |

**PA= Producer accuracy and UA= User accuracy.*

*4.3. Accuracy Assessment for the LULC Classification*
To determine the classification's accuracy, all LULCs were contrasted with reference data [31]. For all categorised LULC maps, the total accuracy was 87.50 % in 1989, 81.00 % in 2001, 89.00 % in 2010, and 88.00 % in 2017. For all categorised images, the kappa statistics are 0.84 in 1989, 0.76 in 2001, 0.86 in 2010, and 0.85 in 2017. This is an indication of better performance in classifying images. Table 6 displays the accuracy assessment's findings.

*4.4. Landscape metrics analysis*
From 1989 to 2017, there were changes in the areas and the geographical distribution of land use, which had an impact on the spatial pattern and form of land use. The patch complexity (represented by NP, LSI, and MSI), aggregation (represented by LPI and CONTAG), and diversity were the four key categories into which the landscape metrics were fitted here to describe the various landscape pattern characteristics (represented by SHDI). Table 7 lists the descriptions of the employed landscape metrics. These indicators have changed values throughout time.

Changes in landscape-level metrics from 1989 to 2017 are shown in Figure 6. Over time, the total number of patches (NP) has grown in the landscape, reflecting a shift in land use and the resulting fragmentation [29]. Patch complexity peaked in 2001 after reaching its lowest point in 1989. Between 1989 and 2001, there was a significant shift in patch number; since then, it has remained mostly constant.

An indication of the patch's relative size is the LPI. LPI grew from 7.60 % to 9.02 % in 2001 as a result of land use fragmentation. LPI felt from 2001 to 2010, which may show that agricultural land predominated in 2001. A comparable outcome was discovered by Xi et al. (2018) in their investigation of spatiotemporal variation analysis of landscape pattern response to land use change from 1985 to 2015 in Xuzhou City, China [32].

**Table 7**. Statistics of landscape metrics used in this study

| Landscape metrics | | NP | LPI | LSI | MSI | CONTAG | SHDI |
|---|---|---|---|---|---|---|---|
| Range | | NP ≥ 1, without limit | 0 < LPI ≤ 100 | LSI ≥1, without limit. | MSI ≥ 1, without limit. | 0 < CONTAG ≤100 | SHDI ≥0, without limit |
| Unit | | None | Percentage | None. | None. | Percentage | None. |
| Year | 1989 | 33905 | 7.60 | 123.88 | 1.27 | 42.43 | 1.15 |
| | 2001 | 51555 | 9.02 | 139.24 | 1.23 | 34.65 | 1.29 |
| | 2010 | 38882 | 4.21 | 120.90 | 1.26 | 35.17 | 1.33 |
| | 2017 | 39145 | 4.57 | 121.41 | 1.26 | 35.17 | 1.33 |

A straightforward indicator of class aggregation or clumpiness is offered by LSI. This index also shows a similar pattern for temporal sequences. There was the highest LSI in 2001, decreasing to a maximum in 2010 and an increase in 2017. Between 1989 and 2001, these indicators had the greatest amount of change. Due to land use fragmentation, these indices grew in 2001, but since then, they have decreased [13].

The MSI compares a patch form's complexity to that of a standard shape (a square or nearly square) of the same size. From 1989 through 2001, MSI dropped, and then it rose. This index demonstrates how the complexity of the form grows as patch numbers rise.

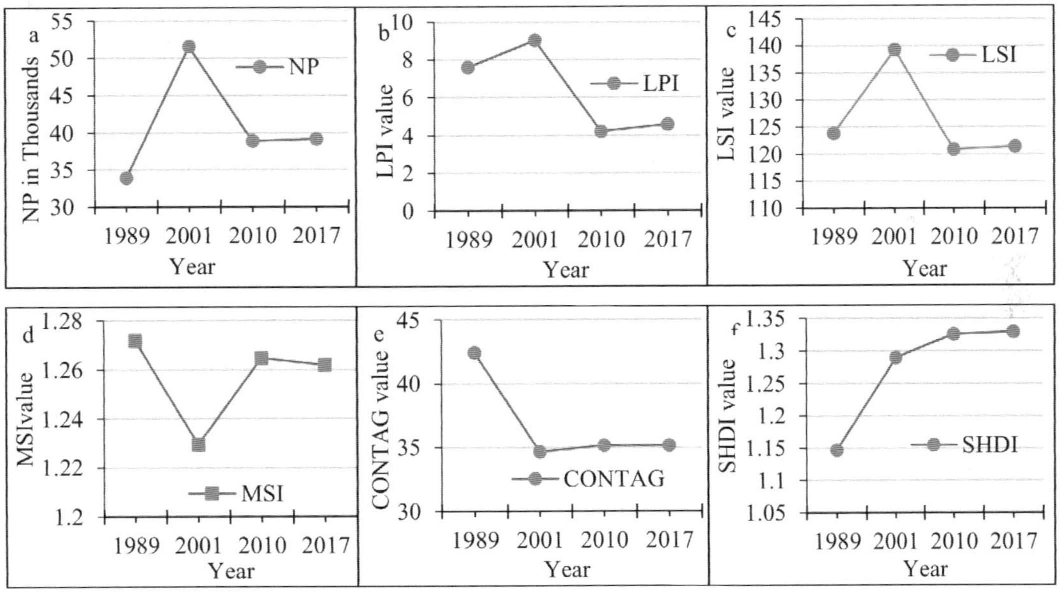


**Figure 7.** Line graphs showing the changes in landscape metrics from 1985 to 2015. (a) NP, (b) LPI, (c) LSI, (d) MSI, (e) CONTAG, and (f) SHDI.

The CONTAG only applies to landscape-level raster images, and it is based on raster "cell" adjacencies rather than "patch" adjacencies. CONTAG is inversely related to NP, LPI, and LSI. The CONTAG fell precipitously after 1989, then fluctuated slightly from 2001 to 2017. This diminished trend supports the notion that the landscape pattern was more diffusely dispersed and less continuous in 2001 [32].

Shannon's Diversity Index (SHDI) measures relative diversity. The diversity metrics showed rising variability of this terrain from 1989 to 2010, although they levelled in that year. This fact is clear from the way that farmland is divided up into smaller areas due to the regular expansion of urban land [32].

This study demonstrates that eutrophication in all lakes and small water bodies, sediment deposition, disposal of solid waste, and filling up of ponds/wetlands for building development cause the water bodies to steadily decrease. Agricultural land has declined due to the development of industries like the brick industry. Throughout this time, there have been varying levels of agricultural fallow. Between 1989 and 2017, the majority of the agricultural fallow land that had been transformed was turned into agricultural land, vegetation, and residential (built-up) regions. Throughout the research period, the vegetation gradually grew. Due to this rise, a significant size of areas of agricultural and agricultural fallow was converted into vegetation and created into various kinds of forest, or "Bagan." Das and Sarkar [24] conducted similar research on their use of the Markov model to forecast changes in LULC: An examination of the Bhagirathi-Hugli River at the catchment level.

A growing interest in the study of LULC change brought on by urbanisation has resulted in the publication of several associated research findings, notably for the extraction of urban areas based on indices [33, 34].

The high rate of population growth and the expansion of infrastructure over the period under review are the causes of the growth of the built-up area. In this research region (adjusting blocks), the population increased from 2047227 in 1991 to 2774123 in 2011 but the population growth rate dropped from 20.39 % between 1991 and 2001 to 12.56 % and 2011 [35, 36].

## 5. Conclusions

The study evaluated landscape metrics based on remote sensing data properties such as spatial and temporal resolution differences and dominant land cover classes. Landscape metrics successfully detected changes in land cover in the study area. The changing pattern is consistent with results previously reported in the region using other methods [24]. These measures identified changes over the study periods in provinces with different types of land use. The method is very simple and efficient for choosing the best landscape parameters to analyse the model change in landscape ecology. This method may be applied in any environment and under all circumstances, and it would be logistically challenging to process hundreds of metrics if they were otherwise chosen from a suitable collection of measurements.

**Acknowledgments**

The authors like to acknowledge USGS for providing the required resources for the study and also like to thank the Department of Geography, Adamas University for providing the necessary facility to conduct the study. Additionally, we would like to thank the editor and anonymous reviewers for their insights and recommendations for strengthening the work.

## References

[1] Dewan, A. M., Yamaguchi Y., 2009. Using remote sensing and GIS to detect and monitor land use and land cover change in Dhaka Metropolitan of Bangladesh during 1960-2005. Environ. Monit. Assess. 150: 237-249. http://dx.doi.org/10.1007/s10661-008-0226-5.

[2] Riebsame, W.E., Meyer, W.B., Turner, B.L. II., 1994. Modelling Land-use and Cover as Part of Global Environmental Change. Climate Change. Vol. 28, 45-46.

[3] Cheruto, M.C., Kauti, M.K., Kisangau, P.D., Kariuki, P., 2016. Assessment of Land Use and Land Cover Change Using GIS and Remote Sensing Techniques: A Case Study of Makueni County, Kenya. J Remote Sensing & GIS, 5: 175. https://doi:10.4175/2469-4134.1000175.

[4] Ayivor, J.S., Gordon, C., 2012. Impact of land use on river systems in Ghana. West African Journal of Applied Ecology, 20 (3): 83-95.

[5] William, F., Lora, R., Kenneth, D.A., Timothy, B.M., Timothy, G.R., Smith, J.L.R., Christian, G.R., 2001. Land Use Change and Effects on Water Quality and Ecosystem Health in the Lake Tahoe Basin, Nevada, and California. USGS Science for a Changing World, 01-29.

[6] Gregory, K.J., 1995. Human activity and palaeohydrology. Chichester: Willy, 151-172.

[7] Cushman, S.A., McGarigal, K., Neel, M.C., 2008. Parsimony in landscape metrics: Strength, universality, and consistency. Ecological Indicators 8: 691-703. http://dx.doi.org/10.1016/j.ecolind.2007.12.002.

[8] Yin, D., Chen, X., Yan, L., Huang, Z., 2007. The research and realization of the land-use change forecasting model in development zones based on RS and GIS. IEEE Proc, IGARSS, 3429–3432.

[9] Sylla, L., Xiong, D., Zhang, H.Y., Bangoura, S.T., 2012. A GIS technology and method to assess environmental problems from land use/cover changes: Conakry, Coyah, and Dubreka region case study, The Egyptian Journal of Remote Sensing and Space Sciences, 15:31–38.

[10] Kumar, S., Radhakrishnan, N., Mathew, S., 2014. Land use change modelingtation using a Markov model and remote sensing, Geomatics, Natural Hazards and Risk, 5(2), 145-156, DOI:10.1080/19475705.2013.795502.

[11] Zhang, K., Yu, Z., Li, Y., Zhou, W., Zhang, D., 2007. "Land use change and land degradation in China from 1991 to 2001," Land Degradation and Development, vol. 18, no. 2, 209–219.

[12] Liu, J., Kuang, W., Zhang, Z., 2014. Spatiotemporal characteristics, patterns, and causes of land use changes in China since the late 1980s, Journal of Geographical Sciences, 24(2), 19 5-210. DOI:10.1007/s11442-014-1082-6.

[13] Li, Y. and Liu, G., 2017. Characterizing Spatiotemporal Pattern of Land Use Change and Its Driving Force Based on GIS and Landscape Analysis Techniques in Tianjin during 2000–2015. Sustainability, 9, 894. https://doi.org/10.3390/su9060894.

[14] Alfiky, A., Kaule, G., Salheen, M., 2010. Agricultural fragmentation of the Nile Delta; a modelling approach to measuring agricultural land deterioration in Egyptian Nile Delta. Procedia Environmental Sciences 14, 79 – 97.

[15] Kadigi, R.M.J., Kashaigili, J.J., Sirima, A., Kamau, F., Sikira, A., Mbungu, W., 2017. Land fragmentation, agricultural productivity, and implications for agricultural investments in the Southern Agricultural Growth Corridor of Tanzania (SAGCOT) region, Tanzania, J. Dev. Agric. Econ, 9(2), 26-36.

[16] Majumder, D., 1978. West Bengal District Gazetteers, Nadia. Government of West Bengal, 5-16.

[17] Biswas, K.R., 2001. Rivers of Bengal. Vol. I, Government of West Bengal, pp. xviii, xxix, 87, plate-18 and 19.

[18] Bagchi, K., 1978. Diagnostic Survey of Deltaic West Bengal, A Research and Development Project. Department of Geography, Calcutta University, 17.

[19] Bandyopadhyay, S., Ghosh, K., Saha, S., Chakravorti, S., De, S.K., 2013. Status and impact of brickfields on the river Haora, West Tripura. Transactions 35 (2), 275-286.

[20] District Census Hand Book, Nadia (2001): Directorate of Census Operations, West Bengal.

[21] Reis, S., (2008): Analyzing Land Use/Land Cover Changes Using Remote Sensing and GIS in Rize, North-East Turkey. Sensors 8, pp. 6188-6202.

[22] Samal, D.R., and Gedam, S.S., 2015. Monitoring land use changes associated with urbanization: An object-based image analysis approach. European Journal of Remote Sensing, vol. 48, pp. 85-99. doi: 10.5721/EuJRS20154806.

[23] Wang, J. F., 2010. Spatial Data Analysis Tutorial; Science Press: Beijing, China.

[24] Das, S., Sarkar, R., 2019. Predicting the Land use and Land cover change using Markov model: A catchment level analysis of the Bhagirathi- Hugli River. Spatial Information Research, 27 (4): 439-452. https://doi.org/10.1007/s41324-019-00251-7.

[25] Ghosh, U., Khan, D.K., 2014. Spatial pattern analysis of two landscapes in North-western parts of Orissa, India. American Journal of Mathematical Analysis, 2, 1-3. doi: 10.12691/ajma-2-1-1.1.

[26] Forman, R. T. T., and Godron, M., 1986. Landscape Ecology. John Wiley & Sons, New York. 619 pp.
[27] McGarigal, K., 2015. Fragstats help. University of Massachusetts, Amherst. http://www.umass.edu/landeco/research/fragstats/documents/fragstats.help.4.2.pdf).
[28] McGarigal, K., and Marks, B.J., 1995. FRAGSTATS: Spatial pattern analysis program for quantifying landscape structure. Gen. Tech. Rep. PNW-GTR-351. Portland, OR: USDA, Forest Service, Pacific Northwest Research Station.
[29] Botequilha Leitão, A., Miller, J., Ahern, J., McGarigal, K., 2006. Measuring Landscapes: A Planner's Handbook. Island Press, Washington DC.
[30] McGarigal, K. and Marks, B.J., 1995. Spatial Pattern Analysis Program for Quantifying Landscape Structure, Gen. Tech. Rep. PNW-GTR-351; US Department of Agriculture, Forest Service, Pacific Northwest Research Station: Portland, OR, USA.
[31] Sundarakumar K, Harika M, Begum A, Yaini S and Balakrishna K (2012): Land use and land cover change detection and urban sprawl analysis of Vijayawada city using multitemporal landsat data. International Journal of Engineering Science and Technology, vol. 4, pp. 170-178.
[32] Xi, Y., Thinh, N. X., Li, C., 2018. Spatio-temporal variation analysis of landscape pattern response to land use change from 1985 to 2015 in Xuzhou City, China. Sustainability, 10 (11), 4287. https://doi.org/10.3390/su10114287.
[33] Zhang, J., Li, P. J., Wang, J. F., 2014. Urban built-up area extraction from Landsat TM/ETM plus images using spectral information and multivariate texture. Remote Sens. 6: 7339–7359.
[34] Zhang, Q. L., Pandey, B., Seto, K. C., 2016. A Robust Method to generate a consistent time series from DMSP/OLS Nighttime light data. IEEE Trans. Geosci. Remote Sens., 54: 5821–5831.
[35] District Census Hand Book, Barddhaman. (2011). Directorate of Census Operations, West Bengal.
[36] District Census Hand Book, Nadia. (2011). Directorate of Census Operations, West Bengal.

# GIS-based Valuation of Ecosystem Services in the Lower Delta Plain of West Bengal with Special Reference to Indian Sundarbans

**S Mitra[1], R Roy Chowdhury[1], A Saha[2], K Mukherjee[1], T Bhadra[1]***

[1] Department of Geography, Adamas University, West Bengal, India
[2] Department of Biotechnology, Adamas University, West Bengal, India

*Corresponding Author Email: tuhinbhadra.au@gmail.com

**Abstract.** The Indian Sundarbans partly hosts the world's largest mangrove forest and support the livelihood of almost five million people. The Holocene delta of Indian Sundarbans was formed by the river Ganges. The decay of the distributaries of the river Ganges in deltaic West Bengal began with the eastward avulsion of the river Ganges. The rivers of Sundarbans now stand disconnected from their sources and therefore do not get a significant amount of freshwater from upstreams. The people of Sundarbans depend on the services of the Sundarbans ecosystem which is directly linked to the upstream freshwater flow, rainfall and groundwater. The scarcity of freshwater during dry seasons, high salinity of surface water and soil, siltation and drainage congestion are major constraints on livelihood security in this area. The biodiversity of Sundarbans is also threatened by water stress. To ensure the sustainable conservation of the rich biodiversity and well-being of the people of Sundarbans, improving ecosystem services in the region is extremely essential. From this perspective, the present study aims to identify major flow-dependent ecosystem services in the region and estimate the value of those ecosystem services using geoinformatics. The land use land cover (LULC) maps have been prepared for the years 1990, 2000, 2010 and 2020 from Landsat images for the valuation of major ecosystem services. Accuracy assessment has been performed using high-resolution satellite images and field data. Change detection analysis has been done to understand the changing land use pattern in the study area. The LULC-based valuation of the ecosystem services has been carried out for the years 1990, 2000, 2010 and 2020. The area under each LULC class has been multiplied by the global Ecosystem Service Value (ESV) of that class to calculate the total ESV of the specific LULC class. The study reveals that the ESVs are the highest for freshwater bodies including rivers, and estuaries followed by forests. With the decreasing upstream freshwater flow and deteriorating mangrove health, the services of flow-dependent ecosystems have changed drastically.

**Key Words:** Ecosystem Service Valuation, Geoinformatics, Land-use Change, Mangrove forest, River discontinuity, Sundarbans.

## 1. Introduction

The Millennium Ecosystem Assessment report (2005), defines ecosystem service as the benefits that people get from ecosystems. Ecosystems not only offer an abundance of products and services but also perform a crucial role in the upbringing and flourishing of the prosperity of the people [1]. Ecosystems are one of the most important elements for survival on earth and are also very important for the continuance and sustenance of human beings. An ecosystem has a direct impact on human life and provides us with different types of environmental services like hydrological services and climatic services and different types of cultural services like cultural values, recreation etc. There is also some indirect impact of the ecosystem on human life like pollination, soil formation etc. which is known as supporting services. A well-developed ecosystem provides us with food and water which are essential for our life [2]. Ecosystem services have not received a proper place in the economic market as compared to economic services, therefore received very little attention in this regard [3].

The ecosystem services are categorised into four classes. These are (i) Provisioning services, the products (i.e. food, water, fibre, fuel) which we directly receive from ecosystems, (ii) Regulating services, the profits which we receive through the controlling power of the ecosystem, like erosion and flood control, water purification, climate regulation etc. (iii) Cultural services, the benefits that people receive as spiritual enhancement, recreation etc. from the ecosystem, (iv) Supporting services, which supports other ecosystems and their services. Primary production, soil formation, provisioning of habitat etc. are examples of supporting services. In the present study, land use land cover (LULC) maps have been used to assess the ESVs.

A global fundamental connection between human and natural systems can be derived from the LULC change map in social and physical prospects. After the beginning of large-scale changes in LULC, socioeconomic advancement has been significantly impacted by the influence of anthropogenic functions and global climate change. These changes address the chief environmental impacts which extended to the reduction of ecosystem services on a global scale. The interaction between the human body and physical properties, as well as the climate and ecology of the planet, are all directly impacted by LULC change, making it one of the key components of global environmental change. On the other hand, LULC change has also been acknowledged as the maximum extensive driver of ecosystem degradation. Among the several factors, human activity has been the main driver of land acquisition for several years. On the other hand, the LULC change has led to great economic development and the welfare of the people including the cost of ecosystem services given by the land.

The Indian Sundarbans is located in the lower delta plain of West Bengal. It partly hosts the world's largest mangrove forest and provides a livelihood to about five million people. Due to the eastward avulsion of the river Ganges, the distributaries in the western part of the delta have decayed with time. Rivers in the Indian Sundarbans area are now disconnected from their upstream sources and do not receive fresh water from upstreams. The people of the Indian Sundarbans depend on the services of the deltaic ecosystem which is directly linked to the upstream freshwater flow, rainfall and groundwater. The Indian Sundarbans is encompassed by various rivers and estuaries that play a major role in different environmental services. LULC has undergone several changes in the region and the remote sensing analysis provides the synoptic fact of the LULC at the various spatio-temporal scales [4].

Remote sensing techniques have been broadly used to measure and map the features of the ecosystem. It is also used for demarcating the activities of ecosystem processes in combination with the available devices and data [5-7]. The main advantages of remote sensing are spatially uninterrupted and intensive monitoring of the earth's surface. The earth observation satellites provide the remotely sensed data of different spatial and temporal resolutions which are used for mapping and modelling the LUCL as well as the ecosystem services of a region [8-14].

There has been a lot of effort to categorize the services of the ecosystem [3, 15]. Meaningful and compatible methods of valuing ecosystem services based on realities have been provided by different existing literature [16-20]. Stated Preference, Revealed Preference and Cost Based Methods are the main three categories of non-GIS-based ecosystem valuation methods. In the case of the Stated Preference Method, a well-structured questionnaire is used to stimulate individuals' choices for any change in given natural resources. The Stated Preference-based Contingent Valuation Method expresses public choices to pay for a transfer in the quantity as well as the quality of natural products or services offered in a speculative market [21]. Choice Modelling is another type of Stated Preference procedure which expresses public choices by providing a series of alternatives and asking them to choose their desired option [22]. The Hedonic Pricing method adopts objects of interest which are perfectly traded through the market demand. For instance, as natural beauty and good environmental conditions are most often implicitly sold, their value can be determined by comparing the costs of two identical homes, one of which is built in an area with a magnificent natural landscape and environmental conditions while the other is built in an area with zero natural beauty [23]. Apart from this, there are many non-GIS-based methods to evaluate ecosystem services. However, in this paper, the GIS-based method of valuing ecosystem services has been used as it is cost-effective and requires less amount of resources and time.

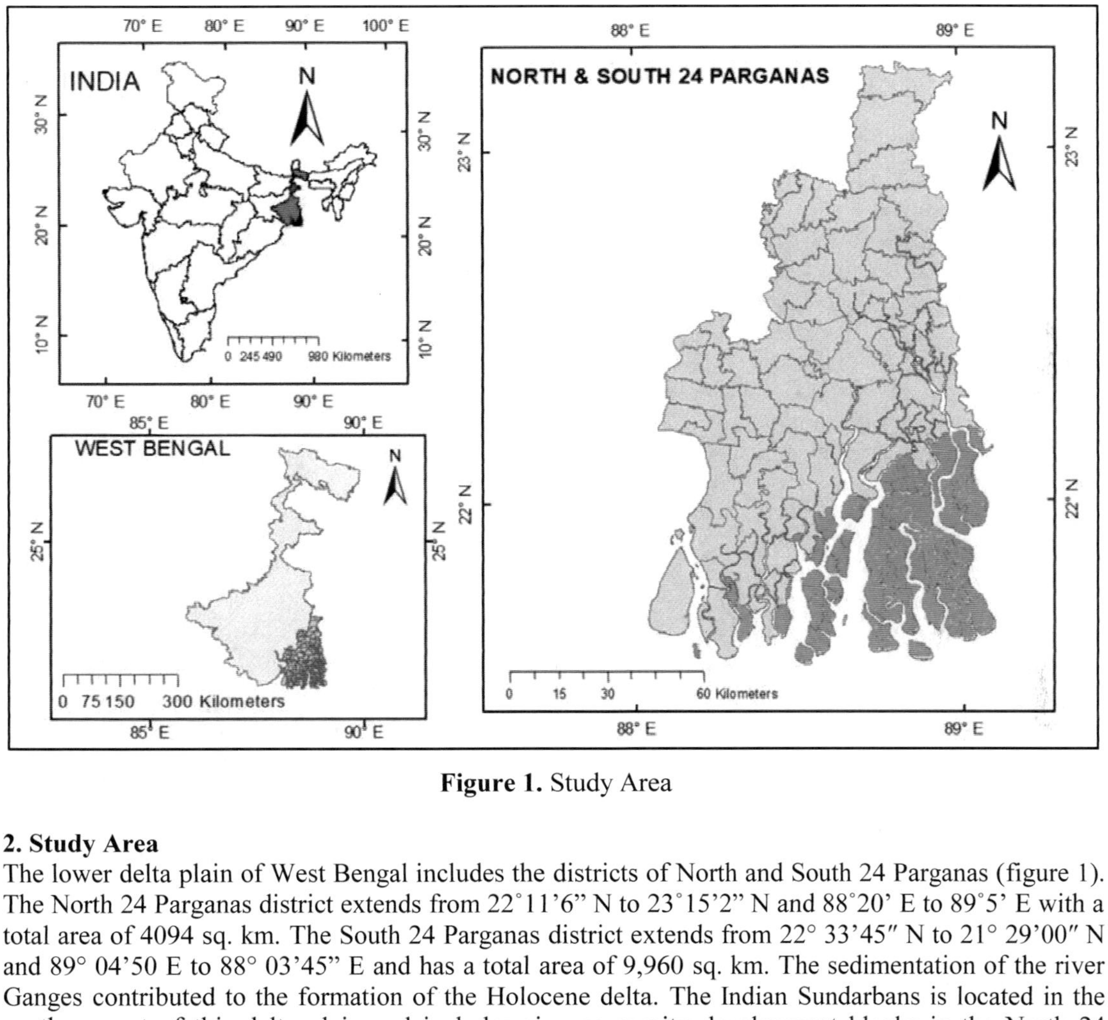


**Figure 1.** Study Area

## 2. Study Area

The lower delta plain of West Bengal includes the districts of North and South 24 Parganas (figure 1). The North 24 Parganas district extends from 22°11'6" N to 23°15'2" N and 88°20' E to 89°5' E with a total area of 4094 sq. km. The South 24 Parganas district extends from 22° 33'45" N to 21° 29'00" N and 89° 04'50 E to 88° 03'45" E and has a total area of 9,960 sq. km. The sedimentation of the river Ganges contributed to the formation of the Holocene delta. The Indian Sundarbans is located in the southern part of this delta plain and includes six community development blocks in the North 24 Parganas district and thirteen community development blocks in the South 24 Parganas district. The total area of the Indian Sundarbans is 9630 sq. km which is subdivided into a block region (i.e. 5367 sq. km) and a reserve forest area (i.e. 4263 sq. km.). The forest is further segmented into a core area of 1700 sq. km which is protected from human intervention and a buffer area of 2563 sq. km in which resources can be used sustainably.

There are four major flow-dependent ecosystems in the study area. These are the mangrove ecosystem, the riverine ecosystem which further may be subdivided into freshwater and saline water tidal blocks, the estuarine ecosystem and the floodplain-freshwater wetland ecosystem. Each of these ecosystems performs a variety of ecosystem services like drainage, navigation, sediment flushing, shoreline stabilization, prevention of salt intrusion, storm protection, recycling of organic nutrients, carbon sequestration, food and other resources supply etc. The provisioning of nursing-breeding-feeding grounds for aquatic animals is a major service of these ecosystems. The ecosystem services increase and decrease depending on flow availability from upstream. The main problems of the region are that most of the rivers of the lower delta plain have lost their link with their sources. The abandoned river channels are being converted into aquaculture farms and agricultural lands. The disconnected rivers are unable to provide ecosystem services due to the unavailability of freshwater flow from upstreams. In this perspective, the valuation of existing ecosystem services in the region is highly essential for their sustainable management. The objective of the study is to monitor LULC change dynamics in the lower delta plain of West Bengal and to estimate the Ecosystem Service Value (ESV) from the LULC information.

## 3. Methodology

To assess the ESV from LULC information, the LULC maps have been prepared through the maximum likelihood method of supervised classification (figure 2). Landsat TM 5 images have been used to prepare LULC maps of the study area for 1990 and 2000. Whereas, the Landsat 7 ETM image and Landsat 8 OLI/ TIRS image have been used for the years 2010 and 2020 respectively. An accuracy assessment has been done to estimate classification accuracy. The study finds that, for the year 1990, the Overall Accuracy (OA) is 93% and the Kappa Coefficient (KC) value is 0.89. For the year 2000, the OA is 94% and the KC value is 0.93. Whereas for the years 2010 and 2020, the OA values are 94% and 90% respectively and the KC values are 0.93 and 0.88 respectively. The accuracy assessment results indicate that the classification is satisfactory. The change detection analysis has been performed on the classified images and the area under each LULC class has been calculated using the GIS platform.

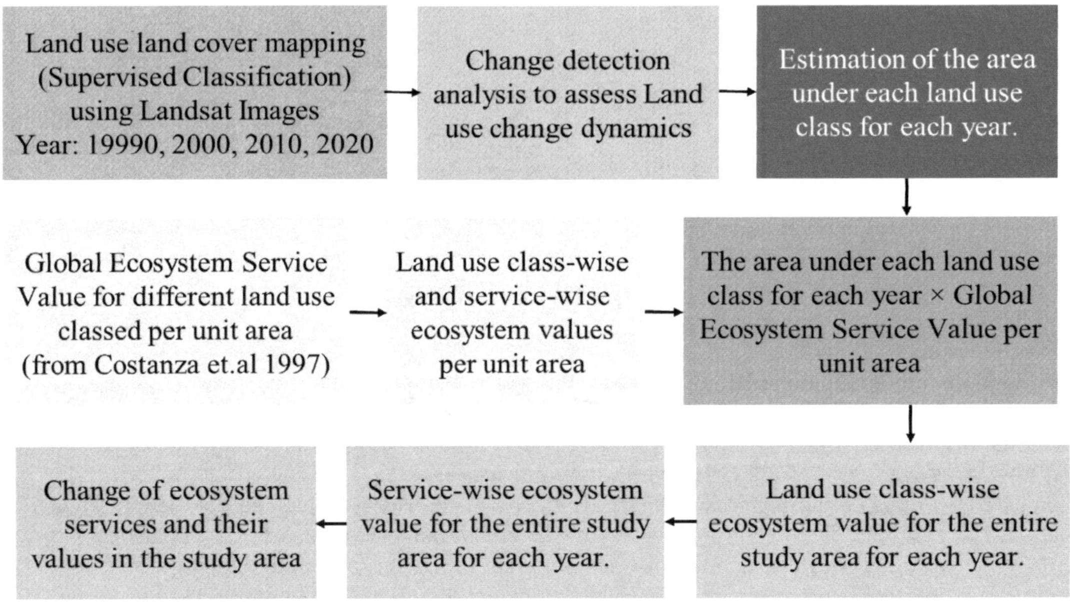


**Figure 2.** Methodology of GIS-based Valuation of Ecosystem Services

Land use class-wise ecosystem service and their values (per unit area) have been taken from Costanza et.al. [3]. The per unit area value has been multiplied by the total area under each class. Change in ESVs has also been estimated under this study using LULC change information. According to Costanza et al. [3], the average global annual ESVs of rivers are as follows, water regulation: 5445 US$ per hectare, water supply: 2117 US$ per hectare, waste management: 665 US$ per hectare, food production: 41 US$ per hectare, recreation: 230 US$ per hectare. The average global annual ESVs of croplands are as follows, pollination: 14 US$ per hectare, biological control: 24 US$ per hectare, and food production: 54 US$ per hectare [3]. Whereas, the average global annual ESVs of forest or vegetation are, climate regulation: 141 US$ per hectare, disturbance regulation and water regulation: 2 US$ per hectare, water supply: 3 US$ per hectare, erosion control: 96 US$ per hectare, soil formation: 10 US$ per hectare, nutrient cycling: 361 US$ per hectare, waste management: 87 US$ per hectare, biological control: 2 US$ per hectare, food production: 43 US$ per hectare, raw materials: 138 US$ per hectare, genetic resources: 16 US$ per hectare, recreation: 66 US$ per hectare, cultural value: 2 US$ per hectare [3]. The average global annual service values of estuaries have also been estimated by Costanza et al. [3]. These values are as follows, disturbance regulation: 567 US$ per hectare, nutrient cycling: 21100 US$ per hectare, biological control: 78 US$ per hectare, habitat provision: 131 US$ per hectare, food production: 521 US$ per hectare, raw material supply: 5 US$ per hectare, recreation: 381 US$ per hectare, cultural values: 29 US$ per hectare [3]. Costanza et al. [3] have also estimated the average global annual ESVs of mangrove forests. These values are as follows, disturbance regulation: 1839 US$ per hectare, waste treatment: 6696 US$ per hectare, habitat provision: 169 US$ per hectare, food production: 466 US$ per hectare, raw material: 162 US$ per hectare, recreation: 658 US$ per hectare [3]. These ecosystem values (in US$) have been multiplied by the total area of each LULC class (i.e. rivers, croplands, vegetation, estuaries and mangroves) to estimate the total ESVs of the study area (figure 2). The ecosystem-specific and service-specific values of the major ecosystems have finally been presented in millions of US$ in the present study. The valuation has been done separately for the upper catchment area of the Sundarbans mangrove forest (which includes the districts of North and South 24 Parganas) and the mangrove forest region of Sundarbans to focus on the world's largest mangrove ecosystem.

## 4. Results and Discussion

The LULC maps, prepared in the geoinformatics platform, show that there are nine major LULC classes in the study area. These are water, agricultural land, fallow land, vegetation, settlement, estuary, wasteland, dense forest and degraded forest. According to the LULC map of 1990 (figure 3a), the area of major LULC classes are as follows, water body: 113200 hectares, agricultural land: 261800 hectares, fallow land: 308500 hectares, vegetation: 113300 hectares, settlement: 86100 hectares, estuary: 144500 hectares, wasteland: 3400 hectares, degraded Forest: 6000 hectares and dense forest: 204200 hectares. Some areal changes in the major LULC classes have been identified in the year 2000. The area of the LULC classes, estimated from the LULC map of 2000 (figure 3b) are as follows, water body: 103100 hectares, agricultural land: 340600 hectares, fallow land: 256100 hectares, vegetation: 110000 hectares, settlement: 101100 hectares, estuary: 145700 hectares, wasteland: 4400 hectares, degraded Forest: 10100 hectares and dense Forest: 197900 hectares. According to the LULC map of 2010 (figure 3c), the area of major LULC classes are as follows, water body: 95700 hectares, agricultural land: 197900 hectares, fallow land: 398900 hectares, vegetation: 106400 hectares, settlement: 122600 hectares, estuary: 149200 hectares, wasteland: 4300 hectares, degraded Forest: 21800 hectares and dense forest: 182900 hectares. A huge change in LULC has been observed in the year 2020. The area of major land use classes, estimated from the LULC map of 2020 (figure 3d) are as follows, water body: 90300 hectares, agricultural land: 260400 hectares, fallow land: 353000 hectares, vegetation: 101000 hectares, settlement: 134600 hectares, estuary: 141900 hectares, wasteland: 11900 hectares, degraded Forest: 24700 hectares and dense forest: 179500 hectares.

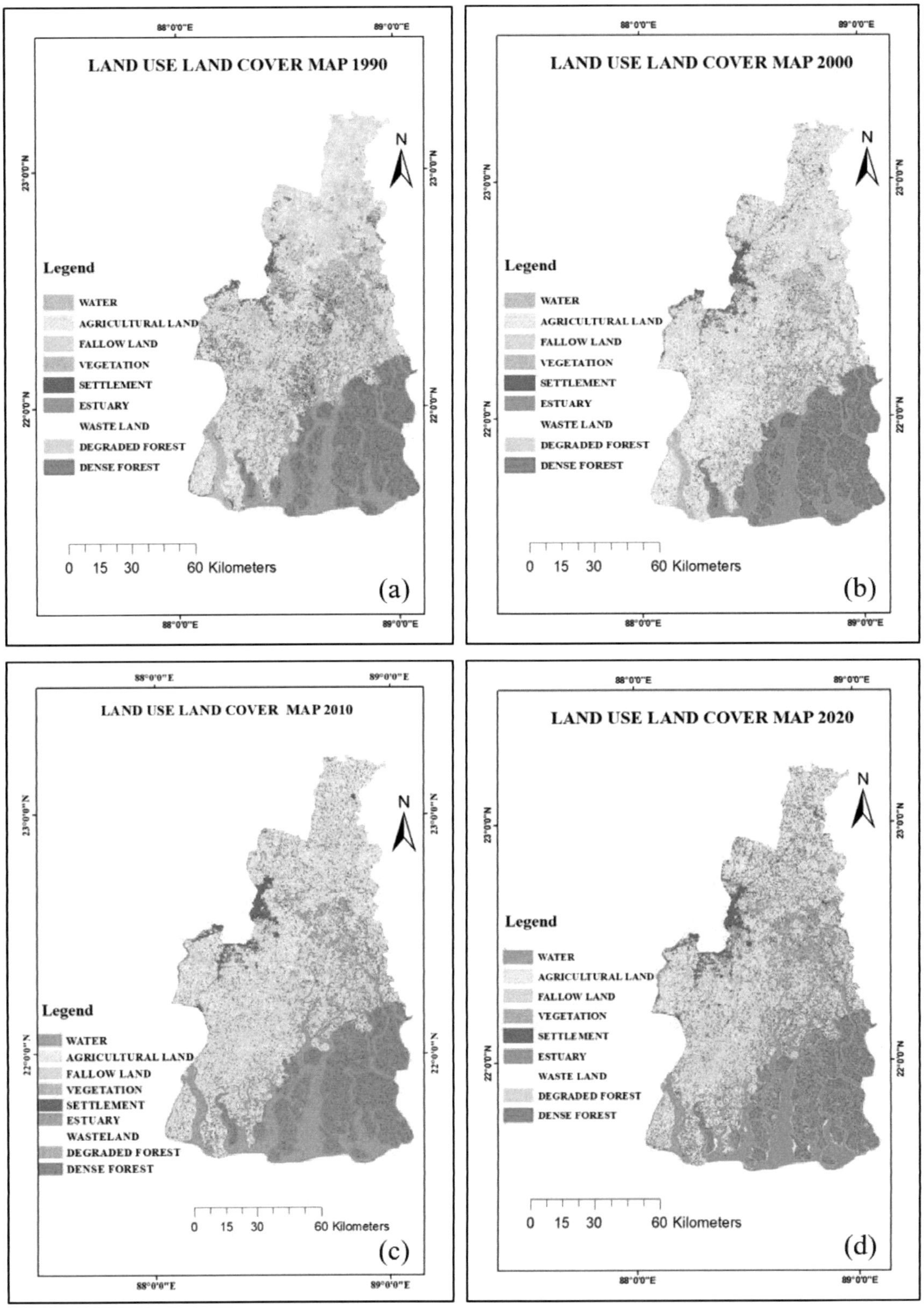


**Figure 3.** LULC map of the years 1990 (a), 2000 (b), 2010 (c) and 2020 (d)

A detailed change detection analysis has been performed to understand the temporal change of LULC in the study area. The study shows that almost 174000 hectares of agricultural land remained agricultural land during the time frame from 1990 to 2000 (figure 4a). Whereas almost 54806 hectares of the agricultural field were transformed into fallow land, 42312 hectares of the agricultural field were transformed into the settlement, 1652 hectares of the agricultural field were transformed into vegetation and around 2117 hectares of agricultural field were transformed into waterbody during the same time frame (figure 4a). The change of fallow land into other land use classes from 1990 to 2000, was as follows, agricultural land: 143700 hectares, fallow land (no change): 134317 hectares, settlement: 23092 hectares, vegetation: 1273 hectares, water body: 1170 hectares. Negligible changes have been observed from the land use class settlement, degraded forest, water body, estuary and wasteland to other land use classes. The transformation of vegetation land into other land use classes from 1990 to 2000, was as follows, agricultural land: 35214 hectares, fallow land: 9794 hectares, settlement: 20314 hectares, vegetation (no change): 97547 hectares and water body: 485 hectares. The study shows that almost 194319 hectares of dense forest remained dense forest during the time frame from 1990 to 2000 (figure 4a). Whereas almost 5756 hectares of dense forest became degraded forest, 3187 hectares of dense forest became an estuary and 1438 hectares of dense forest became wasteland during the time frame (figure 4a).

The temporal change of LULC classes has also been estimated during the time frame from 2000 to 2010 (figure 4b). The study shows that almost 204170 hectares of agricultural land remained agricultural land during the time frame from 2000 to 2010 (figure 4b). Whereas, almost 73752 hectares, 54925 hectares, 1335 hectares and 1485 hectares of agricultural land were converted into fallow land, settlement, vegetation and waterbody respectively. The change of fallow land into other land use classes from 2000 to 2010, was as follows, agricultural land: 56345 hectares, fallow land (no change): 187611 hectares, settlement: 12578 hectares, vegetation: 966 hectares, water body: 537 hectares. Around 24724 hectares of vegetation area were converted into agricultural land, 1001 hectares of vegetation area were converted into fallow land and 4888 hectares and 326 hectares of vegetation area were converted into settlement and water bodies respectively. Whereas, almost 83263 hectares of vegetation area remained unchanged during this time frame. Negligible changes have been observed from the land use class settlement, degraded forest, water body, estuary and wasteland to other land use classes. The study shows that almost 182371 hectares of dense forest remained dense forest during the time frame from 2000 to 2010 (figure 4b). Whereas almost 13075 hectares of dense forest became degraded forest, 2481 hectares of dense forest became an estuary and 779 hectares of dense forest became wasteland during the time frame (figure 4b).

The LULC changes between the year 2010 and 2020 (figure 4c) shows major changes in agricultural land, fallow land, vegetation and dense forest. Whereas the changes were negligible for the land use class settlement, degraded forest, water body, estuary and wasteland to other land use classes. Almost 150435 hectares of agricultural land remained agricultural land during the time frame from 2010 to 2020 (figure 4c). Whereas almost 86001 hectares of the agricultural field were transformed into fallow land, 40081 hectares of the agricultural field were transformed into the settlement, 1135 hectares of the agricultural field were transformed into vegetation and around 1197 hectares of the agricultural field were transformed into waterbody during the same time frame (figure 4c). The change of fallow land into other land use classes from 2010 to 2020, was as follows, agricultural land: 82141 hectares, fallow land (no change): 157107 hectares, settlement: 35085 hectares, vegetation: 1187 hectares, water body: 563 hectares. The change of vegetation land into other land use classes from 2010 to 2020, was as follows, agricultural land: 24571 hectares, fallow land: 5142 hectares, settlement: 20483 hectares, vegetation (no change): 82353 hectares and water body: 318 hectares. The study shows that almost 174181 hectares of dense forest remained dense forest during the time frame from 2010 to 2020 (figure 4c). Whereas almost 8097 hectares of dense forest became degraded forest, 982 hectares of dense forest became estuary and 993 hectares of dense forest became wasteland during the time frame (figure 4c).

IOP Conf. Series: Earth and Environmental Science 1164 (2023) 012004 doi:10.1088/1755-1315/1164/1/012004

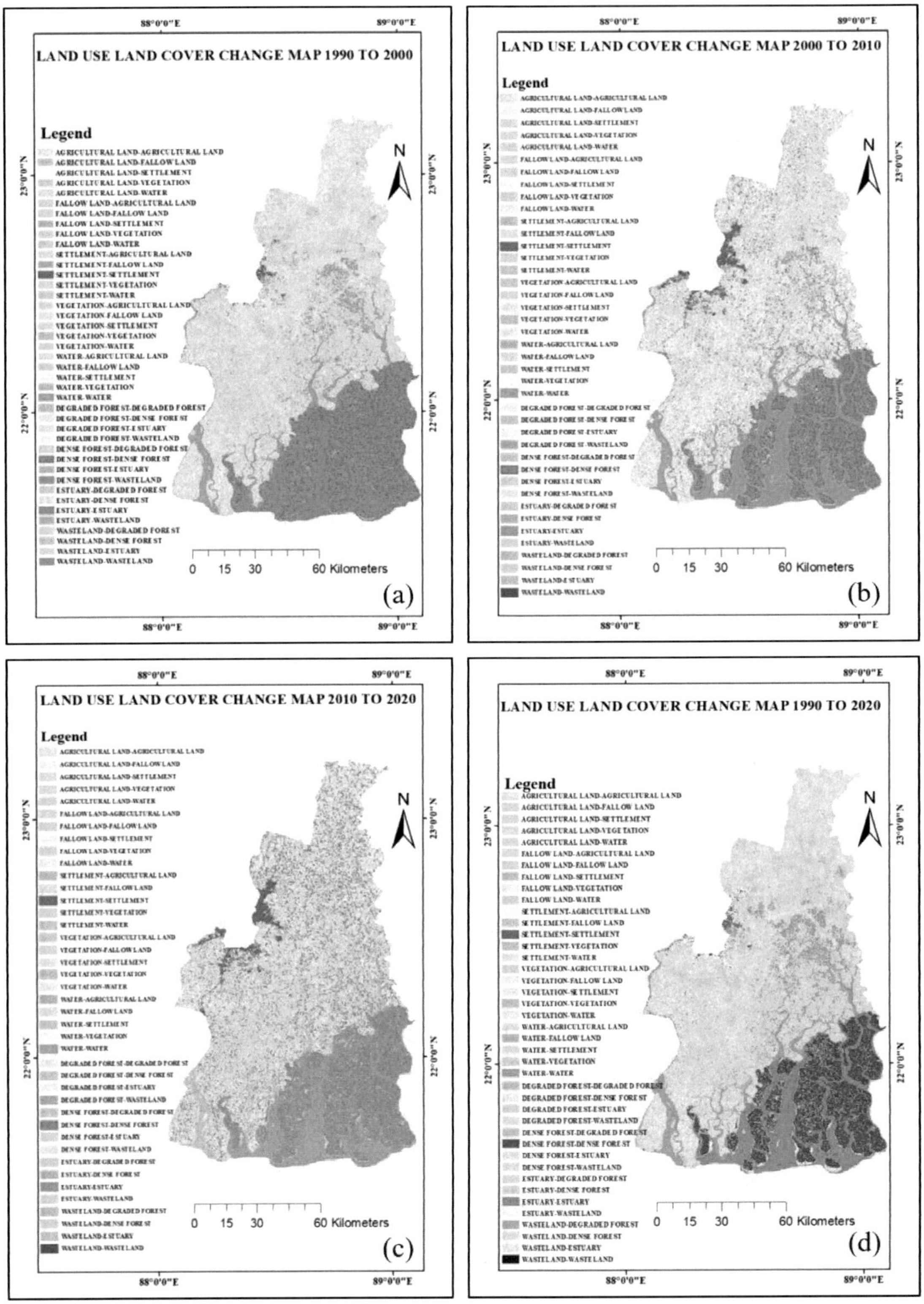


**Figure 4.** LULC change map 1990-2000 (a), 2000-2010 (b), 2010-2020 (c) and 1990-2020 (d)

The temporal change of LULC area from 1990 to 2020 (figure 4d) has been estimated under the study. During these thirty years, almost 4653 hectares, 39033 hectares, 1995 hectares and 2976 hectares of agricultural land were converted into fallow land, settlement, vegetation and waterbody respectively. The change of fallow land into other land use classes from 1990 to 2020, was as follows, agricultural land: 2477 hectares, settlement: 55664 hectares, vegetation: 266 hectares, and water body: 591 hectares. Around 34608 hectares of vegetation area were converted into agricultural land, 21005 hectares of vegetation area were converted into fallow land and 32427 hectares and 898 hectares of vegetation area were converted into settlement and water bodies respectively. Negligible changes have been observed from the land use class settlement, degraded forest, water body, estuary and wasteland to other land use classes. The study shows that almost 17694 hectares of dense forest became degraded forest, 7435 hectares of dense forest became estuary and 3393s hectares of dense forest became wasteland during the time frame (figure 4d). The study reveals that, over the last thirty years, the area of the water body and vegetation in the upper catchment region of Sundarbans have decreased, whereas, the area of agricultural land and settlement has increased (figure 5a). In the mangrove forest region, a decreasing trend has been observed in the dense forest area, whereas the wasteland (i.e. saline blank) and degraded forest area have significantly increased (figure 5b).

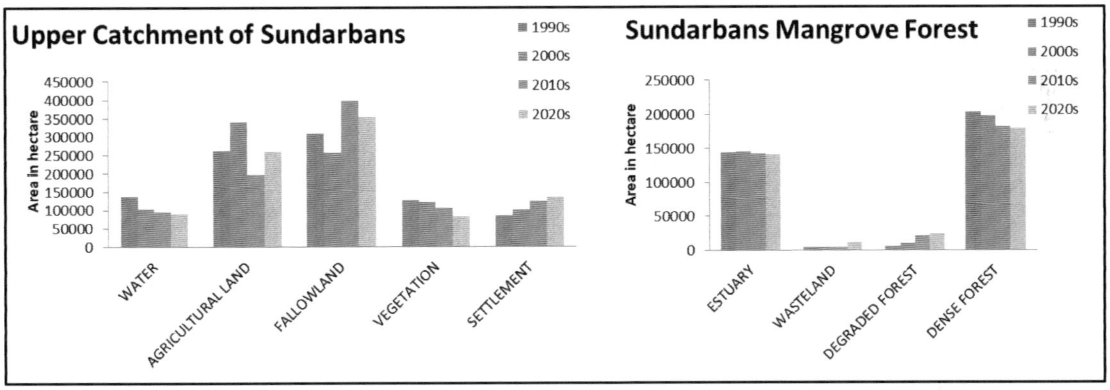


**Figure 5.** Temporal change of LULC area from 1990 to 2020 (a) the upper catchment area of Sundarbans and (b) the mangrove forest region.

The ESVs have been estimated for the years 1990, 2000, 2010 and 2020 in the study (figure 6 & 7). The study shows that, in the upper catchment area of the Sundarbans mangrove forest, the ESV of water is maximum. The ESV of water was 962 million US$ in the year 1990, 876 million US$ in the year 2000, 813 million US$ in the year 2010 and 767 million US$ in the year 2020 (figure 6). A decreasing trend in the ESVs of water has been observed in the present study. The ESVs of vegetation in the region is the second highest. It was 109 million US$ in the year 1990, 107 million US$ in the year 2000, 103 million US$ in the year 2010 and 98 million US$ in the year 2020 (figure 6). The decreasing trend has also been observed in the ESVs of vegetation. For agricultural land, the ecosystem value was 24 million US$ in 1990, 31 million US$ in 2000, 18 million US$ in 2010 and 24 million US$ in 2020 (figure 6). The ESV of fallow land was 28 million US$ in 1990, 33 million US$ in 2000, 37 million US$ in 2010 and 33 million US$ in 2020 (figure 6).

In the Sundarbans mangrove forest region, the maximum ecosystem services are provided by the estuary and dense forest and the values were around 3300 and 2040 million US$ respectively for the year 1990 (figure 7). The ESV of estuaries was 3327 million US$ in 2000, 3407 million US$ in 2010 and 3240 million US$ in 2020 (figure 7). Whereas, the ESV of dense forest (i.e. mangrove) was 1977 million US$ in 2000, 1827 million US$ in 2010 and 1793 million US$ in 2020 (figure 7). A decreasing trend in the ESVs of the dense forest has been observed in the present study. This is due to the reduction in the dense forest area with time. The ecosystem valuation of the degraded forest was around 60 million

US$ in the year 1990, 101 million US$ in 2000, 218 million US$ in 2010 and 247 million US$ in 2020. An increasing trend in the ESVs of the degraded forest has been observed in the present study as the dense forest is being replaced by degraded forest in the study area.

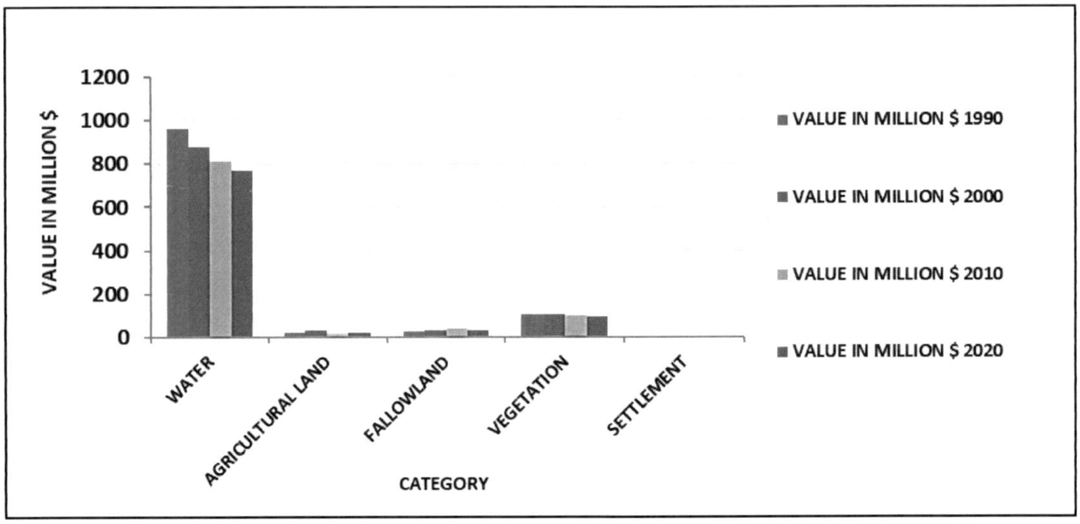


**Figure 6.** Temporal change of land use class-wise ecosystem services values in the upper catchment area of Indian Sundarbans from 1990 to 2020.

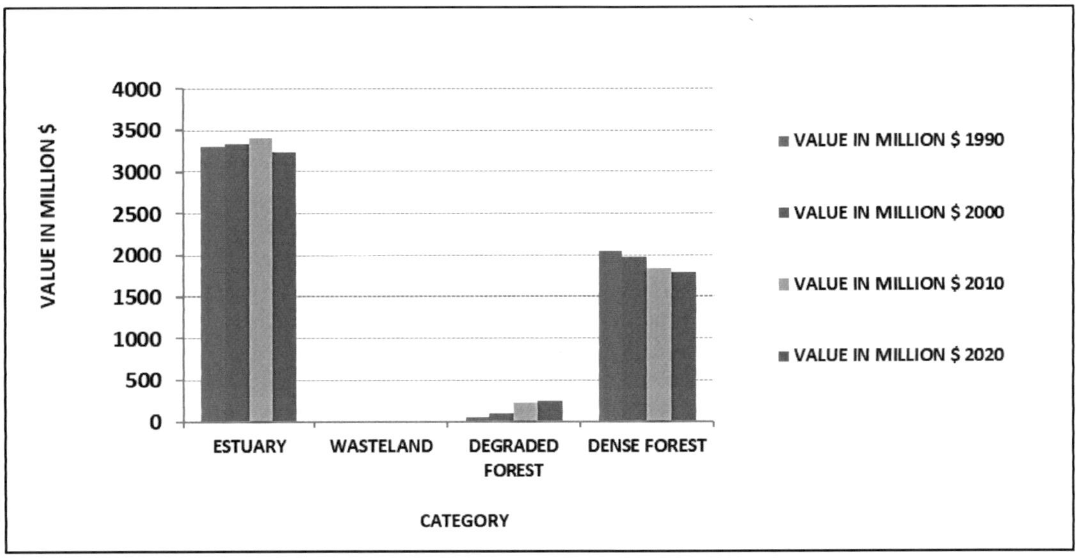


**Figure 7.** Temporal change of LULC class-wise ecosystem services values in the mangrove forest region of Indian Sundarbans from 1990 to 2020.

Ecosystem service-wise values have also been estimated for the years 1990, 2000, 2010 and 2020 under the present study. Figure 8 shows the ESVs in the upper catchment of Sundarbans. It reveals that the value is maximum for water regulation. The ESV of water regulation was around 617 million US$ in 1990, 562 million US$ in 2000, 521 million US$ in 2010 and 492 million US$ in 2020 in the upper catchment area of Indian Sundarbans. A decreasing trend has been observed in the ESVs of water regulation in the study area. The values of water supply are second highest. It was around 239.98 million

IOP Conf. Series: Earth and Environmental Science 1164 (2023) 012004 doi:10.1088/1755-1315/1164/1/012004

US$ in 1990, 218.59 million US$ in 2000, 202.92 million US$ in 2010 and 191.47 million US$ in 2020. The ecosystem values of waste treatment range from 85 million US$ to 69 million US$. The values of Nutrient cycling, food production and recreation range between 25 to 50 million US$. The ESVs of climate regulation, erosion control, pollination, biological control and raw material range between 10 to 20 million US$. The values of disturbance regulation, soil formation, genetic resources and cultural services are negligible, i.e. around 0.2 million US$, 1.0 million US$, 2.0 million US$, and 0.2 million US$ respectively. A decreasing trend has been observed in most of the ESVs.

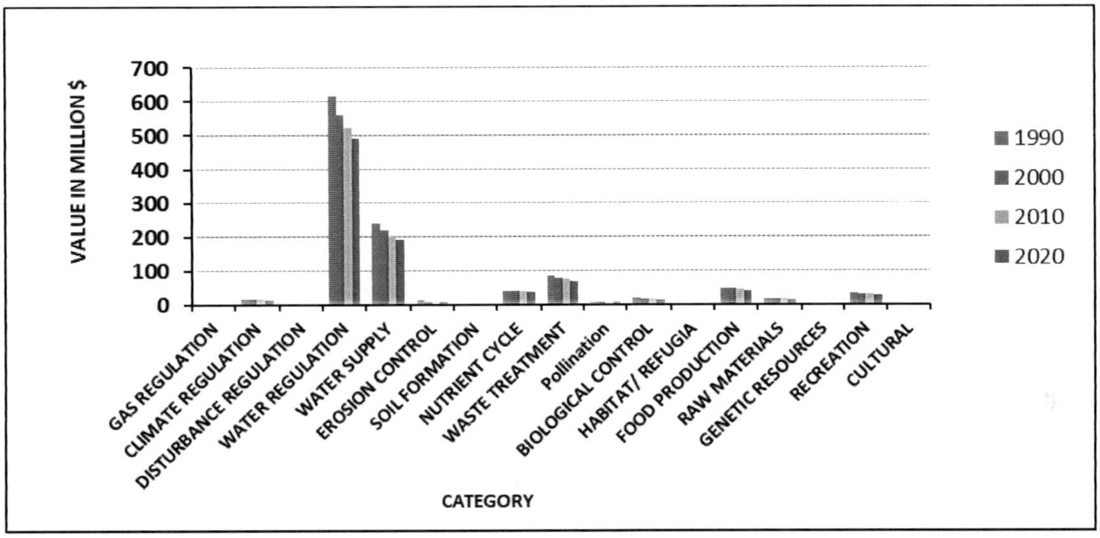


**Figure 8.** Temporal change of ecosystem services values in the upper catchment area of Indian Sundarbans from 1990 to 2020.

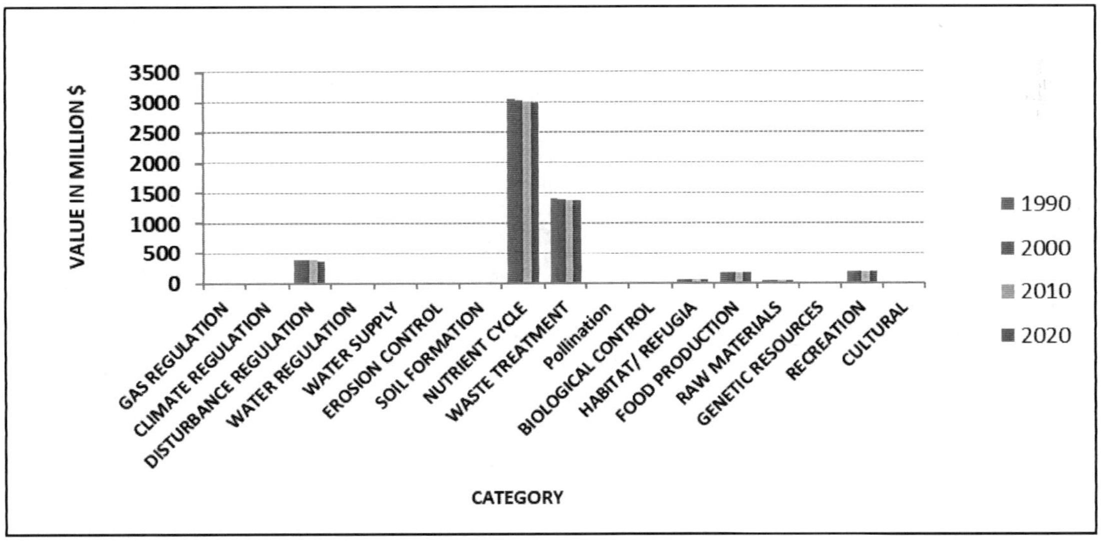


**Figure 9.** Temporal change of ecosystem services values in the mangrove forest region of Indian Sundarbans from 1990 to 2020.

The economic values of different services provided by the Sundarbans mangrove forest have also been estimated under the study. Figure 9 shows that the ESV is the maximum for nutrient cycling and waste treatment in the reserve forest region. The value of nutrient cycling was 3048.95 million US$ in 1990, 3044.27 million US$ in 2000, 3018.12 million US$ in 2010 and 2994.09 million US$ in 2020. Whereas, the value of waste treatment was 1407.50 million US$ in 1990, 1392.77 million US$ in 2000, 1370.67 million US$ in 2010 and 1367.32 million US$ in 2020. A decreasing trend has been observed in the ESVs of nutrient cycling and waste treatment in the study area. The value of disturbance regulation ranges from 375 million US$ to 395 million US$. The ESVs of food production and recreation range from 170 million US$ to 195 million US$. The values of biological control, habitat provisioning, raw materials and cultural services are below 50 million US$.

## 5. Conclusion

The study finds that the area of dense forest has decreased with the passage of time and the area of settlement has increased. It has an adverse effect on the ecosystem services of the study area. The ESVs are the highest for freshwater bodies including rivers, estuaries, mangrove and non-mangrove forests. ESVs of water regulation and water supply are very high in the upper part of the study area. ESVs of nutrient cycling and waste treatment are very high in Sundarbans estuary and mangrove forest areas. The total ESV of the forest in the upper part has changed from 109.7877 million US$ in 1990 to 97.869 million US$ in 2020. Similarly, the value of the mangrove forest ecosystem has changed from 2039.958 million U$ in 1990 to 1793.205 million U$ in 2020. The total ESV of the freshwater body in the upper part has changed from 1050.6704 million U$ in 1990 to 767.3694 million U$ in 2020. Similarly, the ESV of the estuary has changed from 3299.224 million $ in 1990 to 3239.861 million $ in 2020. The health of the mangrove forest is deteriorating with the increasing salinity level of the estuarine rivers and the area of the forest is also decreasing. Therefore, the shrinking forest is unable to provide services like in the past.

## Reference

[1] Zhang X, Niu J, Buyantuev A, Zhang Q, Dong J, Kang S and Zhang J 2016 Understanding grassland degradation and restoration from the perspective of ecosystem services: a case study of the Xilin river basin in inner Mongolia, China *Sustainability* **8** 1-14

[2] Lin X, Xu M, Cao C, Singh RP, Chen W and Ju H 2018 Land-use/ land-cover changes and their influence on the ecosystem in Chengdu city, china during the period of 1992–2018 *Sustainability* **10** 1-17

[3] Costanza R, d'Arge R, Groot de R, Farber S, Grasso M, Hannon B, Limburg K, Naeem S, O'Neill RV, Paruelo J, Raskin RG, Sutton P and Belt MVD 1997 The value of the world's ecosystem services and natural capital *Nature* **387** 253-259

[4] Rai R, Zhang Y, Paudel B, Acharya K B and Basnet L 2010 Land use and land cover dynamics and assessing the ESVs in the trans-boundary Gandaki river basin, central Himalayas Sustainability 10 1-17

[5] Chopra R, Verma VK and Sharma PK 2001 Mapping, monitoring and conservation of harika wetland ecosystem, Punjab, India, through remote sensing *Int J Remote Sens* **22** 89-98

[6] Ustin SL, Darling D, Kefauver S, Greenberg J, Cheng YB and Whiting ML 2004 Remotely sensed estimates of crop water demand *Remote Sens Model Ecosyst Sustain* **5544** 230-240

[7] Chambers JQ, Asner GP, Morton DC, Anderson LO, Saatchi SS, Espirito- Santo FDB, Palace M and Souza Jr C 2007 Regional ecosystem structure and function: ecological insights from remote sensing of tropical forests *Trends Ecol Evol* **22** 414-423

[8] Estreguil C and Lambin EF 1996 Mapping forest-cover disturbances in Papua New Guinea with AVHRR data *Journal of Biogeography* **23** 757-773

[9] Knaeps E, Sterckx S and Raymaekers D 2010 A seasonally robust empirical algorithm to retrieve suspended sediment concentrations in the scheldt river *Remote sens.* **2** 2040-59

[10] Lewis D, Phinn S and Arroyo L 2013 Cost- effectiveness of seven approaches to map vegetation communities- a case study from northern Australia's tropical savanas *Remote Sens.* **5** 377-414

[11] Nemani R, Hashimoto H, Votava P, Melton F, Wang W, Michaelis A, Mutch L, Milesi C, Hiatt S and White M 2009 Monitoring and forecasting ecosystem dynamics using the terrestrial observation and prediction system (TOPS) *Remote Sens Environ* **113** 1497-1509

[12] Verburg PH, Steeg JVD, Veldkamp A and Willemen L 2009 From land cover change to land function dynamics: a major challenge to improve land characterization *J Environ Manage* **90** 1327-35

[13] Zinnert JC, Shiflett SA, Vick JK and Young DR 2011 Woody vegetative cover dynamics in response to recent climate change on an Atlantic coast barrier island: a remote sensing approach *Geocarto Int* **26** 595-612

[14] Marghany M and Hashim M 2010 MODIS satellite data for modelling chlorophyll- a concentrations in Malaysian coastal waters *Int J Phys Sci* **5** 1489-95

[15] De Groot RS, Wilson MA and Boumans RMJ 2002 A typology for the classification, description and valuation of ecosystem functions, goods and services *Ecol Econ* **41** 393-408

[16] Faber JH 2007 European experience on application of site-specific ecological risk assessment in terrestrial ecosystems *Hum Ecol Risk Assess* **12** 39-50

[17] Boyd J and Banzhaf S 2007 What are ecosystem services? the need for standardized environmental accounting units *Ecol Econ* **63** 616-626

[18] Wallace KJ 2007 Classification of ecosystem services: problems and solutions *Biol Conserv* **139** 235-246

[19] Schmitt K, Albers T, Pham TT and Dinh SC 2013 Site-specific and integrated adaptation to climate change in the coastal mangrove zone of soc trang province *Viet Nam J Coast Conserv* **17** 545-558

[20] Barbosa C, Atkinson PM and Dearing JA 2015 Remote sensing of ecosystem services: a systematic review *Ecological Indicators* **52** 430-443

[21] Mitchell R and Carson RT 1989 *Using Surveys to Value Public Goods: The Contingent Valuation Method* (Resources for the Future Washington, DC, USA) p 484

[22] Turner RK, Morse-Jones S and Fisher B 2010 Ecosystem valuation A sequential decision support system and quality assessment issues *Ann NY Acad Sci* **1185** 79-101

[23] Day B 2000 The Valuation of Environmental Goods (Imperial College London Mimeo)

# Forest Fire Risk Zone Mapping in Mizoram Using RS and GIS

**N Pramanick[1*], B Kundu[1], R Acharyya[2], A Mukhopadhyay[3]**

[1]School of Oceanographic Studies, Jadavpur University, India
[2]Department of Geography, School of Earth Sciences, Central University of Karnataka, India
[3]Centre for Earth Observation Science (CEOS), University of Manitoba, Canada

*Corresponding Author Email: niloy.pramanick@yahoo.com

**Abstract.** In recent years, India's Northeastern territory has been plagued by wildfires. Mizoram, as per FSI, has been among the most affected regions, with approximately 20,744-recorded wildfires. Forest fire directly or indirectly affects human health, climate change, and the environment. The forest fire risk zonation is generated utilizing remote sensing data, and Geographic Information System based on specified physical and socioeconomic factors. As part of the current study, a Geographical Information System is employed to determine forest fire risk based on predetermined physical and socioeconomic parameters. This study used three distinct models of fire risk zonation index namely FRI (Fire Risk Index), HFI (Hybrid Fire Index), and SFI (Structural Fire Index) to identify the zones in the study area under the risk of forest fires. The Indian state of Mizoram is categorized into five distinct hazard zones based on the probability of wildfire incidents. Fire alerts generated using risk models, and real-time hotspot datasets (forest fire spots) received from MODIS and USGS have been validated. According to the study's findings, 18.84 $km^2$ of the study area is at low risk of forest fire, 11.072 $km^2$ is at moderate risk, and 5.38 $km^2$ is at high risk. The probability from each metric varies since the input and weightage of the different parameters vary from one another. SFI, therefore, therefore predicts a lower frequency of high-risk wildfire-prone zones than HFI and FRI. In this research, the coefficient of discrimination ($R^2$) is employed to assess the reliability of the projected fire indices being compared with real-time hot spots. Here, FRI possesses the highest accuracy ($R^2$=0.892), HFI has moderate accuracy ($R^2$=0.676), and FRI has the lowest accuracy ($R^2$=0.629).

**Keywords**: Forest Fire-Mizoram, Forest fire, Fire risk zone, HFI, DFI, SFI, MODIS.

## 1. Introduction

Forests constitute a valuable food source, shelter, habitat for wildlife, fuel, and essential supplies [1]. Forests occupy about one-third of the total land area on the earth [2]. India's entire forest and woodland area is 80.73 million hectares, accounting for 24.56 % of the nation's geographical area [3]. As per the status of the Forest Report (2009) by FSI, the documented forest area is grouped into three categories reserved forest (51.6%), protected forest (26.79%), and unclassified forest (17.24%) [4]. Madhya Pradesh has the maximum forest cover in India, followed by Arunachal Pradesh, Chhattisgarh, and Odisha [5]. Mizoram (85.4%), Arunachal Pradesh (79.61%), Meghalaya (76.32% ), Manipur (75.46 %), and Nagaland (75.3%) are the five major states in terms of forest cover as a portion of total geographical area [3]. Because of their reliance on fuelwood, food, minor forest resources, and lumber, forests are intertwined with the social system [6]. Fuelwood is the primary source of energy for 70% of the Indian population, and 125 million tonnes are harvested each year [7]. Furthermore, more than half of the cattle population (270 million) is dependent on the forest for grazing resources, and the forest provides NTFP of Rs. 6.5-20 billion yearly [8].

Forest fires are the most common threat currently occurring in forests worldwide [9]
. Forest fires are complicated events driven by both natural and anthropogenic influences [10,11]. Forest fires are regarded as a catastrophe distressing land ecosystems and causing economic devastation for human beings, such as a lack of land use revenue, disruption to crops, and depletion of biodiversity [11-13]. There are three phases to forest fires: grass fires, undergrowth fires, and crown fires. Grass and non-wet undergrowth usually ignite in the absence of moisture and can be controlled easily. It becomes much more difficult to monitor if the fire reaches the top of the trees, particularly conifers, resulting in canopy or crown fire [14]. When this is transformed into a wildfire, the treetops, particularly in the specific instance of conifers, can rapidly spread flames, leading to the canopy fire. Canopy fires are much harder to control and, therefore, can quickly become a conflagration, which basically cannot be extinguished [15]. It leads to an imbalance in the environment, depletion of wildlife, biodiversity, and forest natural resources. Forest fire is primarily considered a natural hazard, although human-induced or anthropogenic influences can also be caused [16]. Initiation and propagation of fire are dependent on many crucial factors such as precipitation volume and frequency, topography, temperature, lightning, fuel availability and distribution, relative moisture concentration, and wind speed all play a role in the presence of inflammatory agents [17].

Wildfires are becoming an increasing concern as the world continues to warm exponentially, burning across international boundaries and significantly raising greenhouse gas emissions in the atmosphere. The fires are also local and national issues and global challenges [18]. The Amazon forest fire has been a recent occurrence, with the deadliest fires that struck the Amazon Basin over more than a decade in 2019. The issue of the Amazon Wildfires has become a serious concern, when environmental concerns become more political [19]. In Brazil, the Amazon rainforest initially occupied 4.1 million sq.km but decreased to 3.4 million sq. km. In the last 50 years, however, 17% of the biodiversity has been irrevocably disrupted. An additional 17% became impaired and deteriorated, primarily associated with the wildfires by human activities [15]. Australia has also experienced similar bushfire incidents recently, and the foremost issue with forest fire is human activities. Bushfires account for about 10% of the total expense of all significant natural hazards in Australia, and they are synonymous with the tremendous loss of wildlife. In Australia, bushfires are an unavoidable phenomenon. In Australia, approximately 50 million forest areas are burnt annually, with about 80 % of the fire-ravaged areas located north of Savanna areas [20]. Fire is most prevalent in the northern tropical savannas, where certain areas of land fire each year. However, the southeast, which is residence to most of the population, is particularly vulnerable to massive wildfires that threaten their lives and property [20].

According to a recent study by FSI (Forest Survey of India), close to 36% of India's forest is vulnerable to fires, with over 10% of that being extremely prone [21]. Forest fire threatens valuable environmental resources and biodiversity and is also one of the leading causes of forest destruction and air quality. According to FSI, in recent years, forest fire occurrences in India have increased in summer by 49,32% (a total of 24,817 wildfire events in 2016, which escalated to 35,888 by 2017 and 37,059 in 2018) [21]. Studies have shown that 90% of forest fires throughout India are human-induced and that around 3.73 million hectares of forest are damaged each year, which leads to losses of US$104 million [22]. Mizoram's forest coverage of 85.41 % has declined from 86.26 % in 2017, according to ISFR ("India State of Forest report"-2019) published by FSI [23]. In Mizoram, as officials stated, about 1,300 wildfires were registered in 2020. Around 1,090 accidents have been caused by Jhum cultivation (slash-and-burn agriculture) and just 210 by natural causes [24]. The forest coverage in Mizoram is 157.05 square kilometers of Very Dense Forests (VDF), 5,800 square kilometers of Moderately Dense Forests (MDF), and 12,047.71 $Km^2$ of Open Forests (OF). Forest cover in the state has decreased by 180.49 $Km^2$ since ISFR 2017.[23]. Forest officials in Mizoram claim that Jhum cultivation, the primitive farming process, explains the large number of wildfire alerts reported in northeastern states [25].

However, because slash-and-burn activities often coincide with dry and windy seasons of the year, wildfires that spread beyond the boundaries of designated zones into surrounding woods are not uncommon. Such flames can entirely destroy entire villages or small communities [26]. Forest fires (i.e., lightning strikes) have unusual natural causes. Forest fires typically refer to actions of human beings, like jhum, where highly flammable flakes of bamboo and charcoal are blown into adjacent regions, leaf litter, and dry grasses are burnt to flames. No firebreaks around the jhum field before combustion occurs. In February and March, annual removal and burning are typically carried out at the driest time of the year. Cattle grazers scorch dry grasslands and forest floors during the dry season to eliminate weeds and promote the growth of fresh grazing shoots. Forest floors are burned to increase visibility for poaching wild animals. Inadequate approved burning, as well as fire line planning and design. Building structures, agriculture, road construction, shifting cultivation, and so on are examples of some other human activities inducing forest fires [26].

To develop a thorough study on forest fire probability, environmental and socioeconomic factors must be considered, and the correlation between various factors and fire risk is established. The term "fire risk," when used in an analysis, refers to the probability of the initiation of a fire or of a region getting burnt, as influenced by the magnitude and occurrence of causative factors [27-29]. Performing fire risk minimization by rebuilding the fire regime and maintaining the fire regime is essential to effectively controlling wildfires. [30]. Since the 1920s, the concepts of hazard and risk have been regularly employed in fire-control matters throughout the United States [28]. According to the MiSRaR project of the European Union, the risk is a combination of hazard and harmful consequences. [31]. Forest fire control, in general, includes of four steps of assessment and evaluation of successful fire response: mapping possible fire risk and danger, detecting hot spots, tracking current fires, and analyzing post-fire depletion [32]. The most critical step is assessing fire risk and deciding where the hazard is greatest; it is necessary to limit harm to people, property, and natural resources. In this work, we focused on determining the spatial range of fire risk across Mizoram. The "Joint Research Center" of the EU ("European Union") has proposed three classes of indices to assess fire risk based on their temporal scale, namely structural indices, dynamic indices, and combined indices [33]. Structural indices are the long-term indices that primarily portray static data at a worldwide scale [34]. They do not alter in short periods. Elevation, vegetation, land use, land cover, aspect, slope, proximity to roads and settlement [35], climate factors, soils [36], and population density are instances of such indices [37]. Dynamic indices are short-term indices that shift somewhat over time due to vegetation or climatic conditions. The indicators are designed to monitor forest fuel's flammability during wildfire incidents. Therefore, the dynamic indexes use variables that change over a brief period and rely on the probability of the ignition and proliferation of forest fires [38]. The combined indices comprise both dynamic and structural variables mentioned above. In this method, the most crucial aspect is how the relevant variables can be easily combined to generate a logical concept [39].However, gathering all accurate details from such a North-eastern state as Mizoram is challenging. As a result, the current study has been dependent on readily available indicator variables. Geographic information systems and Remote sensing are by far the most effective and necessary methods for developing a wildfire susceptibility model to determine and distinguish high-risk and fire risk zones in today's world. The knowledge available on the Internet was also highly beneficial in the development of our project. The purpose of this study is to illustrate the risk of the forest fire scenario in Mizoram. This study aims to establish the correlation between fire risk and various biophysical and socioeconomic variables (elevation, slope, radiation appearance, soil moisture, distance to roads, proximity to settlements, etc.) The aim has been divided into objectives to create a fire risk index and characterize the zones based on fire risk.

## 2.Study Area

Mizoram is among the seven states that make up the region of India's northeast renowned as the "Land of 7 Sisters." [40]. Compared to other states of India, Mizoram is a small state (fifth smallest state) covering 21,081km$^2$ [41]. The forests cover 91 % (15, 94,000 hectares) of Mizoram's total area, putting it third among Indian states in total forest cover [42]. According to the 2011 census, Mizoram's total population was 1091,014, India's second inhabited state [43].The latitudinal extension of the state is 21° 58'N to 24° 35' N, and the longitudinal extension is 92° 15' E to 93° 29' E [44] (figure 1). The neighbouring states of Mizoram are Assam, Tripura, and Manipur and share a 722-kilometre border with the neighbouring countries of Bangladesh and Myanmar in the southeast [45]. Mizoram does have the most varied hilly terrain in India's eastern region. The steeply sloped hills are segregated by rivers that flow either north or south, forming deep gorges across the mountain ranges. The hills have an average elevation of around 1000 meters. Phawngpui (Blue Mountain) is Mizoram's highest elevation, standing at 2210 meters [46]. Mizoram does have a moderate climate, of summer temperatures ranging between 20°C - 29°C, while temperatures are steadily rising due to climate change. The region is dominated by monsoons, with heavy rain during the dry (cold) season between May and September. With total rainfall of 254 cm/year, the climatic condition is tropical humid subtropical. The rainfall is approximately 215cm in Aizawl and around 350 cm in Lunglei [47].

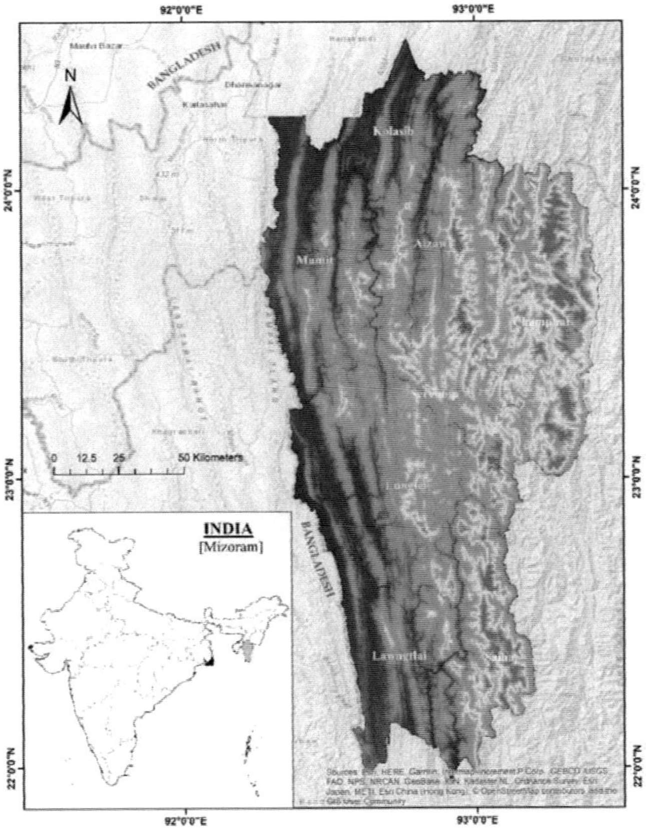


**Figure 1.** Study area (Mizoram) map selected for the current study.

*2.1. Land use and Land Cover*

Mizoram has the most extensive forest cover in the state based on the total land area[23]. The land use/land cover types (figure 2) found in this region are primarily vegetation agricultural lands, current shifting cultivation lands, mining sites, wetlands settlement and wastelands etc. The vegetation in the region is divided into four types evergreen or semi-evergreen, deciduous, forest plantations and scrublands. Deciduous trees are the most prominent and widespread however they are highly flammable

IOP Conf. Series: Earth and Environmental Science 1164 (2023) 012005 doi:10.1088/1755-1315/1164/1/012005

[21]. The evergreen forest has been presently concentrated in the study area's eastern section, whereas the forest plantation is primarily found in the southwest region. Scrubland covers the majority of the north and northwestern portion. The general settlement pattern is scattered, and linear forms of settlement can be found alongside roads and cultivated lands. Figure 3 shows that most agricultural fields are found accross the north-eastern portion of the state. Within the study area, fields used for shifting cultivation are mostly found in the northeastern and eastern sections. These regions are very vulnerable to fire caused by human activity. Various physical and socioeconomic causes linked to the fire have been chosen. This section creates a fire risk map with various variables to determine how each factor affects the fire itself.

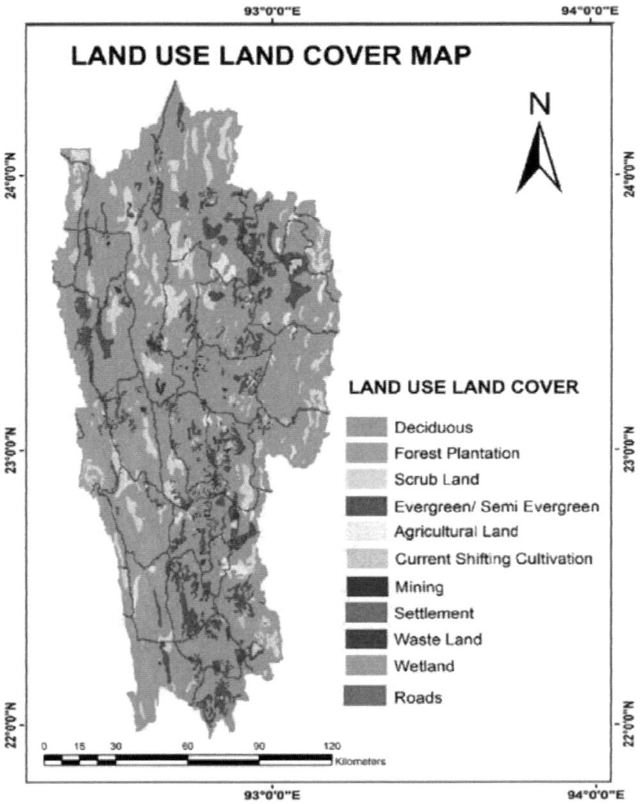


**Figure 2.** Land Use and Land Cover Map of the study area

## 3. Materials and Methods

### *3.1. Data*

The current study is based on remotely sensed data that is currently available. The datasets described in table 1 provide insight into the inputs (figure 2 and figure 3) used to accomplish the objectives mentioned earlier regarding the current study to delineate the study area's forest fire risk zones. The remotely sensed LANDSAT 8 OLI/TIRS [48] datasets have been utilized to acquire the scenarios of the land cover, settlement pattern, and soil moisture across Mizoram. Line vector data sets have been acquired from OpenStreetMap and are used for road layer analysis [49], shown in figure 3. The region's height, slope, and aspect have been then estimated using the SRTM DEM dataset of 30 meter spatial resolution [50] as inputs for the current investigation, as shown in figure 3. The datasets of real-time

IOP Conf. Series: Earth and Environmental Science 1164 (2023) 012005 doi:10.1088/1755-1315/1164/1/012005

fireplaces in Mizoram were then obtained from MODIS [51] for validation purposes. ArcGIS 10.3.1 has been employed to evaluate geospatial and remotely sensed datasets in a GIS framework. Excel is being used for specific quantitative data processing and assessment regarding the current study.

**Table 1.** Datasets used for the current study.

| Data Type | Description | Location/Extension | Time | Resolution | Source |
|---|---|---|---|---|---|
| **Land-cover** | Multispectral Satellite Imagery (Landsat) | Mizoram (21° 58' N-24° 35' N 92° 15' E -93° 29' E) | 2019 | 30 metres | https://earthexplorer.usgs.gov/ |
| **Settlement (BUI)** | Multispectral Satellite Imagery (Landsat) | Mizoram (21° 58' N-24° 35' N 92° 15' E -93° 29' E) | 2019 | 30 metres | https://earthexplorer.usgs.gov/ |
| **Roads** | Line Shapefile | Mizoram (21° 58' N-24° 35' N 92° 15' E -93° 29' E | | | https://openstreetmap.in/#4.37/22.82/82 |
| **Soil Moisture (NDMI)** | Multispectral Satellite Imagery (Landsat)\ | Mizoram (21° 58' N-24° 35' N 92° 15' E -93° 29' E) | 2019 | 30 metres | https://earthexplorer.usgs.gov/ |
| **Real-time Fire points** | MODIS | Mizoram (21° 58' N-24° 35' N 92° 15' E -93° 29' E) | 2019 | 1km | https://earthdata.nasa.gov/active-fire-data |
| **Elevation** | Digital Elevation Model (DEM) | Mizoram (21° 58' N-24° 35' N 92° 15' E -93° 29' E) | 2019 | 30 metres | https://srtm.csi.cgiar.org/ |
| **Slope** | Digital Elevation Model (DEM) | Mizoram (21° 58' N-24° 35' N 92° 15' E -93° 29' E) | 2019 | 30 metres | https://srtm.csi.cgiar.org/ |
| **Aspect** | Digital Elevation Model (DEM) | Mizoram (21° 58' N-24° 35' N 92° 15' E -93° 29' E) | 2019 | 30 metres | https://srtm.csi.cgiar.org/ |

IOP Conf. Series: Earth and Environmental Science 1164 (2023) 012005 doi:10.1088/1755-1315/1164/1/012005

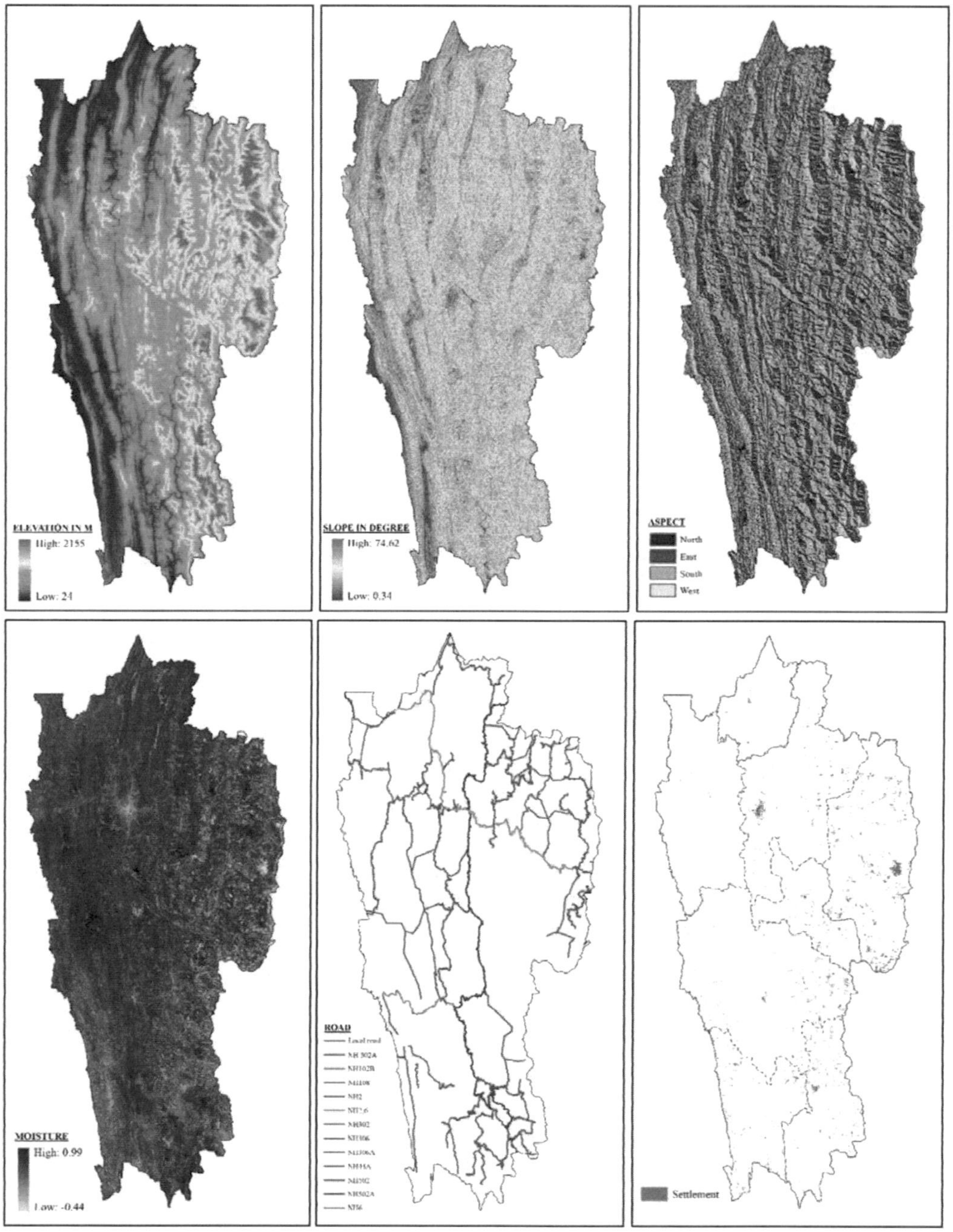


**Figure 3.** The current study requires the above-illustrated datasets: Elevation, Slope, Aspect, Moisture, Road, and Settlement

### *3.2. Methods*

Figure 4 displays a flow chart that outlines all the measures used with the datasets incorporated to generate the current study results. The steps have been executed in the application of ArcGIS 10.3.1 has been employed to evaluate geospatial and remotely sensed datasets in a GIS framework. Excel is being used for specific quantitative data processing and assessment regarding the current study.

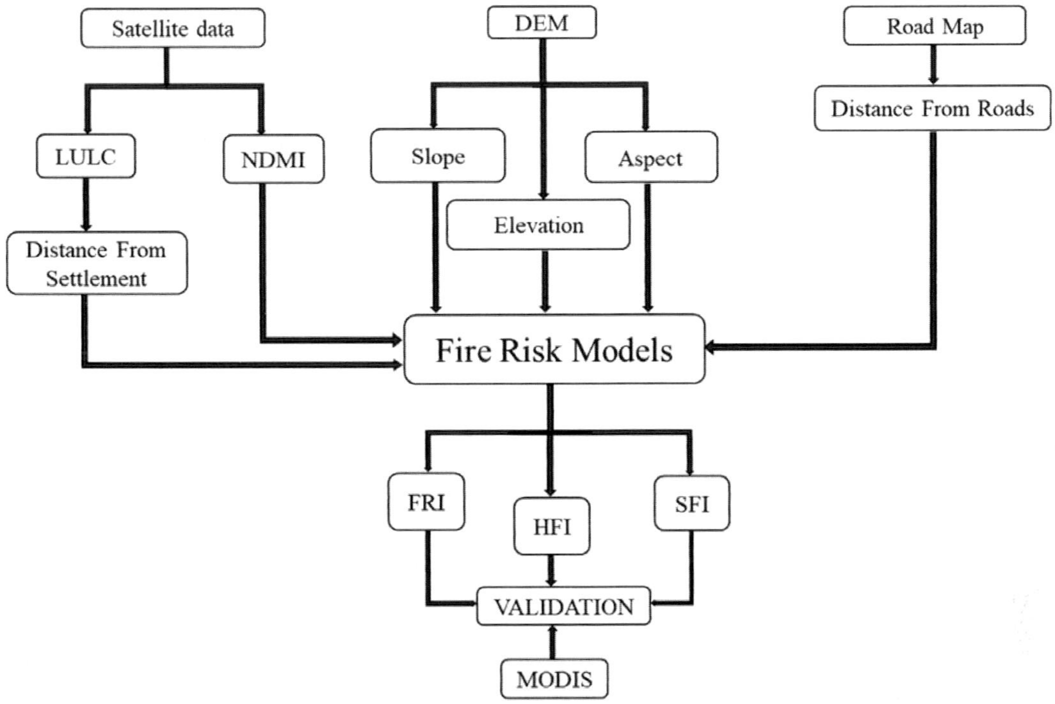


**Figure 4.** Flow chart showing the Methodology of the current study.

*3.2.1. Elevation.* The elevation is a critical topographical factor that plays a significant role in the rapid flame spread. It is a significant physiographic parameter related to temperature, humidity, and wind [52]. It also affects vegetation structure, moisture content fuel, and air humidity [53]. Thus, higher elevations are more susceptible to fire because the winds are more effective. The study area has been categorized into five categories based on its elevation (in meters): High (960-2155 m), Moderate (479 - 960 m), and Low (<479 m), depicted in figure 5(a) and table 2. The classified elevation pattern of the study area is illustrated in figure 5(a), with the north-eastern and south-eastern regions being more vulnerable to wildfire due to their higher elevation. On the other hand, the western portion has lower elevation zones and therefore bears a lower fire risk.

*3.2.2. Slope.* Among the factors that influence the spread of fire is slope [54]. Consequently, calculating the region's steepness and inclination is essential. The slope of the study area, depicted in figure 5(b), has been calculated using the SRTM DEM. The potential of fire's spread rises when the fire moves closer to the land surface, which is further boosted by winds [55]. The fire spread is quicker in the uphill direction than in the downhill direction [56]. The slope in the study area was measured in degrees and classified into five separate groups based on steepness, as shown in figure5(b). They are classified as Low (< 15%), Moderate (15 – 28%), and High (>35 %). From east to west, the terrain slopes downward. The south-eastern portion of the region has a steeper slope and is more prone to forest fires.

*3.2.3. Aspect.* The amount of solar energy received by a region is proportional to its aspect (figure 5 c). As per figure 5(c), the classified part of the area indicates that the east-facing slopes get more sunshine, have higher temperatures, and are steeper. As a result, vegetation on east-facing slopes is usually drier, deciduous, and less dense. There is much-shifting cultivation and wasteland; thus, the east-facing slope is high fire-risk zones. The north-facing slopes are moderately prone to forest fires. Then, the west and south-facing slopes of the research area are the least vulnerable to forest fires.

*Soil Moisture.* Soil moisture content has a direct influence on the occurrence and spread of forest fires [11,57,58]. As a result, developing a soil moisture map (figure 5d) of the study area is critical to determining the correlation between soil moisture and fire risk. Soil moisture and fire risk are inversely related, with wet soil indicating a lower risk of wildfire and dry soil indicating a higher risk of wildfire. SWIR (short-wave infrared) bands absorb more moisture than NIR (near-infrared) bands because of the increased reflection of NIR (near-infrared) bands in thin canopy layers. In the current investigation, the NDMI (Normalized Difference Moisture Index) has been computed utilizing the bands of NIR and SWIR to assess the level of moisture in the vegetation [59]. This index is responsive to fluid water composition variations in vegetative canopies.

$$\mathrm{NDMI} = (\mathrm{NIR} - \mathrm{SWIR}) / (\mathrm{NIR} + \mathrm{SWIR}) \tag{1}$$

Lower NDMI values imply that excessively dry vegetation ignites more readily than lush and wet vegetation having a higher value of NDMI [60]. As per figure 5(d), the region has been categorized into three categories based on NDMI values representing soil moisture content, namely high (<0), moderate (0 – 0.02), and low (>0.2), as mentioned in table 2.

**Table 2.** Categories of input variables as per the level of fire risk

| Feature | High | Moderate | Low |
|---|---|---|---|
| Elevation(m) | 960–2155 | 479 - 960 | 24-479 |
| Slope(°) | >28 | 15 - 28 | <15 |
| Aspect(°) | East | North | West, South |
| Moisture (NDMI) | <0 | 0 - 0.2 | >0.2 |
| Distance from Road (m) | <1000 | 1000 - 2000 | >2000 |
| Distance from Settlement (m) | <1000 | 1000 - 2000 | >2000 |

*3.2.4. Distance from Roads.* Road density and proximity are potentially significant criteria because roads enable people to infiltrate forested regions and trigger ignitions. Furthermore, forested or grassland regions surrounding communities are more vulnerable to firebreaks since residents within the forest could accidentally trigger fires [61,11]. Roads are essential because there is a direct correlation between the distances across roads to the forest and the accessibility generated by the road, which impacts the capability of firefighters to manage major fires [62]. Furthermore, road structures raise the danger of human-induced fire, so it is crucial to keep a safe distance from the road diagram, as seen in figure 5 (e). As per figure 5(e), the region was divided into three distinct categories (in meters): high (<1000 meters), moderate (1000-2000 meters), and low (>2000 meters), depicted in table 2. Figure 5 (e) shows the longitudinal segments of pink adjacent to the forest and are most vulnerable to forest fires. Although zones highlighted in blue are farther away from the trees, they are less vulnerable to forest fires. Forest regions that are less than 1000 meters from a road face a greater risk of human-induced fire.

*3.2.5. Vicinity to settlements.* Anthropogenic variables, such as the vicinity to settlements and roads, are significant variables that affect fire incidents [63]. Human settlements surrounding forests are another critical aspect that could pose a problem to the environment. The tribal communities within the forests can also be a significant cause of forest fires [16]. Sometimes, tribal communities set fire to the forest for "*Jhum*" cultivation, clear pathways, gather cooking fuel, and so on, but it can also happen unintentionally due to carelessly tossing cigarette butts, a campfire, or the gathering of non-timber forest items. Figure 5(f) depicts the proximity of the forest fire danger zone from the settlements. To further assess the zones at high risk from human populations, the study area has been categorized into 3 distinct classes of vulnerability based on distance (in meters), namely highly vulnerable (1000 metres), moderately vulnerable (1000 – 2000 metres), and least vulnerable (>2000 metres) to forest fire risk, as shown in figure 5 (f) and table 2.

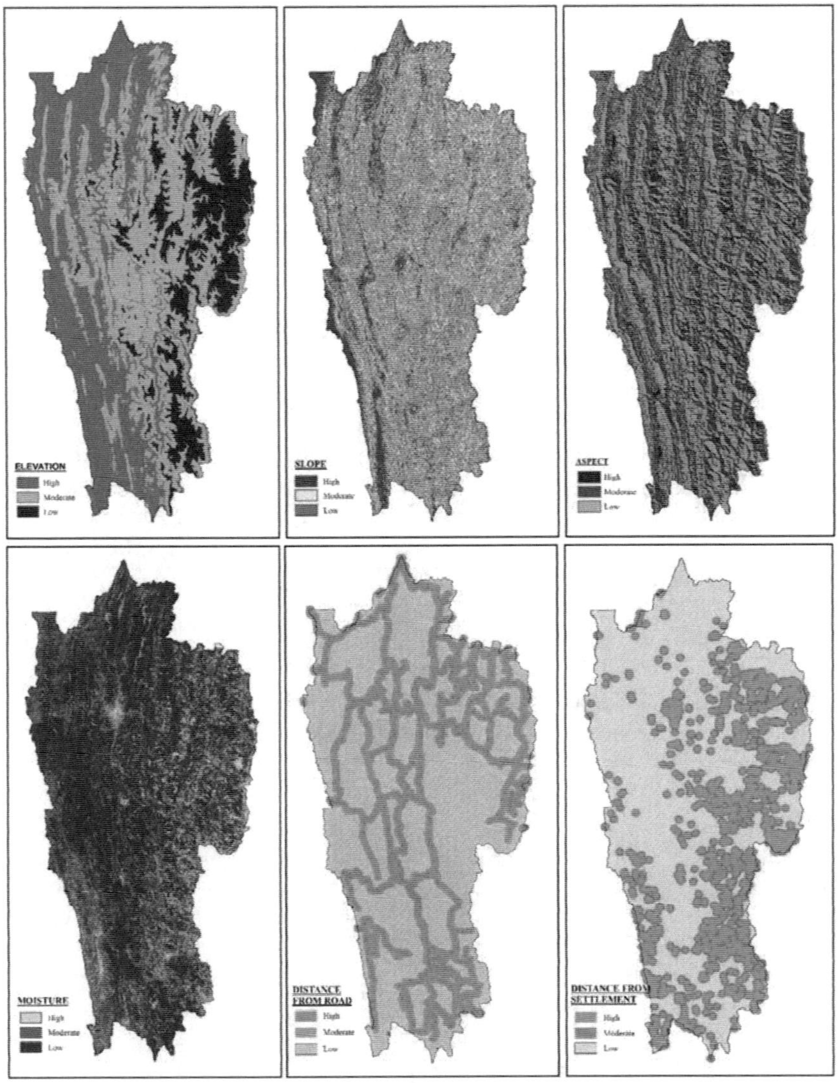


**Figure 5.** Classified input datasets for the above-illustrated datasets: Elevation, Slope, Aspect, Moisture, Road, and Settlement

*3.2.6. Fire risk indices.* In different regions, structural, dynamic, as well as combined indices have been widely used. Since they are designed for a specific region, they have varying levels of potential to monitor forest fire risk [11]. However, no distinctions have been made among the indices to determine their effectiveness in the study area.Different fire risk indices for the current study have beenestimatedto create an accurate fire risk index. The following indices are the methods for determining potentially vulnerable fire risk zones of the state:

- Fire Risk Index: FRI is an experimental index implemented by Erten *et al.*[64] in Turkey and by Siachalou *et al.*[64,65] in Greece.FRI stands for fire risk index, generated using the parameters. The following equation (2) [11] has been used to estimate SFI to determine fire risk :

$$\text{FRI} = 7\ x\ vt + 5\ (s + a) + 3\ (Dr + Ds) \tag{2}$$

Where the parameters like moisture (vt), slope (s), aspect (a), distance from road (Dr), distance from the settlement (Ds, have been used to estimate the index of FRI.

- Structural Fire Index: The Structural Fire Index (SFI) is a quantitative index focused on a compilation of five significant parameters on fire risk: elevation, slope, vegetation moisture, and vicinity to roads [66]. The soil's moisture has been given more weightage than the other variables in this fire index [11].In this study, the following equation (3) has been used to estimate SFI in order to determine fire risk :

$$\text{SFI} = 1 + 100v + 30s + 10a + 5r + 2e \tag{3}$$

where it have the factors such as moisture (v), aspect (a), elevation (e), slope (s), distance from road (r) has been used to estimate this particular index.

- Hybrid Fire Index: The indices FRI and SFI do not apply the factors of proximity to settlements and altitude. Since each variable does affect fire risk, the omitted parameters could impede the indices' capability to identify fire risk [11]. Consequently, it is believed that integrating all variables from the FRI and SFI indices will improve the study's outcomes. HFI, has been estimated by using the following equation (4).

$$\text{HFI} = \{100v + 50s + 25a + 10(r + c) + 10e\} / 10 \tag{4}$$

Where the index has been calculated by using the parameters of moisture in vegetation ($v$), aspect ($a$), slope ($s$), distance from roads ($r$), proximity to settlements ($c$), elevation ($e$). Each of the variables has a different weightage

## 4. Results and Discussion

In the current study, Mizoram's forest fire risk zonation has been carried out based on physical parameters including elevation, slope, aspect, soil moisture, and human-induced factors like proximity from road and settlement. The risk zone analysis is applied to all variables dependent on their susceptibility to flames. The three indices of SFI, FRI, and HFI have been estimated (figure 6) to determine the research area's forest fire risk zones. As a consequence, the study region has been classified into three forest fire risk categories, namely high, low, and moderate, relying on their influence on forest fire occurrence.

According to the HFI index for the study area (figure 6 c), 18.14 $km^2$of the region are within the low-risk zone, 14.53 $km^2$are in moderate risk, and 7.15 $km^2$are under the high-risk zone of the forest fire. The area with a high fire risk is highlighted in red in figure 6(c), mainly across the northeastern and southern regions of the research area. As per figure 6(a), FRI index shows that the region of 21.14 $km^2$ is in a low-risk area. The 16.64 $km^2$ is then encompassed within a zone of moderate fire risk. Them an area of 8.43 $km^2$comes under the high forest fire risk zone. According to the FRI index, figure 6(a) shows that the northeastern, southeastern portion of the state bears a high forest fire risk. The SFI index, depicted in figure 6(b), indicates that 18.84 $km^2$ of the study area is at low risk. Then, 11.076 $km^2$ of the study area is at moderate risk, and 5.38 $km^2$ is highly susceptible to forest fires. SFI implies a lesser concentration of high-risk areas than other indices such as HFI and FRI.

IOP Conf. Series: Earth and Environmental Science 1164 (2023) 012005 doi:10.1088/1755-1315/1164/1/012005

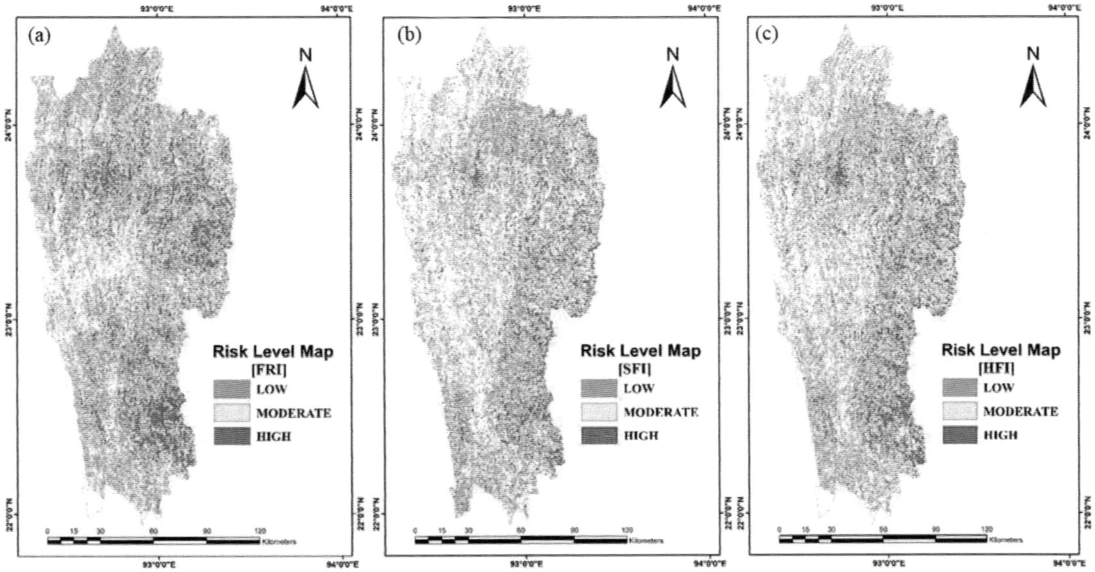


**Figure 6.** Fire risk indices: a) FRI (Fire Risk Index) b) SFI (Structural Fire Index) c) HFI (Hybrid Fire Index)

### *4.1. Probability Distribution:*

The level of fire risk in the study area has been illustrated by using probability distributions (figure 7) by quantifying the risks of forest fires in a specific region, based on the afore-mentioned estimations of FRI, SFI, and HFI. Thus, the level of fire risk has been shown in line graphs, which are estimated for three respective fire index models (SFI, FRI, and HFI) used in this study. In the current study, forest fire risk has been categorised by utilizing probability distributions, determined by each index into three fire risk categories: low, moderate, and high (figure 7). Hence, probability distributions of forest fire risk have been estimated, based on the SFI, FRI, and HFI to assess the frequency of fire risk over the study area. Then the probability from each index differs since the weightage and input of the various parameters vary. Forest fire risk levels in the line graphs indicate low, moderate, and high levels according to the risk levels indicated by the indices. Values between 1 - 10 indicate low risk, 11 - 30 indicate moderate risk, 60 - 100 indicate high risk. According to the current study's estimates, FRI project a higher probability of forest fire risk in the study area than SFI and HFI (figure 7) in all three categories of forest fire risk (low, moderate, and high).

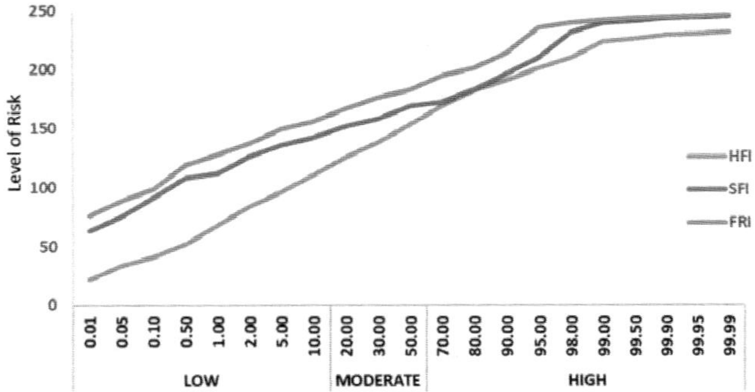


**Figure 7.** Three indices (HFI, SFI, and FRI) derived from the probability distribution, encompassing levels of the minimal, medium, and severe fire risk.

### 4.2. *Estimation of Accuracy of Fire Indices*

Due to the varying weightage assigned to the different factors used to measure the fire indices. It is now necessary to determine the consistency of the models. Real-time hot spots (fire points) data have been used to determine the precision of three distinct models. . The statistical meaure, Coefficient of Determination (r-squared or $R^2$) represents the degree of accuracy to real-time results. In this study, $R^2$ has been used to assess how accurate the results of our study's estimated fire risk indices (FRI, SFI, HFI) are the dependent variable, with real-time fire hotspots functioning as independent variable. $R^2$ quantifies the amount of variation within the dependent parameter (fire risk indices FRI, SFI, HFI) that can be characterized by the considered independent variable (actual fire hotspots). As a result, by correlating the estimated fire indices to real-time hot locations, $R^2$ has been used to validate the results. Figure 8 shows that FRI has the highest accuracy ($R^2$=0.892), HFI has moderate accuracy ($R^2$=0.676), and FRI has the least accuracy ($R^2$=0.629), shown in the graphs(figure 8) below.

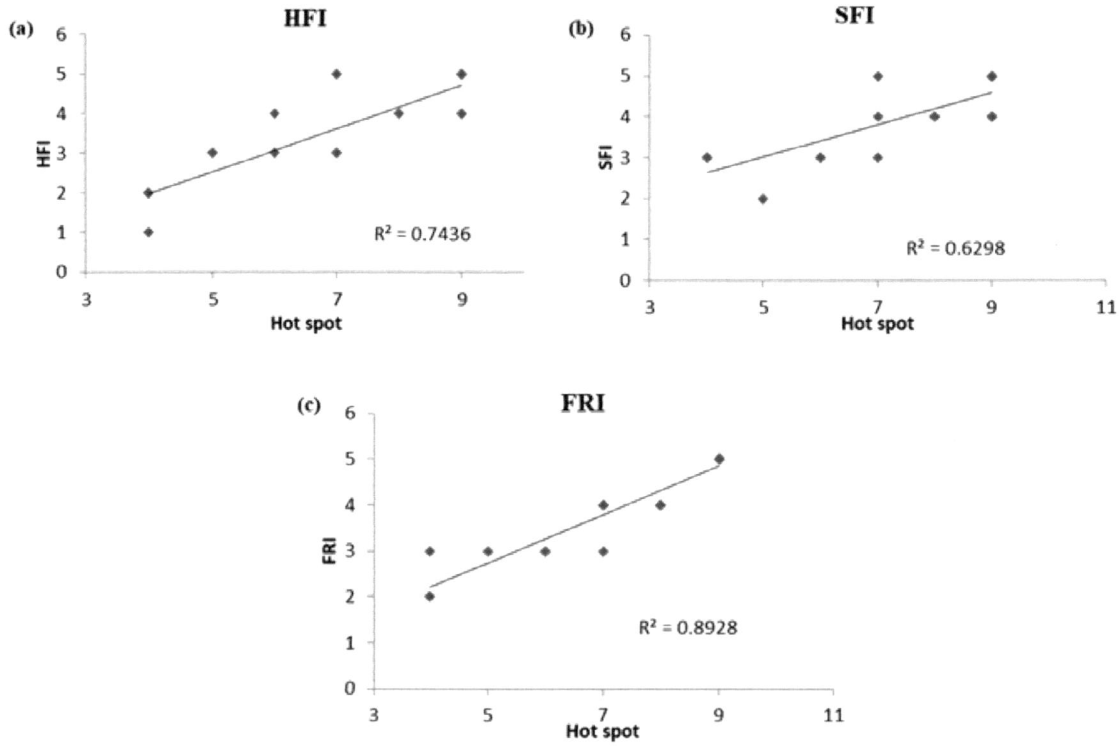


**Figure 8.** Accuracy estimation of the forest fire risk indices: (a) HFI, (b) SFI and (c) FRI

### 4.3. *Model Validation*

Model validation is a critical aspect of risk zone mapping. It evaluates the projected model's precision or predictive potential by comparing it to a real-time dataset [67]. In this study, the fire risk was deemed to be the most credible finding (figure 9)—all of the real-time hotspot data derived from MODIS. Figure 8 depicts the highest number of hot spots in the high-risk region and a few hot spots clustered in the moderate risk zone, with very few hot spots found in the low-risk zone. It indicates that the data obtained from this fire risk index map is correct in locating specific regions vulnerable to fire risk.

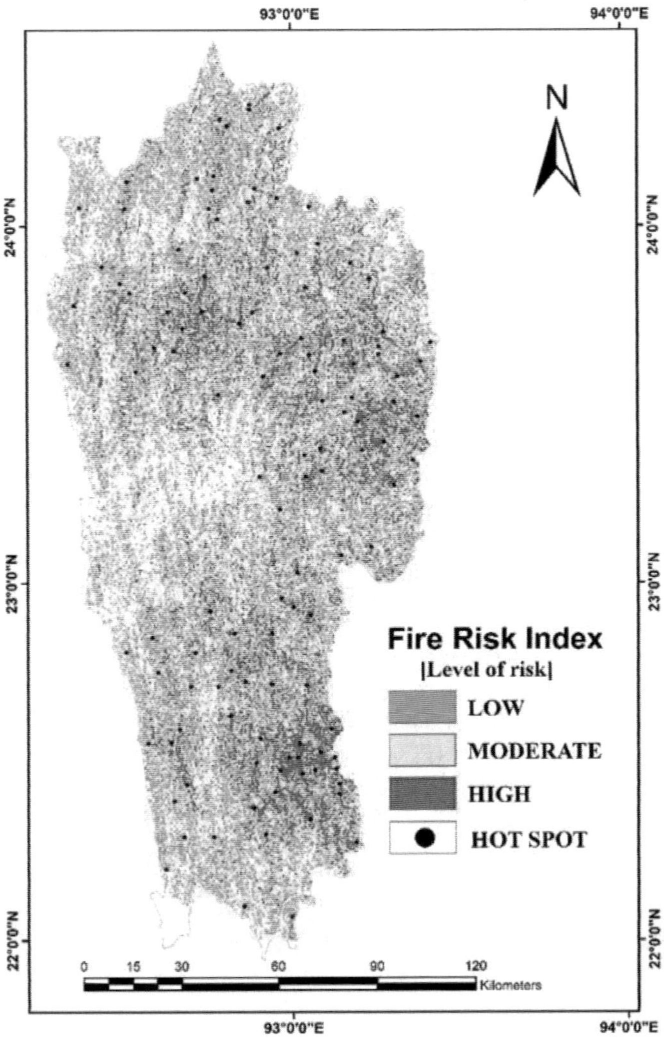


**Figure 9.** Model validation map

## 5. Conclusions:

As part of this study GIS has been used to integrate multiple layers of datasets for use in the modeling of forest fire risk, and the use of remote sensing enabled fire risk information to be obtained. Our main aim is to find out the risk zone area of forest fire, and in the result, we get the accurate risk zone area by using FRI, SFI, HFI, model, but according to model validation and accuracy, FRI model gives us more accurate result to finding risk zone map. In FRI, the result shows that many fire points fall around the shifting cultivated land, roads, and settlement. These are the origin of fires that indicates the increasing number of fire incidents are mostly affected by human activities. Its affects environmental balance, climatic condition, and loss of biodiversity. Sometimes human activities become most vulnerable. As a result, the recent Amazon forest fires have had an impact on the ecology, climatic conditions, and biodiversity loss. As satellite sensors capture images of wildfires over a large area, RS and GIS applications were used to map forest fire risk zones in the research area, and the results have the potential to improve disaster management in mitigating forest fires.

**References**

[1] Altieri M A, Anderson MK, ,and Merrick L C 1987. Peasant agriculture and the conservation of crop and wild plant resources. *Conservation biology*, 1(1), 49-58.
[2] Kumar R 2013 Forest, Categories, Types, Functions and Institutional Framework for protection 61
[3] Sahana M, Areendran G, Raj K, Sivadas A , Abhijitha C S , and Ranjan K 2022 Introduction to Forest Resources in India: Conservation, Management and Monitoring Perspectives. *In Conservation, Management and Monitoring of Forest Resources in India* (pp. 3-31). Cham: Springer International Publishing.
[4] Ashutosh D K and Satendra 2014 *Forest Fire Disaster Management* (New Delhi)
[5] Forest Survey of India 2011 *India State of Forest Report 2011* vol I
[6] Biswas P K 1993 Forest, People ,and Livelihoods: The Need for Participatory Management *FAO* 6
[7] MSSRF 2010 Status Report on use of fuelwood in India 1–12
[8] Roy P S and Giriraj A 2008 Land Use and Land Cover Analysis in Indian Context *J. Appl. Sci.***8** 1346–53
[9] Miles L, Newton A C, DeFries R S, Ravilious C , May I, Blyth S, and Gordon J E 2006. A global overview of the conservation status of tropical dry forests. *Journal of biogeography*, ***33***(3), 491-505.
[10]Vasilakos C, Kalabokidis K, Hatzopoulos J and Matsinos I 2009 Identifying wildland fire ignition factors through sensitivity analysis of a neural network *Nat. Hazards***50** 125–43
[11]Adab H, Kanniah K D and Solaimani K 2013 Modeling forest fire risk in the northeast of Iran using remote sensing and GIS techniques *Nat. Hazards***65** 1723–43
[12]Liu W, Wang S, Zhou Y, Wang L and Zhang S 2010 Analysis of forest potential fire environment based on GIS and RS *2010 18th Int. Conf. Geoinformatics, Geoinformatics 2010*
[13]Ayanz J S, Gitas I, Camia A and Oliveira S 2011 *Advances in Remote Sensing and GIS applications in Forest Fire Management From local to global assessments*
[14]Disaster N and Authority M 2020 Forest Fire Management Global Best Practices
[15]Hirschberger P and 4con forestconsulting Www.forestconsulting.de 2016 *Forests ablaze Causes and effects of global forest fires* | ed WWF Deutschland (Berlin)
[16]Ajin R, Loghin A-M, Vinod P and Jacob M 2016 Forest Fire Risk Zone Mapping Using RS and GIS Techniques: A Study in Achankovil Forest Division, Kerala, India *J. Earth, Environ. Heal. Sci.***2** 109
[17]Dale V H, Joyce L A, McNulty S, Neilson R P, Ayres M P, Flannigan M D, Hanson P J, Irland L C, Lugo A E, Peterson C J, Simberloff D, Swanson F J, Stocks B J, Wotton B M C-A C C N-P 813564, P R and P M 2001 Climate change and forest disturbances: Climate change can affect forests by altering the frequency, intensity, duration, and timing of fire, drought, introduced species, insect and pathogen outbreaks, hurricanes, windstorms, ice storms, or landslides *Bioscience***51** 723–34
[18]Sundt N 2021 Climate Change is a Burning Global Issue *World Wildl. Fund* 1–2
[19]Stewart M 2019 An Analysis of Amazonian Forest Fires *Towardsdatascience* 1–33
[20]Booth D T H 2009 *Bushfires in Australia Prepared for the 2009 Senate Inquiry into Bushfires in*
[21]Forest Survey of India 2019 *Identification of Fire Prone Forest Areas Based on GIS Analysis of Archived Forest Fire Points Detected in the Last Thirteen Years* (Dehradun: Forest Survey of India)
[22]Srivastava P 2019 Forest fire—Impeding effects on the air quality
[23]Forest Survey of India 2019 *Ministry of Environment Forest and Climate Change* vol II
[24]PTI 2020 1,300 forest fires reported in Mizoram in 2020 *Indian Express* 1–3
[25]Azad S 2019 Forest fire incidents in country rise by 49% in three years *The Times Of India* 1–2

IOP Conf. Series: Earth and Environmental Science 1164 (2023) 012005 doi:10.1088/1755-1315/1164/1/012005

[26]Darlong V T 1998 Traditional community-based fire management among the Mizo shifting cultivators of Mizoram in northeast India *Food Agric. Organ. United Nations* 1–6
[27]Bachmann A and Allgower B 2001 A consistent wildland fire risk terminology is needed *Fire Management Today* ed M Dillon (Washington, DC: U.S. Department of Agriculture) pp 28–33
[28]Hardy C C 2005 Wildland fire hazard and risk: Problems, definitions, and context *For. Ecol. Manage.* **211** 73–82
[29]Dlamini W M 2011 Application of Bayesian networks for fire risk mapping using GIS and remote sensing data *GeoJournal* **76** 283–96
[30]Keifer M, Caprio A, Lineback P, Coordinator G I S, Folger K and Technician G I S 1999 Incorporating a GIS model Of ecological need into fire management plannin *Natl. Park.*
[31]MiSRaR 2010 Methodology forest fire risk map *1st MiSRaR thematic seminar* (Dordrecht)
[32]Roy P S 2003 Forest Fire and Degradation Assessment Using Satellite Remote Sensing and Geographic Information System *Satell. Remote Sens. GIS Appl. Agric. Meteorol.* 361–400
[33]Joint Research Center (JRC) 2002 *Pilot projects on forest fires*
[34]Chéret V and Denux J P 2011 Analysis of MODIS NDVI time series to calculate indicators of Mediterranean forest fire susceptibility *GIScience Remote Sens.* **48** 171–94
[35]Puri K, Areendran G, Raj K, Mazumdar S and Joshi P K 2011 Forest fire risk assessment in parts of Northeast India using geospatial tools *J. For. Res.* **22** 641–7
[36]Sebastián-López A, Salvador-Civil R, Gonzalo-Jiménez J and SanMiguel-Ayanz J 2008 Integration of socio-economic and environmental variables for modelling long-term fire danger in Southern Europe *Eur. J. For. Res.* **127** 149–63
[37]Li L M, Song W G, Ma J and Satoh K 2009 Artificial neural network approach for modeling the impact of population density and weather parameters on forest fire risk *Int. J. Wildl. Fire* **18** 640–7
[38]Salinero E C 2003 *Wildland fire danger : estimation and mapping : the role of remote sensing data* (Singapore: River Edge, N.J. : World Scientific)
[39]Grana M and Duro R J 2008 *Computational Intelligence for Remote Sensing* (Berlin: Springer)
[40]North East Zone Cultural Center 2021 Introduction Of North East *Minist. Cult. Gov. India* 1–4
[41]Census2011 2021 Top 10 Smallest States of India by Total Area *Census2011* 1–3
[42]IAS INSIGHTS 2020 *Mizoram Current Affairs General Knowledge Yearbook 2020* (New Era Publication)
[43]Kant K 2020 Mizoram – State Overview, Geography, Transportation, Education, Language *Embibe* 1–17
[44]Ministry of Communication & Information Technology, National Informatics Centre and Mizoram State Centre 2021 Mizoram at a glance *Minist. Commun. Inf. Technol. Natl. Informatics Cent. Mizoram State Cent.* 1–2
[45]DPpedia 2021 About: Mizoram *DPpedia* 1–35
[46]Mizoram StateE-Governance S 2021 History of Mizoram *A Gov. Mizoram Undert.* 1–14
[47]Geological Survey of India 2011 *Geology and mineral resources of Manipur, Mizoram, Nagaland and Tripura*
[48]USGS 2019 EarthExplorer
[49]OpenStreetMap contributors 2017 Planet dump retrieved from https://planet.osm.org
[50]Jarvis, A.; Reuter, H; Nelson, A.; Guevara E 2008 SRTM 90m DEM Digital Elevation Database
[51]NASA Firms 2019 MODIS Collection 6 NRT Hotspot / Active Fire Detections
[52]Gao X, Fei X and Xie H 2011 Forest fire risk zone evaluation based on high spatial resolution RS image in Liangyungang Huaguo Mountain Scenic Spot *ICSDM 2011 - Proc. 2011 IEEE Int. Conf. Spat. Data Min. Geogr. Knowl. Serv.* 593–6
[53]Castro R, Chuvieco E, Castro R and Chuvieco E 2008 Modeling forest fire danger from geographic information systems Modeling Forest Fire Danger from Geographic Information Systems *Geocarto Int.* 37–41
[54]Weise D R and Biging G S 1997 A qualitative comparison of fire spread models incorporating wind and slope effects *For. Sci.* **43** 170—180

[55]DeBano L F, Neary D G and Ffolliott P F 1998 *Fire Effects on Ecosystems* (Wiley)
[56]Kushla J D and Ripple W J 1997 The role of terrain in a fire mosaic of a temperate coniferous forest *For. Ecol. Manage.***95** 97–107
[57]Hemmleb M, Weritz F, Schiemenz A, Grote A and Maierhofer C 2006 Multi-spectral data acquisition and processing techniques for damage detection on building surfaces *Proc. ISPRS Comm. V Symp. "Image Eng. Vis. Metrol.*
[58]Lin M L, Chen C W, Shih J Y, Lee Y T, Tsai C H, Hu Y T, Sun F and Wang C Y 2009 Using MODIS-based vegetation and moisture indices for oasis landscape monitoring in an arid environment *Int. Geosci. Remote Sens. Symp.***4** 338–41
[59]Sader S A and Jin S 2006 Feasibility and accuracy of modis 250m imagery for forest disturbance monitoring *ASPRS. Annual Conference, Reno* (Citeseer)
[60]Bo-Cai G 1996 NDWI A Normalized Difference Water Index for Remote Sensing of Vegetation Liquid Water From Space *Remote Sens Env.***58** 257–66
[61]Jaiswal R K, Krishnamurthy J and Mukherjee S 2005 Regional study for mapping the natural resources prospect & problem zones using remote sensing and GIS *Geocarto Int.***20** 21–31
[62]Morrison P H 2007 Roads and Wildfires 40
[63]Avila-Flores D, Pompa-Garcia M, Antonio-Nemiga X, Rodriguez-Trejo D A, Vargas-Perez E and Santillan-Perez J 2010 Driving factors for forest fire occurrence in Durango State of Mexico: A geospatial perspective *Chinese Geogr. Sci.***20** 491–7
[64]Erten E, Kurgun V and Musaoglu N 2004 Forest Fire Risk Zone Mapping From Satellite Imagery and GIS A Case Study *Geo-imagery bridging continents, Turkey* (Istanbul,: ISPRS)
[65]Siachalou S, Doxani G and Tsakiri-Strati M 2009 Integrating Remote Sensing Processing and GIS to Fire Risk Zone Mapping: a Case Study for the Seih-Sou Forest of Thessaloniki *Proc. 24th Int. Cartogr. Conf.* 1–10
[66]Chuvieco E and Congalton R G 1989 Application of remote sensing and geographic information systems to hydrology *Remote Sens Env.* 147–59
[67]Beguería S 2006 Validation and evaluation of predictive models in hazard assessment and risk management *Nat. Hazards***37** 315–29

# Assessing the Water Quality of Indian Sundarban Estuaries using Remote Sensing Techniques

**S Chatterjee[1]*, T Bhadra[1*]**

[1] Department of Geography, Adamas University, Kolkata, India

*Corresponding Author Email:
suman.chatterjee27@yahoo.in, tuhinbhadra.au@gmail.com

**Abstract.** Remote Sensing for water quality assessment is a newly emerging field. Although numerous studies have been conducted to evaluate the correlation between the Digital Number (DN) of bands and water quality parameters (WQP), further studies are required to make this methodology a robust low-cost technique for water quality assessment. With a brief case study of spatio-temporal analysis of WQPs of the Sundarban estuary, the current work explores the possibility of multispectral remote sensing for large-scale water quality monitoring. The study is about establishing empirical relationships between DN values (single, multiple or combination of many bands) with a limited number of in-situ measurements of WQPs such as Chlorophyll algae, Turbidity, pH, Salinity and Euphotic Depth. The study also assesses seasonal variation of WQPs for the period 2013-2014. Stepwise regressions have been performed to select the best predictors of each WQPs; afterwards, simple or multiple regression has been performed according to the result of stepwise regression. Temporal variation has been assessed for summer and winter using those predicted maps. The predictors of Chlorophyll-a, Euphotic Depth, pH, Salinity, Turbidity are B4/B5; B4/B6; B5/B7; B5/B7; B6/B7 for summer and B3/B5; B2/B3, B4; B1, B3/B6; (B2/B5); (B2/B5) for the winter. Three best model estimates are pH and chlorophyll-a of both the seasons with RMSEs of 0.202865, 0.059793, 0.477288568, and 0.224603275905043 respectively. Higher chlorophyll-a, pH and turbidity found in Hooghly estuary in both seasons, at Matla, Thakuran, Raimangal and Harinbhanga higher salinity found in both season and higher euphotic depth found in summer.

**Keywords:** Chlorophyll-a, Estuary, Euphotic Depth, pH, Remote Sensing, Salinity, Sundarban, Turbidity, Water Quality.

## 1. Introduction

Sundarban Estuary is one of the most vulnerable estuaries with respect to climate change [1–3]. The resources of the rich mangrove and the livelihood of the inhabitants highly depend on freshwater source and quality of the water in the region [4]. And the water quality is depends on the fresh water river joining the Bay of Bengal and forming this deltaic region [5, 6]. Therefore, it is mandatory to monitor the water quality of those rivers and the estuaries frequently. Water quality monitoring with dense monitoring stations are sometimes difficult to establish due to cost factor. Therefore, there should be an alternative way to estimate the water quality with less number of stations. Such methodologies have been emerging in last decades with the help of remotely sensed data where more accurate mapping is possible with comparatively less available monitored data. Water quality monitoring is a complicated and expensive task to carry out [7]. Comprehensive spatio-temporal assessment is crucial for the assessment of ecological degradation and its management [8]. Number of samples, selection of water quality variables and spatio-temporal scale plays an important role in determining the costs of the program [9]. The task of sample designing becomes daunting for large scale monitoring since the number of variables tends to increase with the expansion of spatio-temporal domain [10]. With large scale monitoring program, we need more resources and it increases exponentially with the increasing number of samples and in this context remote sensing techniques comes handy because it offers great possibilities with improved spectral and spatial resolution sensors.

Numerous studies regarding application of remote sensing in water quality assessment have been conducted since the advancement in moderate resolution sensors e.g. Landsat Thematic Mapper, Enhanced Thematic mapper, Operational land imager based on the empirical relationship between spectral response of the water and the physico-chemical characteristics of water [11–15]. WQPs such as suspended solids (Turbidity), algae, DOM, Oils, few aquatic plants and heat releases from the water surface are able to change the reflected or emitted solar energy spectrum [16]. Relationship between water quality parameters and spectral reflectance is generally established using statistical empirical models [17]. However the model needed to be changed with multi-temporal aspect due to dynamic characteristics of those WQPs with the change of sun elevation, angle and atmospheric parameters [18]. Cost effectiveness, large area cover and open access moderate to high resolution satellite data made this technique very efficient and felicitous for spatio-temporal analysis of WQPs than the conventional techniques [15, 16, 19–21].

Current Study is about exploring the possibilities of multispectral remote sensing for large scale water quality monitoring with a short case study of spatio-temporal analysis of WQPs of the Sundarban estuarine. The study is about establishing empirical relationship between DN values (single, multiple or combination of many bands) with limited number of in-situ measurements of WQPs such as Chlorophyll algae, Turbidity, pH, Salinity and Euphotic Depth. The study also assesses seasonal variation of WQPs for the period 2013-2014.

## 2. Study Area

The study area comprises estuaries of Hugli, Muriganga, Saptamukhi, Thakuran, Matla, Gosaba, Harinbhanga and Raimangal and a part of Bay of Bengal covering a total area of 5758 sq km. The study area is demarcated by the River Hooghly in the West, Rivers Harinbhanga and Raimangal in the East, six administrative blocks of the adjacent 24 Parganas (N) and 13 blocks of 24 Parganas (S) of west Bengal in the north and Bay of Bengal in the south. The study area is restricted within the latitudes of 21°31' N and 22°32' N and longitude between 87°34' E and 89°33' E (figure1).

## 3. Data Collection

Two types of data used in the current study i.e. remotely sensed data and water quality data. Satellite data has been retrieved from USGS Earth explorer (table 1).

**Table 1.** Data Acquisition from USGS Earth Explorer

| Sensors | Path and Row | Acquiring date and Time |
|---|---|---|
| Landsat TM data | Path: 138, Row: 45 | 31/12/2013, 04:32:48.2312149Z |
| Landsat OLI data | Path: 138, Row: 45 | 22/04/2014, 04:31:17.1198760Z |

***Source*:** Determined by the author.

### *3.1. Water Quality data*

Water quality data has been collected from two sources primary and secondary sources. Water samples were collected hourly for two dates i.e. 31/12/2013 and 22/04/2014 (winter and summer) from 12 well distributed sampling stations (figure 1), in-situ measurement (pH, Salinity, turbidity) were carried out using probe during the sampling. Due to limited resources and fund the data of 22/04/2014 for the stations H1, H2, H3, H4, M1, M2 and M3 were taken from Secondary Sources (table 2). For chlorophyll-a water samples poured into 1 L light blocking container and transported to the laboratory as soon as possible. Turbidity meter was used to measure turbidity. Euphotic depths were calculated from secchi disk depth.

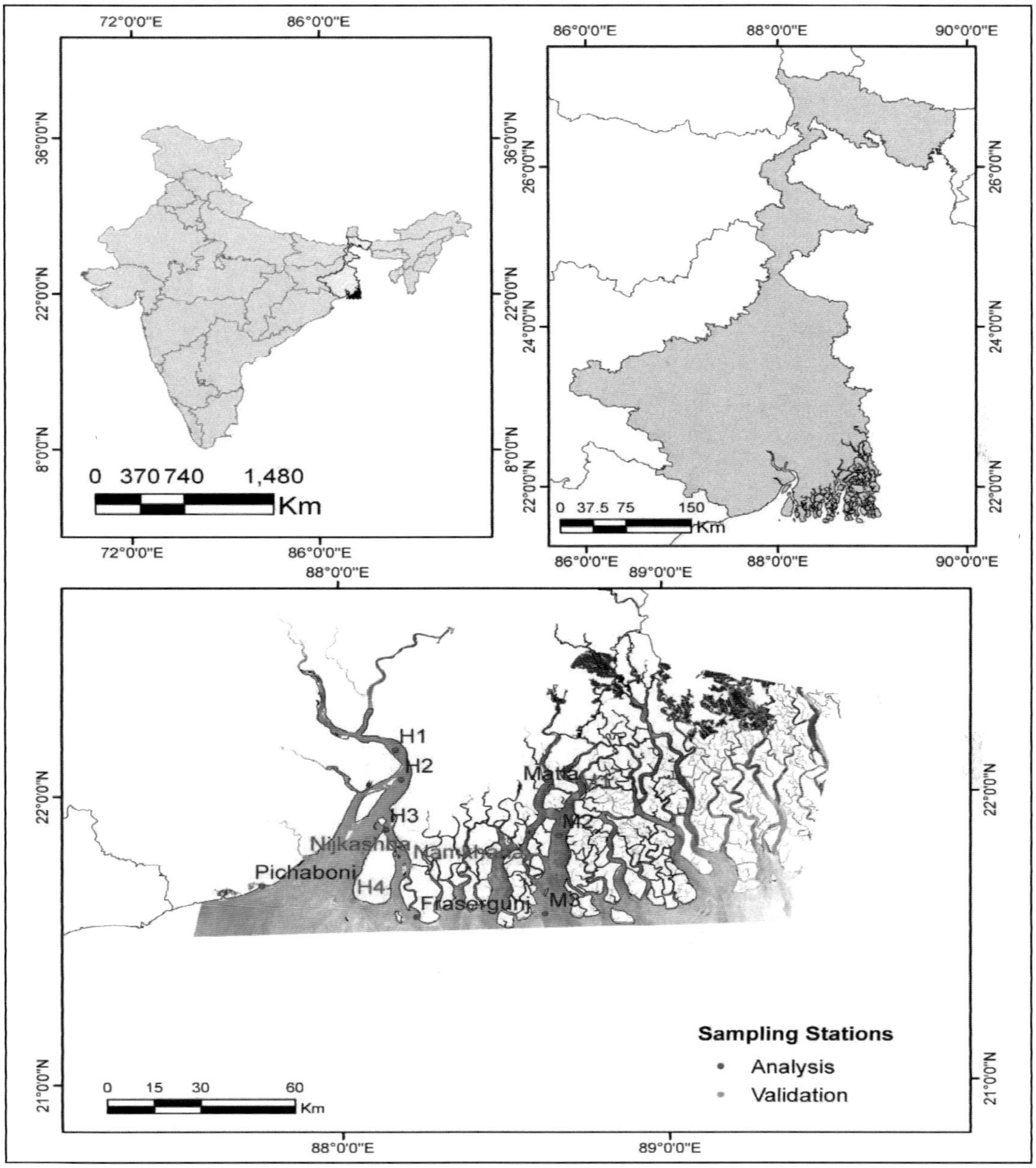


**Figure 1.** Location and Extent Map of the Study Area.

**Table 2.** Sources of water quality data

| WQPs | Dates | Data Sources | Stations | Season |
|---|---|---|---|---|
| pH, Salinity, | 31/12/2013 | Primary | All 12 Stations | Winter |
| Chlorophyll-a, | 22/04/2014 | Primary | Nijkashba H5 | Summer |
| Turbidity, | | | Namkhana H6 | |
| Euphotic | | | Pichaboni H7 | |
| depth | | | Frasergunj H8 | |
| | | | Matla M0 | |
| | | Secondary [22] | Diamond Harbour H1 | |
| | | | Haldia H2 | |
| | | | Tip of Sagar Island H3 | |
| | | | River Mouth H4 | |
| | | | Jharkhali M1 | |
| | | | Bonnie Camp M2 | |
| | | | Dhanchi M3 | |

## 4. Methodology

Relationship between DN value and WQPs measurement have been tried to established using stepwise regression method later the equations have been retrieved using simple/multiple regression method (figure 2). Finally, spatial maps of WQPs have been generated using Erdas Imagine Model maker. Among total 12 sampling stations, eight have been used for analysis and four stations for validation purpose (figure 1). Water sampling for two dates have been conducted hourly and for 24-hour cycle. But the data used for regression is the data taken in the same hour in which satellite image were taken.

### *4.1. Atmospheric Correction*

Success in establishment of relationship between WQPs and Digital band values depend on the accurate noise free satellite data and data should be corrected excluding atmospheric effects [23]. Therefore, it is necessary to remove the scattering and absorption effects of the atmosphere from the radiance received by the sensors. In this study atmospheric correction has been performed in add on tools of Erdas Imagine named ATCOR 2014 for each of the images.

### *4.2. Extraction of Water body*

Water body of the image has been extracted using Normalised Difference water Index then converted in binary NDWI Image using conditional statement. Subseting water body from the satellite images: Binary NDWI images have been converted to Polygon then rectified it by excluding the unnecessary polygons. Those vector files of water body boundary have been used to subset the satellite images to area of interest.

### *4.3. Extraction of DN values*

Water and its properties are dynamic in nature. To address the problem and to avoid sampling error an averaged DN value have been used instead of single DN value of the respective coordinates of in situ water sampling. Each image was geographically controlled (to give latitude and longitude) and then each sea-truth data collection site was located on the image using the GPS positions. Means and standard errors of radiance values for each wavelength band have been extracted using Cubic convolution methods where: The closest 4 x 4 block (representing an area 120 m x 120 m) of input cells is used to compute each output cell value (Sampling location). The weighting factors for the average of the input cells have been computed using a cubic (third-order) function of distance (figure 3).

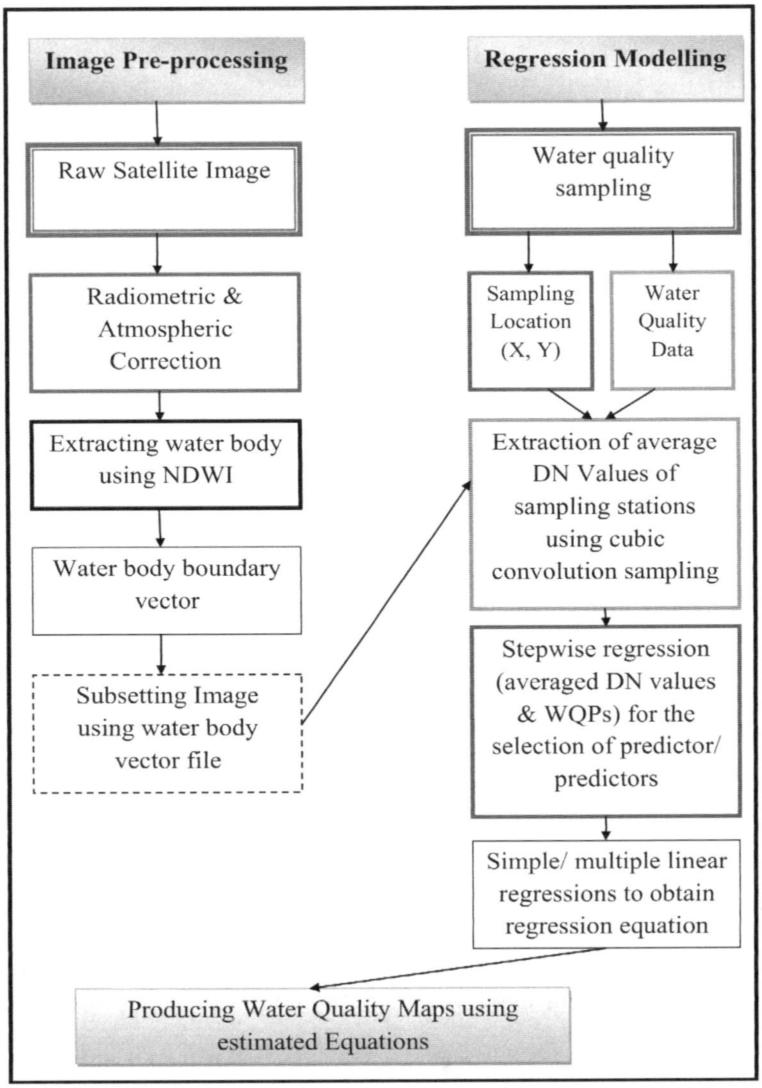


**Figure 2**. Methodology adopted for Water Quality Mapping

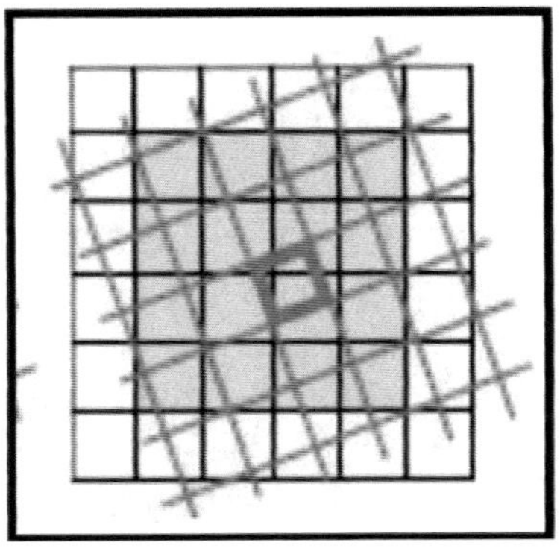

**Figure 3.** Depicting Cubic Convolution Sampling

IOP Conf. Series: Earth and Environmental Science 1164 (2023) 012006 doi:10.1088/1755-1315/1164/1/012006

*4.4. Regression modelling*

Stepwise regressions have been performed in IBM SPSS 21 software to predict the best predictor/predictors for each specific WQPs. WQPs have been taken as the dependent variable and the averaged digital values of various bands (and ratios of those values; except Thermal band) as the independent variables (table 3). To retrieve the equation after stepwise regression simple linear for single significant predictor and multiple linear regression for multiple significant predictors have been performed in NCSS 11 software. Finally, those estimated equations have been used in the model maker to generate the water quality maps.

**Table 3.** Sources of water quality data

| **Dependent variable** | **Independent variables (30)** |
|---|---|
| WQPs (pH, Salinity, Chlorophyll-a, Turbidity and Euphotic depth) | OLI and TM Bands 1 to 7, B1/B2, B1/B3, B1/B4, B1/B5, B1/B6, B1/B7, B2/B3, B2/B4, B2/B5, B2/B6, B2/B7, B3/B4, B3/B5, B3/B6, B3/B7, B4/B5, B4/B6, B4/B7, B5/B6, B5/B7, B6/B7,(B3+B4)/2, $B3^2 \times B4^2$ |
| B1 to B7 denoting Band 1 to Band 7 | |

***Source*:** Determined by the author.

*4.5. Validation of the method*

The method used in the study has been validated using the data of four sampling stations i.e. Nijkashba, Namkhana, H4 and M1. Root mean square error has been calculated with the actual measurement and predicted measurements of those stations.

## 5. Result and Discussion

These sections has been discussed in two parts Result of regression modelling and Spatio-temporal assessment of WQPs.

*5.1. Result of regression modelling*

All the predictors and estimated equations have been displayed in table 4 and 5. Single predictor proved to be significant for all the parameters except euphotic depth and pH in winter (table 4 and 5). All the independent variables of the models explained well the variation in the dependent variable at 5% significant levels.

**Table 4.** Result of Regression modelling (Summer)

| Model No. | Dependent Variables | Best Predictors | Estimated Equation | R | N |
|---|---|---|---|---|---|
| 1. | Chlorophyll-a | B4/B5 | Chlorophyll-a = (3.56190085919674) + (-0.893942334752662) * (B4_B5) | 0. 808 | 8 |
| 2. | Euphotic Depth | B4/B6 | Euphotic-depth = (50.2914547388159) + (-3.3163191989554) * (B5_B7) | 0. 734 | 8 |
| 3. | pH | B5/B7 | pH = (8.12259133379721) + (-0.0234404151408061) × (B5_B7) | 0.803 | 8 |
| 4. | Salinity | B5/B7 | Salinity = (26.8558879218168) + (-2.07430326441807) * (B5_B7) | 0.771 | 8 |
| 5. | Turbidity | B6/B7 | Turbidity = (3.21139688543917) + (178.16226932608) * (B6_B7) | 0.778 | 8 |

**Table 5.** Result of Regression modelling (Winter)

| Model No. | Dependent Variables | Best Predictors | Estimated Equation | R | N |
|---|---|---|---|---|---|
| 1. | Chlorophyll-a | B3/B5 | Chlorophyll-a= (3.50997170962726) + (-0.462555493759846) × (B3/B5) | 0.874 | 8 |
| 2. | Euphotic Depth | B2/B3, B4 | Euphotic depth = -175.984730974756 + 356.108565919633 × B2/B3 - 0.00181240873432809 × B4 | 0.981 | 8 |
| 3. | pH | B1, B3/B6 | pH = 9.04815528499263 - 0.00111443710606523 × B1 + 0.0210166922306594 × B3/B6 | 0.995 | 8 |
| 4. | Salinity | B2/B5 | Salinity = (-13.269836840472) + (13.6646420202985) × (B2/B5) | 0.924 | 8 |
| 5. | Turbidity | B2/B5 | Turbidity = (246.233344131179) + (-58.73777462879) × (B2/B5) | 0.870 | 8 |

*5.1.1. Validation.* For winter RMSE of chlorophyll-a, Euphotic depth, pH, salinity and turbidity computed as 0.477288568, 3.548656, 0.202865, 6.499463 and 7.505418 respectively. For summer RMSE of chlorophyll-a, Euphotic depth, pH, salinity and turbidity computed as 0.224603275905043, 3.418713, 0.059793, 3.799318, 19.59952 respectively.

*5.2 Spatio-temporal analysis*

Considering two seasons' summer and winter the WQPs have been reviewed using the spatial maps.

*5.2.1. Chlorophyll-a.* In the study although highest concentration of Chlorophyll-a (3.50 mg/m$^3$) found during winter yet it was more spatially abundant and well distributed in summer especially in Hooghly, Harinbhanga and Raimangal estuaries.

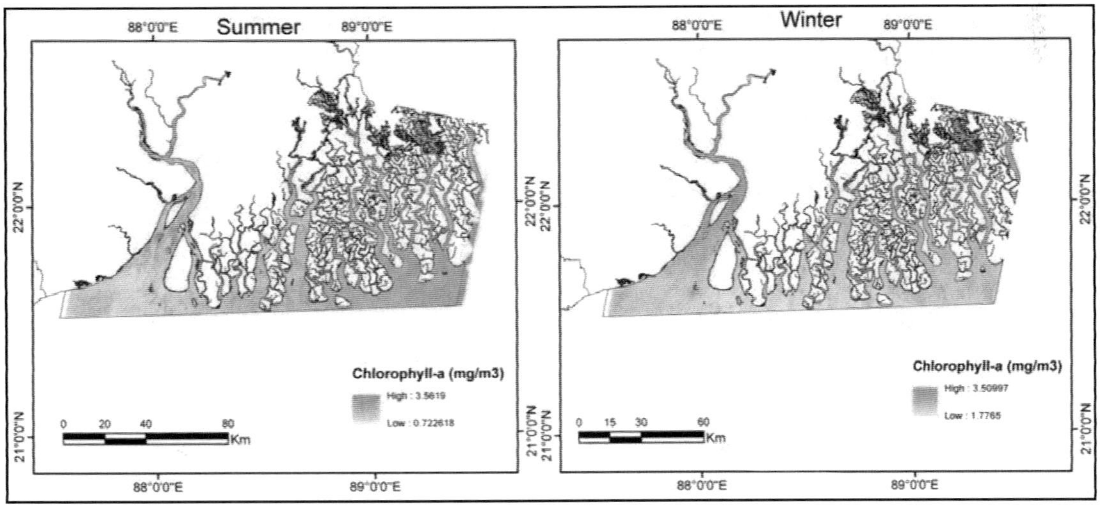


**Figure 4.** Spatio-temporal variation of Chlorophyll-a in Indian Sundarbans

During winter, it was scarce and dispersed; high concentration is found only in Hooghly estuary. In both seasons Chlorophyll-a found higher near shore and streams (figure 4). Among in-situ measurements the highest measured at Pichaboni (2.755483 mg/m$^3$) and the lowest measured at M2 stations (0.98 mg/m$^3$) in summer and in winter the highest at H2 (2.925 mg/m$^3$) and the lowest again at M2 station (1.945 mg/m$^3$).

*5.2.2. Euphotic Depth.* The highest euphoric depth found in winter season (221.135 cm) in the distant parts of the sea. During summer most of the water body found to be clear except Hooghly estuary. Majority of the sea and all the estuaries found ambiguous during winter (figure 5). From in-situ measurements the highest found at Pichaboni (51.25508443 cm), the lowest at H2 (24 cm) in summer and the highest found at Frasergunj (77.06736116 cm), the lowest at H4 (41cm) in winter.

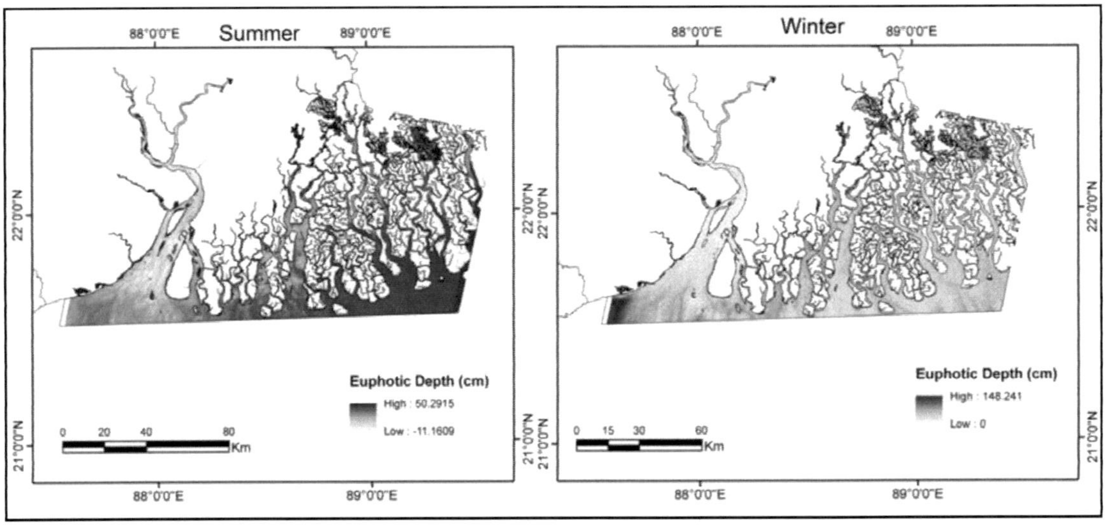


**Figure 5.** Spatio-temporal variation of Euphotic depth in Indian Sundarbans

*5.2.3. Salinity.* Among all estuaries Matla, Thakuran, Raimangal and Harinbhanga estuaries found comparative more saline than Hooghly estuary in both seasons. From both spatial and intensity perspectives in winter surface water found more saline than summer (figure 6).

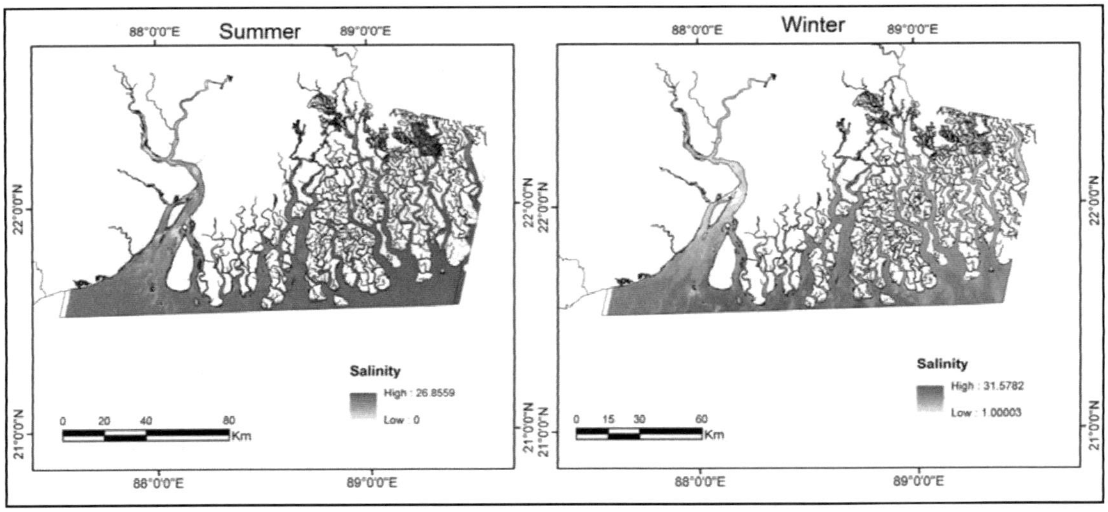


**Figure 6.** Spatio-temporal variation of Salinity in Indian Sundarbans

The highest salinity measured in summer through in-situ measurement was 28.88696 at M1 station and the lowest at H1 station with salinity of 5.533333. In winter the highest salinity measured through in-situ measurement was 23.95417 at M3 and the lowest again at H1 stations with salinity of 1.2125.

*5.2.4. pH.* Hooghly estuary found slightly acidic in both seasons. Most of the sea parts and all the estuaries except Hooghly found to be basic during summer. During winter all estuaries found slightly acidic with low to moderate pH (figure 7). In summer during in-situ measurements the highest pH found at M1 (8.14425) stations and the lowest at H3 (7.947375). In winter the highest pH found at Frasergunj (8.282949) station and the lowest at Nijkashba (7.853767).

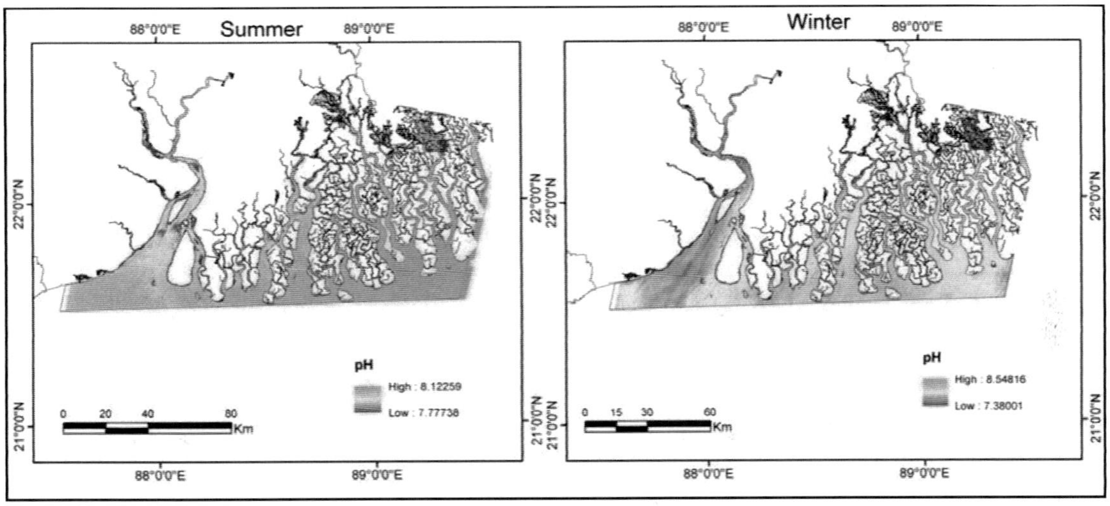


**Figure 7.** Spatio-temporal variation of pH in Indian Sundarbans

*5.2.5. Turbidity.* Water found to be more turbid in summer**.** Turbidity at Hooghly estuary found higher than all the other eastern estuaries in both seasons. The water towards inland i.e. small streams found more turbid than the sea parts in both seasons (figure 8). Among in-situ measurements the highest measured at H4 (285.4167 NTU) and the lowest measured at M3 stations (215.1667 NTU) in summer and in winter the highest at H4 (180.5455 NTU) and the lowest at M2 and M3 stations (83 NTU).

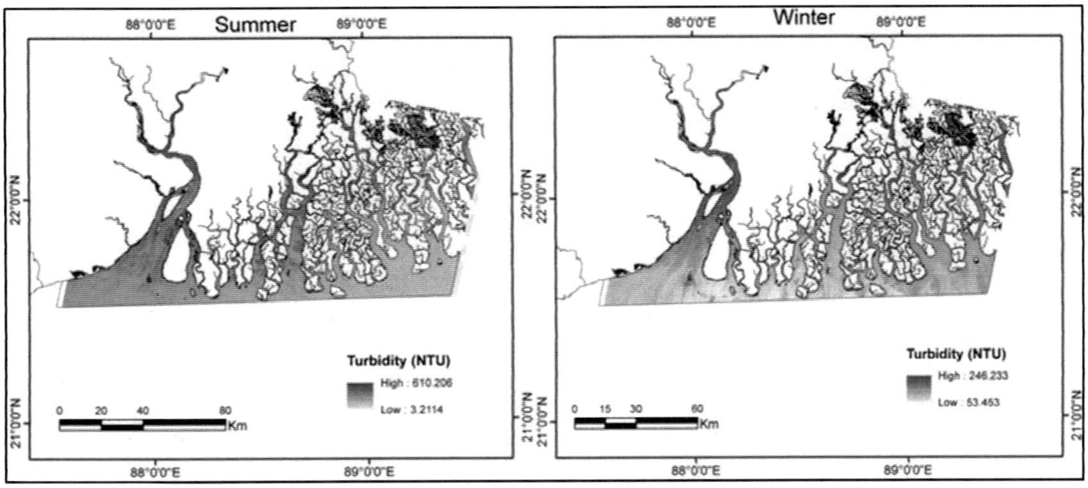


**Figure 8.** Spatio-temporal variation of Turbidity in Indian Sundarbans

### *5.3. Discussion*
We have discussed the findings of our study in following three sections namely regression modelling, Validation and spatio-temporal analysis.

5.3.1. *Regression modelling.* Instead of atmospheric correction the relationship between DN values and WQP measurements change with season with the changes of sun angle, sun elevation. The study revealed that the ratio of bands is the dominant predictor among all other combinations or a single band. Green, red and short-wave infrared-one band are the most used predictors for the models.

*5.3.2. Validation.* Being extremely dynamic entities water and its composition changes every second both spatially and magnitude wise, hence producing error during predictive modelling is inevitable. Apart from atmospheric error, improper extraction of water boundary, error during in-situ measurement, insufficient in-situ stations are other causes. In the study model of turbidity, produced highest error for both seasons might be due to insufficient in-situ measurements and even insufficient numbers of validation stations.

*5.3.3. Spatio-temporal analysis.* Chlorophyll-a is the green pigment present in plants and an indicator of presence of phytoplankton in the coastal and estuarine waters used as a WQP [24]. Chlorophyll concentration fluctuate in different season even in the same season due to several factors such as sea surface temperature, wind forcing, rainfall, supply of nutrients, upwelling cold water etc. Highest concentration found in winter season as some phytoplankton such as diatom and Skeletonema costatum prospers in cooler water [25, 26]. Spatial abundance in summer may be due to the circulation and upwelling of cold water with nutrients [27]. Higher concentration of Hooghly estuary is due to the nutrients supply from inlands, which is also the reason of higher concentration in narrow streams towards inlands. Higher concentration in the distant part at sea (south-east) is observed may be due to high euphotic depth or clear marine water as amount of phytoplankton depends the depth of the clear photic layer. Euphotic depth can be defined as the uppermost layer of water column above the compensation points that supports the net primary productivity. It is a WQP indicating clarity of surface seawater and important in terms of primary production of oceanic ecosystem. In our study, Higher Euphotic depth found at the distant part of the sea and low depth at the turbid water of estuaries as depth of photic zone strongly depends upon the suspended particles present in the water. Hence, it varies from 100 meters in clean ocean to few cm in turbid inland streams. Fresh water supply from inland made Hooghly estuary less saline than the other estuary in both seasons. Salinity is dominant mainly over sea part than the estuaries some mixing can be seen in winter due to upwelling of more saline cold water from bottom of the sea. The acidic nature of Hooghly and other estuaries is may be due to excess nutrients supply and algal blooms or the contribution of acidic compounds from rivers as pH of rivers is generally lesser than the sea water. As salt ions helps in aggregation and settlement of suspended particles higher salinity generally brings higher clarity therefore small rivers found more turbid than the sea part. Higher sediments and suspended particles in Hooghly River also caused highest turbidity.

## 6. Conclusion
Any environmental monitoring program will usually need a substantial investment of time and money. Funding shortages impose restrictions, particularly on individual researchers who lack sufficient funding and assistance. Expenses in every monitoring program come in phases, such as equipment costs, surveying costs, laboratory analysis costs, and so on, but they all revolve around two key factors: the parameters we pick and the quantity of samples we collect. Hence, evaluating the efficiency of moderate resolution remotely sensed data along with fewer numbers of ground truth measurements to assess the different WQPs will be helpful. The method is extremely beneficial and cost effective as it can cover such a large area, which is not possible for conventional in-situ measurements and interpolation techniques. Although establishing relation between DN and WQPs is

comprehensive in terms of data production of both spectral and water quality data in which there is a tendency to accumulate error during the process yet it can produce better results than the conventional methods. Although it produces high RMSEs for turbidity and euphotic depth, the errors were negligible for other parameters. Due to lack of available resources in our study, we used eight in-situ measurement stations, which certainly are not sufficient for such a large area, the error can be reduced using more in-situ measurement, and high resolution remotely sensed data. Although fewer numbers of in-situ measurements can produce satisfactory results, incorporating more ground truths will surely increase the accuracy than conventional interpolation methods like IDW or Spline. This technique requires more studies using more in-situ measurements to make it a robust technique.

**References**

[1] Hazra S, Ghosh T, DasGupta R and Sen G 2002 Sea Level and associated changes in the Sundarbans *Science and Culture* **68** 309–21

[2] Chand B K, Trivedi R K, Dubey S K and Beg M M 2012 *Aquaculture in changing climate of Sundarban, Survey Report on Climate Change Vulnerabilities, Aquaculture Practices and Coping Measures in Sagar and Basanti Blocks of Indian Sundarbans* (Kolkata)

[3] Neogi S B, Dey M, Kabir S L, Masum S J H, Kopprio G, Yamasaki S and Lara R 2017 Sundarban mangroves: diversity, ecosystem services and climate change impacts *Asian Journal of Medical and Biological Research* **2** 488–507

[4] Chowdhury M T . 2010 Resource-dependent livelihoods in the Sundarbans *Center for River Basin Organizations and Management, Solo, Central Java, Indonesia Government of Bangladesh (Gob) National Tourism Policy*

[5] Rahman M M, Rahman M T, Rahaman M S, Rahman F, Ahmad J U, Shakera B and Halim M A 2013 Water Quality of the World's Largest Mangrove Forest *Canadian Chemical Transactions* **1** 141–56

[6] Banerjee K 2013 Decadal Change in the Surface Water Salinity Profile of Indian Sundarbans: A Potential Indicator of Climate Change *Journal of Marine Science: Research and Development* **01**

[7] Hunter S and Waterman R W 2016 *Enforcing the Law: Case of the Clean Water Acts*

[8] Nadmitov B, Hong S, In Kang S, Chu J M, Gomboev B, Janchivdorj L, Lee C H and Khim J S 2015 Large-scale monitoring and assessment of metal contamination in surface water of the Selenga River Basin (2007–2009) *Environmental Science and Pollution Research*

[9] Alilou H, Moghaddam Nia A, Keshtkar H, Han D and Bray M 2018 A cost-effective and efficient framework to determine water quality monitoring network locations *Science of the Total Environment*

[10] McDonald T L 2003 Review of environmental monitoring methods: Survey designs *Environmental Monitoring and Assessment*

[11] Gitelson A, Garbuzov G, Szilagyi F, Mittenzwey K, Karnieli A and Kaiser A 1993 Quantitative remote sensing methods for real-time monitoring of inland waters quality *International Journal of Remote Sensing* **14** 1269–1295

[12] Dekker A G and Peters S W M 1993 The use of the Thematic Mapper for the analysis of eutrophic lakes: a case study in the Netherlands. *International Journal of Remote Sensing* **14** 799–821

[13] Alparslan E, Aydöner C, Tufekci V and Tüfekci H 2007 Water quality assessment at Ömerli Dam using remote sensing techniques *Environmental Monitoring and Assessment* **135** 391–8

[14] Bustamante J, Pacios F, Díaz-Delgado R and Aragonés D 2009 Predictive models of turbidity and water depth in the Doñana marshes using Landsat TM and ETM+ images *Journal of Environmental Management* **90** 2219–25

[15] Nas B, Ekercin S, Karabörk H, Berktay A and Mulla D J 2010 An Application of Landsat-5TM Image Data for Water Quality Mapping in Lake Beysehir, Turkey *Water, Air, and Soil Pollution* **212** 183–97

[16] Ritchie J C, Zimba P V. and Everitt J H 2003 Remote Sensing Techniques to Assess Water

Quality *Photogrammetric Engineering and Remote Sensing* **69** 695–704
[17] Dekker A G, Zamurović-Nenad Ž, Hoogenboom H J And Peters S W M 1996 Remote sensing, ecological water quality modelling and in situ measurements: a case study in shallow lakes *Hydrological Sciences Journal* **41** 531–47
[18] Bagchi D and Bussa R 2016 Application of Remote Sensing in Water Quality and Water Resources Management – An Overview *Bhu-Jal News, Silver Jubilee Publication-II* **25** 39
[19] Haddad K D and Harris B A 1985 Assessment and trends of Florida's marine fisheries habitat: an integration of aerial photography and thematic mapper imagery *Eleventh international symposium. Machine processing of remotely sensed data with special emphasis on quantifying global process: Models, sensor systems, and analytical methods* ed S K M and D B Morrison (West Lafayette)
[20] Bhatti A M, Nasu S, Takagi M and Nojiri Y 2008 Assessing the Potential of Remotely Sensed Data for Water Quality Monitoring of Coastal and Inland Waters *Bulletin of Kochi University of Technology* **5** 201–207
[21] Miller H M, Sexton N R, Koontz L, Loomis J, Koontz S R and Hermans C 2011 *The Users, Uses, and Value of Landsat and Other Moderate-Resolution Satellite Imagery in the United States—Executive Report* (New York: USGS)
[22] Akhand A, Chanda A, Manna S, Das S, Hazra S, Roy R, Choudhury S B, Rao K H, Dadhwal V K, Chakraborty K, Mostofa K M G, Tokoro T, Kuwae T and Wanninkhof R 2016 A comparison of CO2 dynamics and air-water fluxes in a river-dominated estuary and a mangrove-dominated marine estuary *Geophysical Research Letters* **43**
[23] Gordon H R and Morel A Y 2012 *Remote assessment of ocean color for interpretation of satellite visible imagery: A review* (Springer Science and Business Media.)
[24] Krantzberg G, Tanik A, do Carmo J S A, Indarto A, Ekdal A, Gurel M and Li G 2010 *Advances in Water Quality Control* (Scientific Research Publishing, Inc. USA.)
[25] Goldman J C 1977 Temperature effects on phytoplankton growth in continuous culture1 *Limnology and Oceanography* **22** 932–6
[26] Smith W O, Steinberg D K, Bronk D A and Tang K W 2009 *Marine plankton food webs and climate change* (Virginia)
[27] Andrews J C and Gentien P 1982 Upwelling as a Source of Nutrients for the Great Barrier Reef Ecosystems: A Solution to Darwin's Question? *Marine Ecology - Progress Series* **8** 257–69

# Use of "Local Climate Zones" for Detecting Urban Heat Island: A Case Study of Kolkata Metropolitan Area, India

**D Bhowmick[1], K Mukherjee[1*], P Dash[2], R Mondal[3]**

[1]Department of Geography, School of Basic and Applied Science, Adamas University, Kolkata-700126, India
[2] Department of Geography, Khejuri College, Baratala, West Bengal-721431, India
[3] Symbiosis Institute of Geoinformatics, Symbiosis International (Deemed University), Pune – 411016, India

*Corresponding Author Email: kasturi.mukherjee@adamasuniversity.ac.in

**Abstract.** The land surface temperature (LST) increases both spatially and in magnitude as a result of rapid urbanization at the expense of the low radiative surface. This process modifies the urban climate which is attributable to a phenomenon recognized as the Urban Heat Island (UHI). By estimating LST, satellite remote sensing offers a cost-effective and time-saving methodology for spatio-temporal analyses of UHI. The limitation, however, is that urban and rural are becoming outmoded conceptions in landscape classification as a consequence of the high population density, particularly in countries like India. In densely populated regions, the countryside and the cities are no longer differentiated by a clear urban-rural divide. Thus, the traditional study of UHI can be enhanced by the adaptation of the newly developed "Local Climate Zones" (LCZ) characterization scheme based on building types and land cover characteristics. This LCZ mapping is an excellent method for categorizing urban areas associated with the most perceptible change in climate in urban areas, namely UHI. This paper aims to identify the LCZs dynamics over a decade from 2010 to 2020 in the Kolkata Metropolitan Area (KMA), the largest urban agglomeration in eastern India. The LCZ classification scheme has divided the landscapes into 17 conventional classes, 10 built-up classes, and 7 land cover classes that affect air temperature at a screen height of 1- 2 meters above the ground. Each LCZ is distinguishable from another by its thermal property. Each of the classes will help to understand the nature of the local climates. It has thus become comparatively simple to estimate Urban Heat Island Intensity (UHII) with the help of the LCZ scheme. The paper explains the typical features of each LCZ of KMA and accomplishes a discussion on the efficiency of mapping UHII from LCZ classes.

**Keywords:** Local climate zones, Land surface temperature, Urban heat island, Kolkata metropolitan, Urban climate, Urban heat island intensity

## 1. Introduction

The accelerated transformation of the Earth's surfaces into urban lands is a consequence of a rising percentage of the world's population residing in urban areas. By 2050, two-thirds of the projected world's population, approximately 6.2 billion people, is anticipated to live in urban areas, up from the current rate of more than 50% (or roughly 3.5 billion people) (Pacione, 2013) [1]. The world's population is expected to touch 9.8 billion by 2050, with 6.2 billion living in cities. The average annual growth of the Asia and African population has become vigorous; by 2050, 2.3 billion of the total population is expected to live mostly in Africa and Asia [2]. Additionally, it has been predicted that by 2025, 16 of the world's largest 29 megacities will be found in Asia [3]. Asia's population growth rate is historically unprecedented and would be much higher than that of any other region of the globe. Such unprecedented changes in earth's surfaces have direct impacts on the environment in many ways and at many scales. At the micro-scale urbanization alters the biophysical parameters of the cities by altering the local climate of the region [4]. The complicated chemistry of the atmospheric influences on radiation balance and greenhouse gas emissions links the local climate that cities produce to the climate at the regional and global scales. This practice produces many artificial heat sources that considerably worsen the heat island effect and lower the quality of urban environments.

When a city experiences a much warmer environment than nearby rural areas the situation is termed an urban heat island (UHI) [5-6]. The temperature gradient between urban and suburban or rural areas is attributed to the differences in land surfaces that absorb and hold heat. The agricultural fields, vegetation, and water bodies become impermeable surfaces because of urbanization, which considerably increases the retention of solar radiation, lessens convective cooling, and decreases the rate of water evaporation in their surroundings. This warmer condition of cities results in greater energy and resource consumption by the inhabitants which in turn makes the urban population more exposed to heatwaves and other extreme climatic conditions [7-9].

Over the last two decades, scientific understanding of urbanization drivers, changing land use (LU) patterns, and their effects on urban climates have advanced significantly. In-depth research on urban heat islands (UHI) has been conducted, either through direct air temperature measurements at ground monitoring stations or through the retrieval of land surface temperatures (LST) using thermal-infrared (TIR) remote sensing data. Additionally, the relationship between various LUs and various temperature zones is analyzed [10-12]. The estimation of the severity of the UHII which is considered the key parameter for assessing the UHI effect used to be calculated by the temperature differential between urban and rural areas. However, the difficulty in choosing temperatures that accurately represent urban and rural areas is a constraint. Urban-rural ideas in the categorization of landscapes are becoming obsolete because of the dense population, particularly in India. In densely populated areas, a distinct urban-rural border no longer exists between the countryside and cities [13].

Therefore, the traditional study of urban heat islands can be enhanced by the adaptation of the newly developed "Local Climate Zones" (LCZ) characterization scheme from Oke and Stewart. For an integrated quantification of the relationship between UHI and urban morphology, the idea of an LCZ was devised [13]. The contextual meaning for landscape classification of "Local" is the horizontal scales of 102 to 104 meters, "Climate" is screen height air temperature and "Zone" is an area of distinctive surface morphology, land cover, and air temperature [13]. This LCZ mapping is a great way to categorize the urban areas associated with the most perceptible change in urban climate, known as the urban heat island.

There are now 17 standard classifications of LCZ, of which the first ten (LCZ 1 to 10) characterize "built-up" forms, and the latter seven (LCZ A to G), "land cover" types [13]. The names of built-up LCZ classes generally reflect the degree of development (compact/open) and building height (high rise/midrise/low rise). Every LCZ has a distinct temperature regime at screen height that is best observable in places with simple topography, quiet, clear nights, and drier surfaces. Screen-height temperatures in LCZs are significantly influenced by the height, spacing, and surface cover (impervious/previous) of buildings, trees, and other objects [13]. While the surface structure has an impact on airflow, atmospheric heat transfer, and the balance of short-wave and long-wave radiation, the surface cover has an impact on albedo, moisture availability, and the ability to make the ground warm and cool. Consequently, a variety of geometric and surface-cover parameters, as well as values for thermal, radiative, and metabolic characteristics, are provided for each LCZ [13].

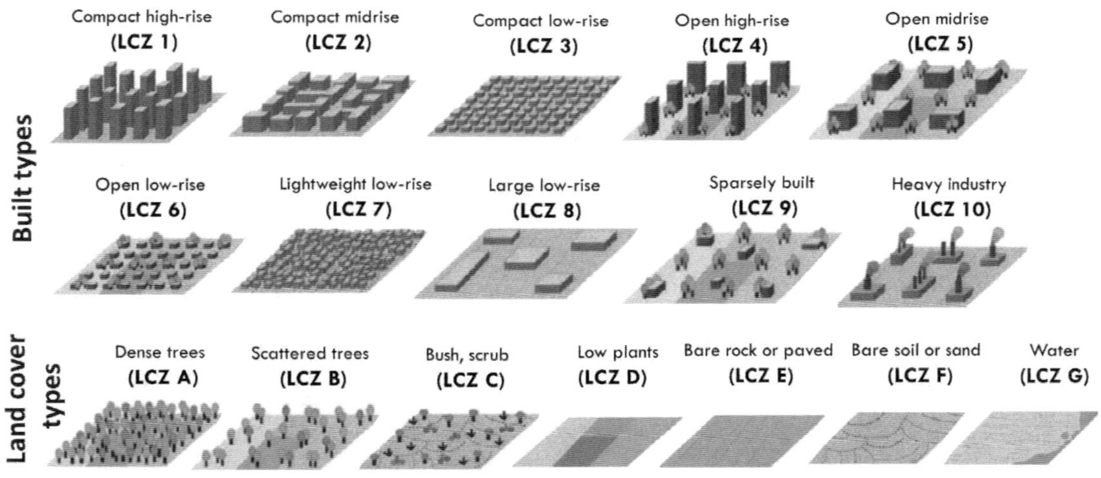


**Figure 1.** LCZ types as given by stwert and Oke, 2011

According to this approach, the landscapes are classified into 17 standard classes, 10 built-up categories, and 7 LC categories (figure 1), all of which have an influence on the air temperature at a height of one to two meters above the ground. Each LCZ is distinguishable from another by its thermal property. Each class will enable us in understanding the nature of the local climates. Thus, it highlights the importance of comparing intra-urban temperatures when investigating the effects of non-homogeneous urban geometry on the development of the local climate. This LCZ model can approximate the UHII rather than an "urban-rural" differential. As LCZ differences draw attention to the comparable surface areas and exposure characteristics of the regions, they are easier to analyze and less prone to obscure. The LCZ will enable the inter-site comparisons and establish the size of the city's urban heat island effect. Heat islands calculated in this method become more reliable indicators of city climate management than those calculated using the urban-rural method [14,15].

Bechtel has compiled the various mapping LCZ schemes that have been proposed, including (i) supervised pixel-based classification [16], (ii) a GIS-based method utilizing building data, (iii) object-based image investigation [17], and (iv) a manually selected sample of each grid cell using Geo-Wiki, followed by the digitization of homogenous LCZs. The classification of diverse urban regions has been done using the universal, robust, basic, and objective supervised pixel-based approach.

It produced significant benefits with excellent overall accuracy [17]. The methodology consists of two steps [17]. The initial step is completed using Google Earth, and it entails digitizing training areas. For the second part geometrical pre-processing and classification, SAGA-GIS has been chosen as a platform [17, 18].

Based on the surface reflectance and emissivity acquired by satellite sensors, various studies on the urban thermal environment (i.e., LST and UHI) have been completed for Indian cities, including Kolkata [19-21]. However, only a few investigations in India have used the recently developed LCZ system. This research has tried to identify the various LCZs of the Kolkata Metropolitan Area (KMA) to detect Urban Heat Island Intensity (UHII). Following this introductory section, the paper is divided into five sections, a description of the KMA area, data and method, result and discussion, and conclusion.

## 2. Study area

With remarkably high population growth from 2001 (14.77 million) to 2021 (20 million), the KMA has been recognized as the third-biggest megacity in India and the biggest urban agglomeration of West Bengal [22]. It encompasses the Kolkata Municipal Corporation and peripheral shares of the other five districts of West Bengal, i.e., North 24 Parganas, South 24 Parganas, Howrah, Hooghly, and Nadia spread on both sides of the Hooghly River. The KMA encompasses an area of 1886.67 km$^2$ and an alarming population expectation of 21.1 million in 2025 [23] has made this site a suitable one for justifying the objectives of this research.

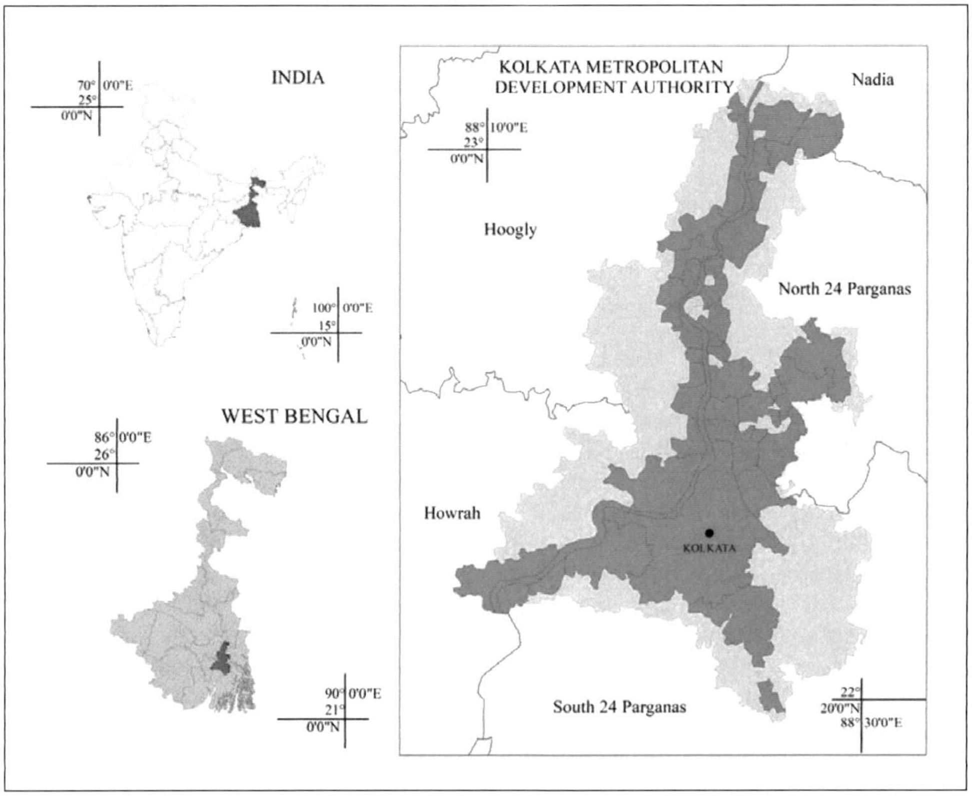


**Figure 2.** Location map of the study area highlighting the municipalities

## 3. Data and methodology

### *3.1. Data*

For path 138 and row 44 for the years 2010 and 2020, the study used Landsat 5 TM and Landsat 8 OLI data from the US Geological Survey's "Earth Explorer" website (table 1). This image was generated using the Universal Transverse Project (UTM) and the World Geodetic System (WGS) - 1984 datum (zone 45 north).

**Table 1:** Description of satellite datasets used in the study

| Data | Period | Purpose of use |
|---|---|---|
| Landsat 8 OLI &TIRS | Summer and winter 2020 | LST calculation |
| | February 2020 | LULC and LCZ classification |
| Landsat 5 TM | Summer and winter 2010 | LST calculation |
| | February 2010 | LULC and LCZ classification |

### *3.2. Classification of Local Climate Zone*
The concept of "regions of consistent surface cover, structure, material, and human activity that extend from hundreds of meters to several kilometers in horizontal scale" refers to Local Climatic Zones (LCZs), which are climate-based categorizations of city structures [13]. In this paper, a GIS-based strategy concentrating on a pixel-based categorization method has been used. For this, researchers connected to the World Urban Database and Access Portal Tools have provided a sample LCZ (.kml) file [24]. The WUDAPT is a global project that uses the LCZ classification framework to assemble data on cities' shape and purpose to improve urban climate categorization. After getting the initial sample files from WUDAPT, the training site classifications have been performed using Google Earth images. These training sites are polygonal borders defined over specific LUs that meet the LCZ classes (figure 3).

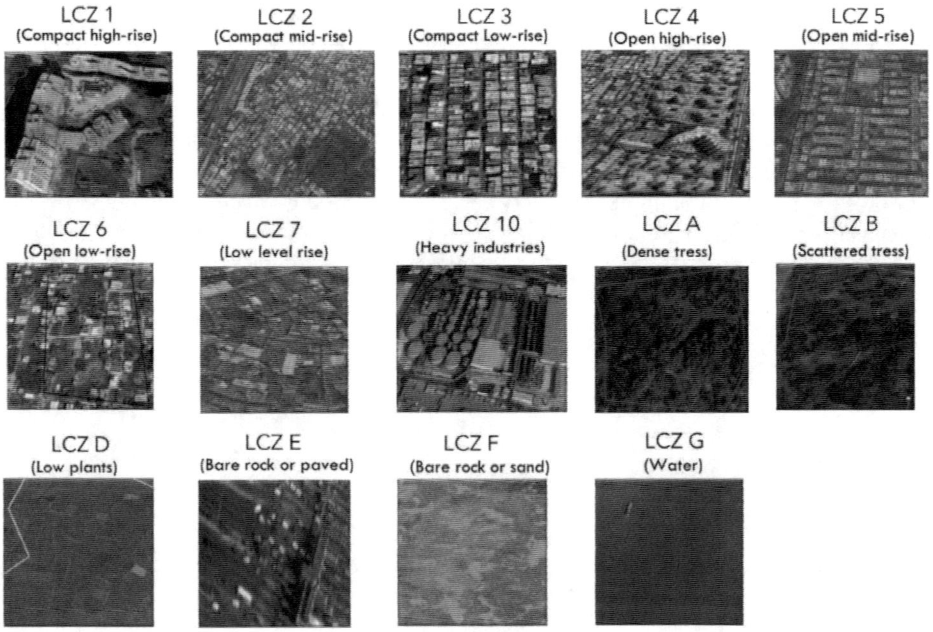


**Figure 3.** Sample images of training areas collected from Google Earth Pro for LCZ classification

The training samples are used as input data for the automated classification performed in SAGA GIS. This software is an open-source GIS program. Landsat 5 and 8 satellite images, masked for the study domain, combined with the vector training sites file are loaded into SAGA for supervised LCZ classification.

The Random Forest (ViGrA) functionality of SAGA was used to finish the categorization. To create a thorough classification of final products, the categorization's findings must be enhanced. The iterative process of the GIS technique used in this study enables us to see our results and run the training locations again until we get the desired precision. The process is iterated until the necessary accuracy is attained because the accuracy of this approach is greatly influenced by the quality of the training sites the researcher uses for categorization.

### *3.3. Extraction of urban built-up*
The built-up areas of 2010 and 2020 were delineated as it is important to evaluate the changing nature of urbanization. The masked images were first classified using the maximum likelihood classifier, into built-up and non-built-up categories [25]. Residential areas, commercial areas, and all infrastructural facilities, such as pavements, transport systems, and artificial structures, are included in the built-up

class of the resulting maps, whereas all other types of LULC observed in the study site, are listed in the non-built-up class. These images were categorized into built-up and non-built-up classes using the supervised maximum likelihood classifier method [26–27]. For this investigation, binary images were created from the categorized LU map using ArcGIS's reclassification approach. Binary images have a built-up class of 1 and a non-built-up class of 0. The vector files were created when the reclassified images were vectorized [28].

### 3.4. *Accuracy Assessment of LCZ and built-up maps*

Accuracy assessment is a method to determine whether the classification for the area of interest was carried out in an unbiased way. The confusion matrix is used extensively as a method of accuracy assessment which also helps us to categorize the errors and refine the work. From the error matrix 'overall accuracy' has been chosen as the most prominent assessment method. It is obtained by dividing classified pixels and the pixels taken as validation points [25]. The overall accuracy is a comparison of the classification conducted on each pixel against the land cover based on the data collected from the ground [29]. It combines all the classes and provides accuracy for the entire image instead of individual classes. Additionally, the kappa coefficient has also been used as a method to determine the accuracy of the classification. The classification accuracy of the image as measured by the Kohen's kappa coefficient in comparison to values selected at random. It ranges from 0 to 1. A coefficient of 0 implies that the agreement between the classified and reference image is null whereas a co-efficient closer to 1 shows higher accuracy. The commission and omission errors are included along with the complete matrix in the kappa coefficient of agreement. The accuracy can thus be calculated by the following formula:

$$k = (\theta_1 - \theta_2)/(1 - \theta_2) \tag{1}$$

where,

$$\theta_1 = (\Sigma^n_{i=1} X_{ij})/N \quad \text{and} \quad \theta_2 = (\Sigma^n_{i=} X_{i+} X_{+i})/N^2 \tag{2}$$

where,

$X_{ij}$ = Count of $ij^{th}$ cell
N = Total pixels in the confusion matrix
$X_{+i}$ = Marginal total of $i^{th}$column
$X_{i+}$ = Marginal total of $i^{th}$ row

For estimating the accuracy of built-up maps, a total of 51 points along with 13 polygons for the year 2010 and 42 points along with 51 polygons for the year 2020 were taken as validation datasets. The accuracy of the LCZ classification is assessed by selecting 300 randomly selected points of reference from the field of study and comparing them to a ground reference image, such as Google Earth. Each satellite image's overall accuracy and Kappa coefficient are calculated by comparing reference pixels to reference data. The reliability of the producer and the user has been established for each LCZ class.

### 3.5. *Estimation of Land Surface Temperature*

For 2010 and 2020, respectively, the Landsat 5 and 8 thermal bands were employed in this analysis. LST is computed using the Mono-Window and Split Window algorithms on Google Earth Engine (GEE). The detailed methodology of LST derivation from Landsat data is described by [28].

### 3.6. *Determining Urban Heat Island Intensity (UHII)*

UHII is referred to the magnitude of temperature by that an urban island is hotter than its neighboring rural areas. In this study, we have made use of LCZ classification to calculate UHII. Generally, the calculation of UHII is done by the equation:

$$UHII = T_{urban} - T_{rural} \tag{3}$$

Where $T_{urban}$ refers to the LST of urban areas and $T_{rural}$ refers to the LST of rural areas. However, determining a clear boundary between urban and rural areas is a difficult task in a constantly developing area. Thus, UHII has been calculated with the help of LCZ land-use class 2 and land cover class D [13, 30].

$$UHII = LCZ\ 2t - LCZ\ Dt \tag{4}$$

Where, LCZ 2t is the mean temperature recorded in LCZ class 2 and LCZ Dt refers to the mean temperature of LCZ D. LCZ 2 is selected to represent the thermal conditions of the urban areas since compact mid-rise buildings are found in both 2010 and 2020 and the building height, the materials used in its construction, building density with the concrete pavements account for the temperature recorded in cities. LCZ D which represents low plants has been selected as the temperature for rural areas as conceptualized by [13]. The brief methodology has been described in figure 4.

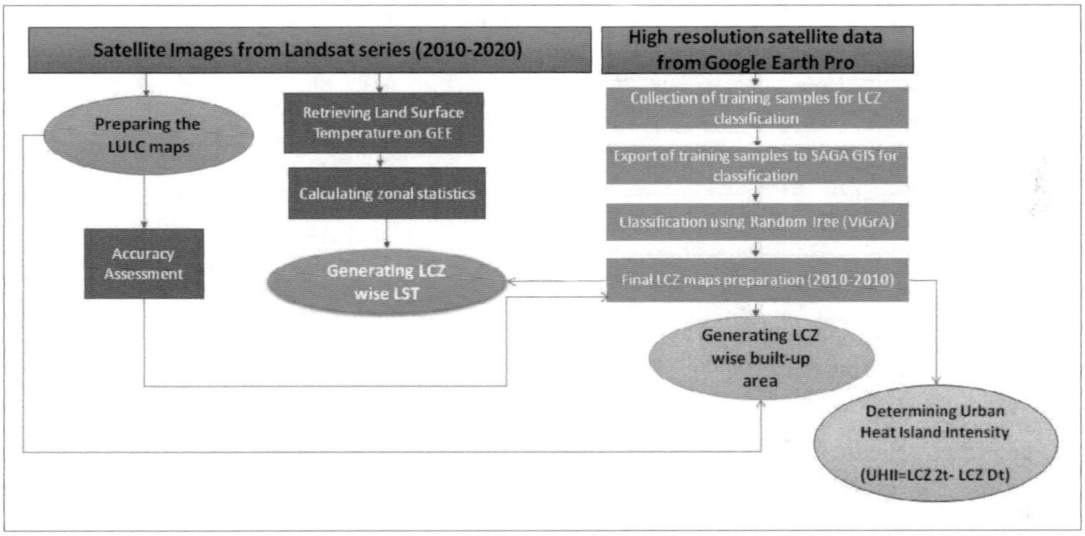


**Figure 4.** Methodological flowchart of the study

## 4. Result and Discussion

### *4.1. Accuracy assessment*

The classification accuracy for built-up maps is furnished by the confusion matrices, and the LCZ classes' classification accuracy is evaluated using the user's accuracy (UA), producer's accuracy (PA), overall accuracy (OA), and Kappa coefficient approaches. The OA and Kappa coefficients are 99% and 0.92 respectively for 2010, and 94% and 0.91 respectively for 2020 LULC maps.

For 2010 and 2020, the OA and Kappa coefficients, as well as the UA and PA of each LCZ class, have been computed. The attained OA are 80.3% and 73.5% and the Kappa coefficient is 0.76 and 0.71 for the LCZ map of 2010 and 2020 respectively. There are variations in the UA and PA of individual classes as it ranges between 50 and 100%. Particularly in 2020, the LCZ 1, 4, 5, and 6 show comparatively poor classification results with the UA lower than or equal to 60%.

The map of 2010 has lower (lower than or equal to 60%) PA in LCZ 2, B, and F, whereas 2020 has lower PA in LCZ 1 and 6. The misclassifications of the built types are due to the deficiency of building geometry information in Landsat images and the misclassifications in vegetation cover like

LCZ B is because of the difficulty to differentiate the vegetation covers with various height and density. Overall, the classifications have met the suggested value and can be used in further research.

### *4.2. LCZ classification based on building type and LU type*

A total of 14 LCZs have been classified in 2010 and 16 LCZ classes have been identified in the year 2020 (figure 5). In 2010, a total of eight LCZs were classified by building types. The compact mid and low rises are represented in LCZ 2 and 3 respectively, open high, mid and low are included in LCZ 4, 5, and 6, the Lightweight and large Low-rises are denoted by LCZ 7 and 8 consecutively and the heavy industries are covered in LCZ 10. Among the six identified land cover classes of 2010 densely vegetated areas and scattered trees are classified as LCZ A and B, the low plants such as agricultural lands are classified as LCZ D. The paved areas such as roads, railway tracks, airports terminal are considered as LCZ E, bare soils or open spaces are assigned to LCZ F and water bodies were assigned to LCZ G. In 2020, in addition to the previous classes of building types, Compact High-rise (LCZ 1) has also been classified. Bush/Shrubs are identified in 2020 and assigned to LCZ C. LCZ 9 the sparsely built buildings with abundant vegetation could not be traced in both years.

### *4.3. Spatial distribution pattern of Local Climatic Zone*

The land surface geometry in KMA displays a substantial variation in land cover and building types, which can be visualized through the LCZ maps (figure 5) and the graph presented in figure 6.

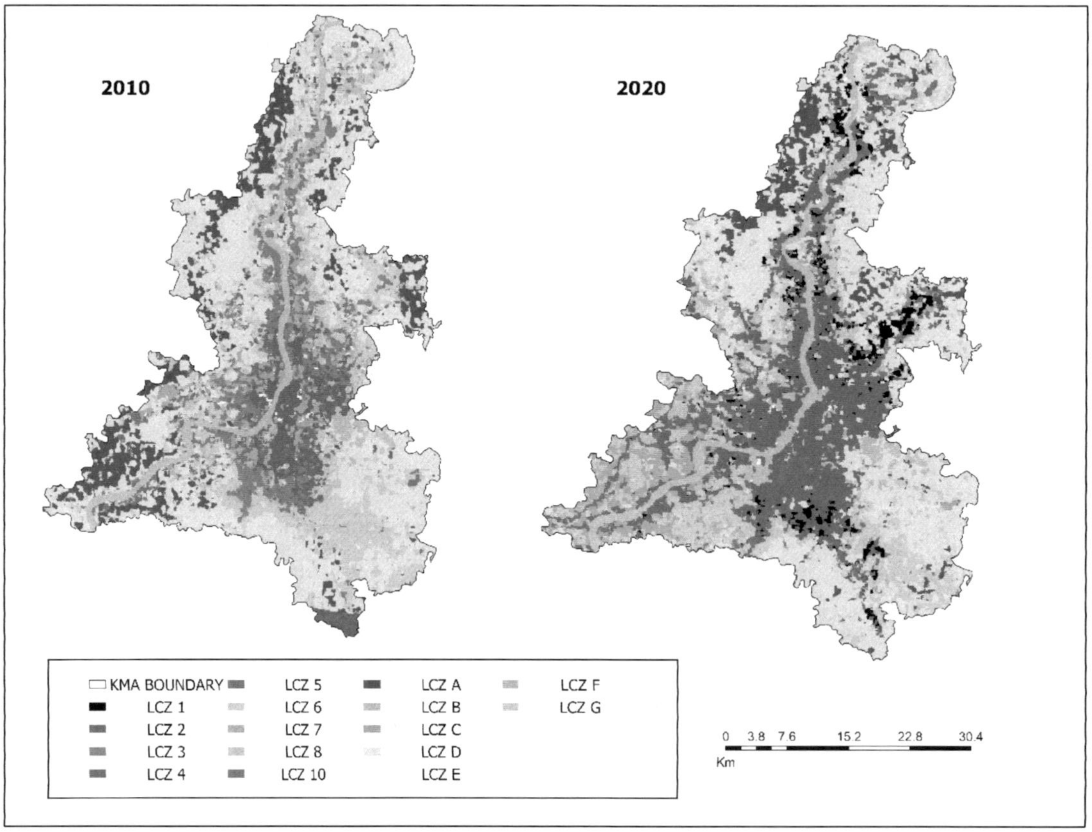


**Figure 5.** Spatial distribution of LCZs in KMA for the years 2010 and 2020

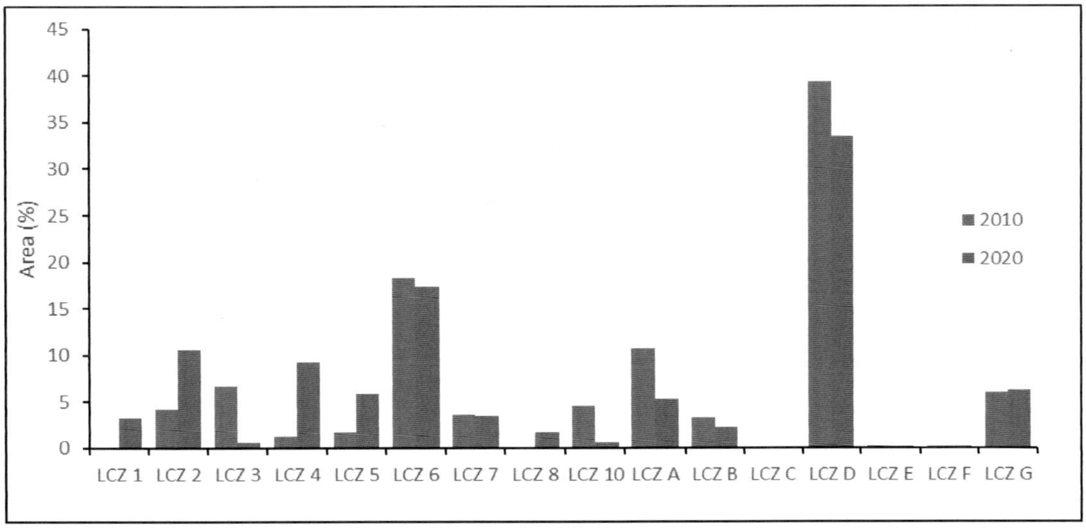


**Figure 6.** Comparison of the spatial extent of various LCZs between 2010 and 2020

The predominant LCZ in the densely populated parts of KMA is LCZ 2, which dominates in the southern portion of this urban agglomeration and includes Kolkata Municipal Corporation, Howrah, Baranagar, and Bally. LCZ 3 has spread over the Hooghly River and is contagious to LCZ 2. This zone is surrounded by a continuous zone of LCZ 6, while LCZ 4 and 5 were distributed sparsely. LCZ 7 and 8 which mostly includes low-cost slum houses occur in between the continuous tracks of LCZ 2 and 3. Heavy industries like jute, textile or artificial weaving, heavy engineering, paper, chemical industries, etc. are located in the Hooghly River belt. LCZ 1 which is prominent in 2020 is located outside of Kolkata Municipal Corporation. In 2010 about 40% of the total area of KMA has been occupied by building types which reached 52% in 2020. Although declining the vegetated LCZs of LCZ A - LCZ D occupies a substantial portion of KMA, mostly surrounding the various building types. In 2010, 53% of the area of this agglomeration was in the account of vegetated LCZs that reduced to 41% in 2020. LCZ E consists of airport terminals, runways, etc., which are consuming significantly lower areas in proportion (0.24% in 2010 and 0.18 % in 2020). Similarly, LCZ F mainly includes sandy shores and bare soils distributed in patches occupying only 0.21% in 2010 and 0.24% of the area in 2020. Hoogly River, East Kolkata Wetland, and other small water bodies are classified as LCZ G.

### *4.4. Change Analysis of LCZs Between 2010 and 2020*

Change detection of LCZ class has been computed to analyze the spatial change of urban geometry over ten years. In 'Built Types' the conversion trend indicates both vertical and horizontal expansions of urban built-ups. In 2010 there was no LCZ 1 present in the area, which emerged in 2020. LCZ 3 is mostly converted to LCZ 1 and 2 from 2010 to 2020 (figure 7). LCZ 4 is partially transformed into LCZ 1, 2, and 3. LCZ 5 has been converted into LCZ 2 and 4. LCZ 6 is changed to LCZ 1, 2, 3, and 5. A vast area of LCZ 7 is transformed into LCZ 2 followed by LCZ 4, 6, and 3. The LCZ 8 of 2010 has been converted mostly into LCZ 6. Half of the LCZ 8 is converted to LCZ 6. LCZ 10 is being converted into all other LCZs under 'Building Types' (LCZ 1 to 9). The gradual decrease of dense vegetation from 2010 to 2020 has been depicted in the conversion of LCZ A into building types classes of LCZ 4, 6, and 7, and also into land cover types of LCZ B and D (figure 7). LCZ B has also changed to mostly building-type classes like LCZ 1, 4, 6, 5, and 2. There is no LCZ C in both years. LCZ D is changed into built-type classes of LCZ 1, 4, 5, and 6 and a land cover class of LCZ B. LCZ E is changed to all different built-type classes like LCZ 1, 2, 3, 4, 5, 7, 8 and 10. Apart from the conversion of LCZ F into built types (LCZ 2, 3, 4, 5, 6, 8, and 10), a few plantation initiatives led to the conversion into LCZ A, C, D, and E. There is little or less change in the percentage of the area LCZ G from 2010 to 2020.

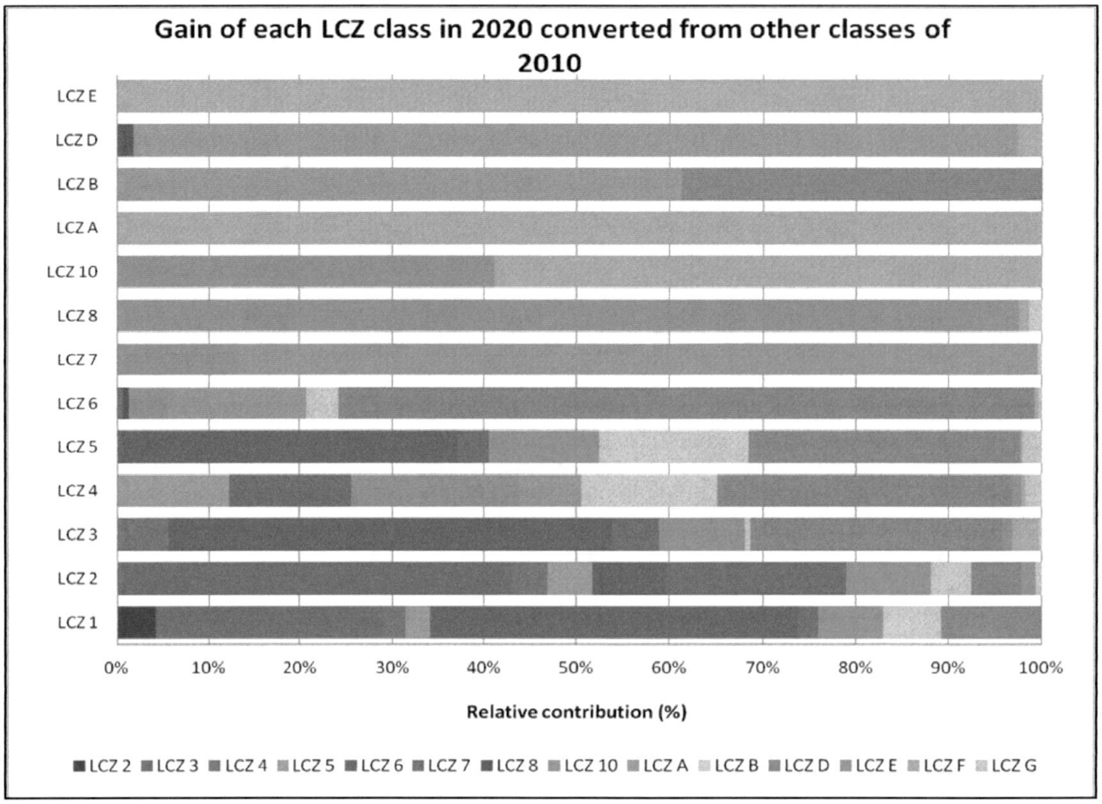


**Figure 7**. Graph showing gain of a LCZ class in 2020 from other classes of 2010. The graph is prepared from the statistics of change analysis between LCZ of 2010 and 2020.

### *4.5. Associations between LST and LCZ classes*

The mean LST of each LCZ class for 2020, for the summer and winter seasons, has been estimated. Substantial seasonal variations of LST within LCZ classes have been observed. Among the building types LCZs, LCZ 8 depicts the highest temperature in both summer (43.96°C) and winter (25.97°C). The low-cost houses that use asbestos—concrete sheets as their roof material gets heated up by the incident solar radiation, thus, high surface reflectance and high LST. Similarly, an industrial area (LCZ 10) also depicts higher average surface temperatures in the summer (40.75°C) and the winter (24.38°C). This is primarily caused due to the extensive use of concrete and iron sheets as building materials, and the removal of vegetation. LST varies between 37.18°C and 38.98°C in LCZ 1- 4. LCZ 5 - 7 exhibit relatively lower LST among the other building-type LCZs. The least LST value has been detected in LU LCZ classes (LCZ A - C) due to the widespread previous landscapes. With mean LST values of 34.05°C and 21.19°C noticed in the summer and winter, successively, LCZ A reveals the second-lowest LST in both seasons. LCZ E which represents bare or paved land recorded the highest mean LST in both summer (41.69°C) and winter (24.71°C). In contrast to other types of LU, bare rocks and paved surfaces often absorb more solar radiation and have higher thermal conductivity. Bare soil also has a higher LST of 37.02°C and 22.08°C in the summer and in the winter respectively due to its low thermal inertia. It warms up quickly during the day and cools down quickly at night.

**Table 2.** Summary of built-up area coverage and LST in each LCZ class

| LCZ Class | Built-up Area ($km^2$) | % of Built-up Area, 2020 | LST(°C) Summer - 2010 | LST(°C) Winter - 2010 | LST(°C) Summer- 2020 | LST(°C) Winter- 2020 |
|---|---|---|---|---|---|---|
| 1 | 32.37 | 7.63 | NA | NA | 37.47 | 22.60 |
| 2 | 109.08 | 25.71 | 35.58 | 22.42 | 38.96 | 23.23 |
| 3 | 3.79 | 0.89 | 34.41 | 21.71 | 37.18 | 22.39 |
| 4 | 77.02 | 18.15 | 33.72 | 21.64 | 37.87 | 23.03 |
| 5 | 29.27 | 6.90 | 33.76 | 21.51 | 36.58 | 22.35 |
| 6 | 56.75 | 13.38 | 33.59 | 21.39 | 35.60 | 22.26 |
| 7 | 6.10 | 1.44 | 36.68 | 23.81 | 35.41 | 21.93 |
| 8 | 19.01 | 4.48 | 34.64 | 21.44 | 43.96 | 25.97 |
| 10 | 6.96 | 1.64 | 35.36 | 22.94 | 40.75 | 24.38 |
| A | 1.89 | 0.45 | 32.26 | 20.57 | 34.05 | 21.20 |
| B | 1.49 | 0.35 | 33.69 | 21.46 | 34.62 | 21.69 |
| C | 0.06 | 0.01 | NA | NA | 39.24 | 24.78 |
| D | 68.14 | 16.06 | 32.90 | 21.29 | 34.70 | 22.09 |
| E | 1.76 | 0.41 | 36.47 | 23.48 | 41.70 | 24.71 |
| F | 1.34 | 0.32 | 36.57 | 23.57 | 37.02 | 22.09 |
| G | 9.22 | 2.17 | 29.33 | 18.77 | 30.78 | 20.42 |

### *4.6. Urban heat island intensity (UHII)*

Table 2 shows the built-up areas of each 2020 LCZ class. It demonstrates that LCZ 2, having the greatest portion of the built-up region, covers an area of 109.08 $km^2$, or 25.71%. $UHI_{Lcz}$ can be expressed as $UHI_{Lcz} = LST_{LCZ\ 2} - LST_{LCZ\ D}$, where $UHI_{Lcz}$ is the result of the disparity in LST of LCZ 2 ($LST_{LCZ\ 2}$) and LST of LCZ D ($LST_{LCZ\ D}$).

The result indicates that the $UHII_{LCZ}$ of KMA is present in both seasons. However, the intensity is as high as 4.26°C in summer and lower during winter (1.14°C).

## 5. Conclusion

The morphology of urban surface characteristics may have complex interactions with urban climate, especially surface temperature. The classical approach of studying the temporal pattern of LST to explore UHI became less significant as the nature of the urban surface cover and built-up morphology become diverse and dynamic. Instead, the classification of the urban area into Local Climate Zones (LCZ) based on building height, density, and land cover types that have a unique interaction with the micro-climate became a robust approach as an alternative. The present study has tried to identify the different LCZs of the Kolkata Metropolitan Area (KMA) and their changing pattern from 2010 to 2020. The study also examined the seasonal variations in LST for each LCZ and calculated the degree to which an urban island is hotter than the rural areas around it. Freely available satellite data and geospatial technology become favorable and important tools for this study.

The study has identified 14 LCZs for 2010 and 16 LCZs for 2020. The dominant increase in the areal coverage has been noticed for LCZ 1, 2, 4, and 5 from 2010 to 2020. The nature of LCZ transformation reveals that the height and compactness of buildings have increased between the periods which may have caused an unprecedented increase in the LST of KMA. The seasonal variations of LST within each LCZ are evident between summer and winter. LCZ 8 showed the highest temperature in both summers (43.96°C) and winter (25.97°C) among all LCZs.

The analysis of UHII has shown that the difference between urban and rural LST is 4.26°C. The overall analysis suggests that the identification of LCZ may provide an accurate estimation and variations of surface temperature within an urban area, especially with heterogeneous surface characteristics.

Since this LCZ categorization approach relies on the effectiveness of urban form on temperature disparity in the city, leading to the creation of more climate-sensitive cities, LCZ maps can depict the spatial variation of the UHI incident. The LCZ map shows geographic elements to assist architects and urban planners in better understanding the thermal environment while making decisions. Furthermore, the LST features of LCZ classes may be used to reduce the negative effects of urbanization through the use of urban vegetation and LU planning. The growth of these cities into megacities is aided by frequent contact between cities in this area of the region. As a result, more cautious planning methods for future urban growth in terms of UHI mitigation are necessary. The present approach may assist planners and urban developers in decision-making for making by-laws towards sustainable development, and researchers for further analysis and research work.

**References**

[1] Punia M, Singh L 2012 Entropy approach for assessment of urban growth: a case study of Jaipur, India *J. Indian Soc. Remote. Sens*. **40**(2) 231-44
[2] UNFPA, State of World Population, 2007: Unleashing the Potential of Urban Growth:United Nations Population Fund, 2007
[3] de Oliveira JA, Doll CN, Kurniawan TA, Geng Y, Kapshe M, Huisingh D2013 Promoting win–win situations in climate change mitigation, local environmental quality and development in Asian cities through co-benefits *J. Clean. Prod.*158:1-6
[4] Weng YC 2007Spatiotemporal changes of landscape pattern in response to urbanization *Landsc. Urban. Plan.***81**(4) 341-353
[5] Grimmond SU 2007 Urbanization and global environmental change: local effects of urban warming *Geogr. J.* **173**(1)83-8
[6] Mirzaei PA, Haghighat F 2010 Approaches to study urban heat island–abilities and limitations *Build. Environ.***45**(10)2192-201
[7] Adamo SB 2010 Environmental migration and cities in the context of global environmental change *Curr.Opin. Environ. Sustain.***2**(3)161-5
[8] Emmanuel R, Loconsole A2015 Green infrastructure as an adaptation approach to tackling urban overheating in the Glasgow Clyde Valley Region, UK *Landsc. Urban. Plan.***138**71-86
[9] Sultana S, Satyanarayana AN 2020 Assessment of urbanisation and urban heat island intensities using landsat imageries during 2000–2018 over a sub-tropical Indian City*Sustain. Cities Soc.***52**101846
[10] Zhou X, Wang YC2011 Dynamics of land surface temperature in response to land-use/cover change. *Geogr. Res*. **49**(1)23-36
[11] Singh P, Kikon N, Verma P2017 Impact of land use change and urbanization on urban heat island in Lucknow city, Central India. A remote sensing based estimate*Sustain. Cities Soc*. **32**100-14
[12] Hu L, Brunsell NA, Monaghan AJ, Barlage M, WilhelmiOV 2014 How can we use MODIS land surface temperature to validate long-term urban model simulations? *J. Geophys. Res. Atmos.***119**(6)3185-201
[13] Stewart ID, Oke TR 2012 Local climate zones for urban temperature studies *Bull. Am. Meteorol. Soc.***93**(12) 1879-900
[14] Bechtel B, Foley M, Mills G, Ching J, See L, Alexander P, O'Connor M, Albuquerque T, de Fatima Andrade M, Brovelli M, Das D 2015 Census of cities: LCZ classification of cities (level 0)–workflow and initial results from various cities *ICUC9* 9th International Conference on Urban Climate jointly with 12th Symposium on the Urban Environment
[15] Lelovics E, Unger J, Gál T, Gál CV 2014 Design of an urban monitoring network based on Local Climate Zone mapping and temperature pattern modelling *Clim. Res.* **60**(1) 51-62
[16] Gamba P, Lisini G, Liu P, Du P, Lin H 2012 Urban climate zone detection and discrimination using object-based analysis of VHR scenes *Proc. 4th GEOBIA* 70-74
[17] Bechtel B, Daneke C 2012 Classification of local climate zones based on multiple earth observation data *IEEE J. Sel. Top. Appl. Earth Obs. Remote Sens*. **5**(4) 1191-202

[18] Bechtel B, Alexander PJ, Böhner J, Ching J, Conrad O, Feddema J, Mills G, See L, Stewart I 2015 Mapping Local Climate Zones for a Worldwide Database of the Form and Function of Cities. *ISPRS Int. J. Geo-Inf.* **4** 199–219

[19] Mohammad P, Goswami A 2021 Spatial variation of surface urban heat island magnitude along the urban-rural gradient of four rapidly growing Indian cities *Geocarto Int.* 1-23

[20] Mathew A, Khandelwal S, Kaul N 2018 Analysis of diurnal surface temperature variations for the assessment of surface urban heat island effect over Indian cities *Energy Build.* **159** 271-95

[21] Ghosh S, Chatterjee ND, Dinda S 2019 Relation between urban biophysical composition and dynamics of land surface temperature in the Kolkata metropolitan area: a GIS and statistical based analysis for sustainable planning *Model. Earth Syst. Environ* **5**(1) 307-29

[22] Census of India, 2011. Primary Census Abstract, Census of India. Gov't. of India

[23] KMDA, 2011 Kolkata Metropolitan Development Authority. http://www.kmdaonline.org/html/about-us (Accessed Dec 2020)

[24] Demuzere M, Kittner J, Bechtel B 2021 LCZ Generator: a web application to create Local Climate Zone maps. *Front. Environ. Sci.* **9** 637455

[25] Bhatta B 2009Analysis of urban growth pattern using remote sensing and GIS: a case study of Kolkata, India *Int. J. Remote. Sens.***30**(18)4733-46

[26] Liu Y, Song W, Deng X 2019 Understanding the spatiotemporal variation of urban land expansion in oasis cities by integrating remote sensing and multi-dimensional DPSIR-based indicators. *Ecol. Indic.***96** 23-37

[27] Xu Y, Qin Z, Lv J 2008 Comparative analysis of urban heat island and associated land cover change based in Suzhou city using landsat data. In2008 International Workshop on Education Technology and Training & 2008 International Workshop on Geoscience and Remote Sensing *IEEE* **2** 316-319

[28] Mukherjee K, Das P 2018Modelling the Relationship between Urban Growth Modes and the Thermal Environment-A Case Study of the Barasat Municipality, West Bengal *J. Geogr. Environ. Earth Sci. Int.*1-9

[29] Rwanga SS, Ndambuki JM 2017 Accuracy assessment of land use/land cover classification using remote sensing and GIS *International Journal of Geosciences* **8**(04) 611

[30] Chieppa J, Bush A, Mitra C2018 Using "Local Climate Zones" to detect urban heat island on two small cities in Alabama *Earth Interact.***22**(16)1-22

# Beach Erosion Assessment and Tourism Viability Analysis at Mandarmani Coast, West Bengal, India

**A Bose** [1*], **S Bardhan** [2]

[1] School of Oceanographic Studies, Jadavpur University, Kolkata, India,
[2] Department of Architecture, Jadavpur University, Kolkata, India

*Corresponding Author Email: monamirsgis@gmail.com

**Abstract:** Mandarmani is a popular seaside tourist destination in Coastal West Bengal, India. The geomorphologically dynamic and ecologically diverse terrain falls within an active coastal belt with strong erosion but limited accretion. It is important to monitor and predict changes in beach morphology and shoreline over a short and long timescale to identify the specific drivers of such changes and their related impacts. While manual ground surveying techniques are relatively more accurate for this purpose, Remote Sensing and GIS techniques are proven to be more applicable as they are relatively accurate, cost-effective, less time-consuming and can cover a large coastal belt. The present paper analysed the beach morphology and shoreline changes along the Digha-Mandarmani coast of West Bengal using RS & GIS as a tool and fish net model to demarcate the erosion and accretion zones, to project the future shoreline position and understand the risk of climatic hazard on the tourism establishments and their sustainability in this region in future. Such a study would help in taking informed policy choices by the planners on such growing tourism destinations developed over time, often violating the existing coastal zone regulations and thus without required environmental clearances. Should be protected with public investment to safeguard the business hub, or will be allowed to face its natural consequences?

**Keywords:** RS-GIS, Shoreline Change, Future Projection, Coastal Tourism, Sustainability

## 1. Introduction

Mandarmani is a well-known seaside tourist destination on the bank of Bay of Bengal in the State of West Bengal, India. Mandarmani is a fairly new addition to the tourism map of the State that has grown largely driven by market forces. However, Mandarmani fall within an active coastal belt where the shoreline is changing with both erosion and accretion due to littoral transport (movement of eroded sand by waves in the near-shore zone) [1]. Natural processes like wave currents, near-shore circulation, sediment characteristics, beach forms, etc. and human interventions like dykes, dams, dredging, sand mining, water extraction, tourism, etc. are responsible for such unintended changes [1]. In the context of coastal vulnerability accreting coastlines are considered less vulnerable, as they result in the addition of land by moving towards the ocean [2]. On the other hand, eroding coastlines are considered highly vulnerable as they result in the loss of land by moving toward the beach as well as result in the loss of natural and man-made resources associated with it [2]. Thus, a healthy coastline and beaches are essential for a successful tourism business and also to protect the critical habitats of many species from any kind of coastal hazard [3]. It is important to monitor and predict shoreline changes over a long period of time to identify the specific causes of erosion and accretion of shoreline of a particular area; and the anthropogenic and human intervention-related impacts, to evaluate geomorphic & ecological changes and accordingly plan for protection measures [4]. To accomplish this, the coastal zone requires an application of Geoinformatics along with site-specific research, due to the complexities associated with the coast even at the regional level [5].

Though Traditional ground surveying techniques are relatively more accurate, it is highly time-consuming and impossible to accomplish for a large coastal belt [6]. On the other hand, Remote Sensing (RS) and Geographical Information System (GIS) based geoinformatics is accurate, cost-effective, less time-consuming and can cover a large coastal belt where the RS provides a source of spatial and multi-temporal data, and the GIS is useful for mapping [7]. Thus, these tools provide an opportunity to develop an information database and support decision-making activities for coastal zone applications [8]. With this perspective, the current study aims to analyse the shoreline changes (erosion and accretion zones) along the Mandarmani coast of West Bengal by using RS & GIS as a tool, project the future shoreline position and understand the vulnerability of the tourism establishments of this region to such changes. The overall objective of this study is to produce shoreline change maps of present and future time that can be used by coastal authorities for making an effective coastal zone management plan for this region.

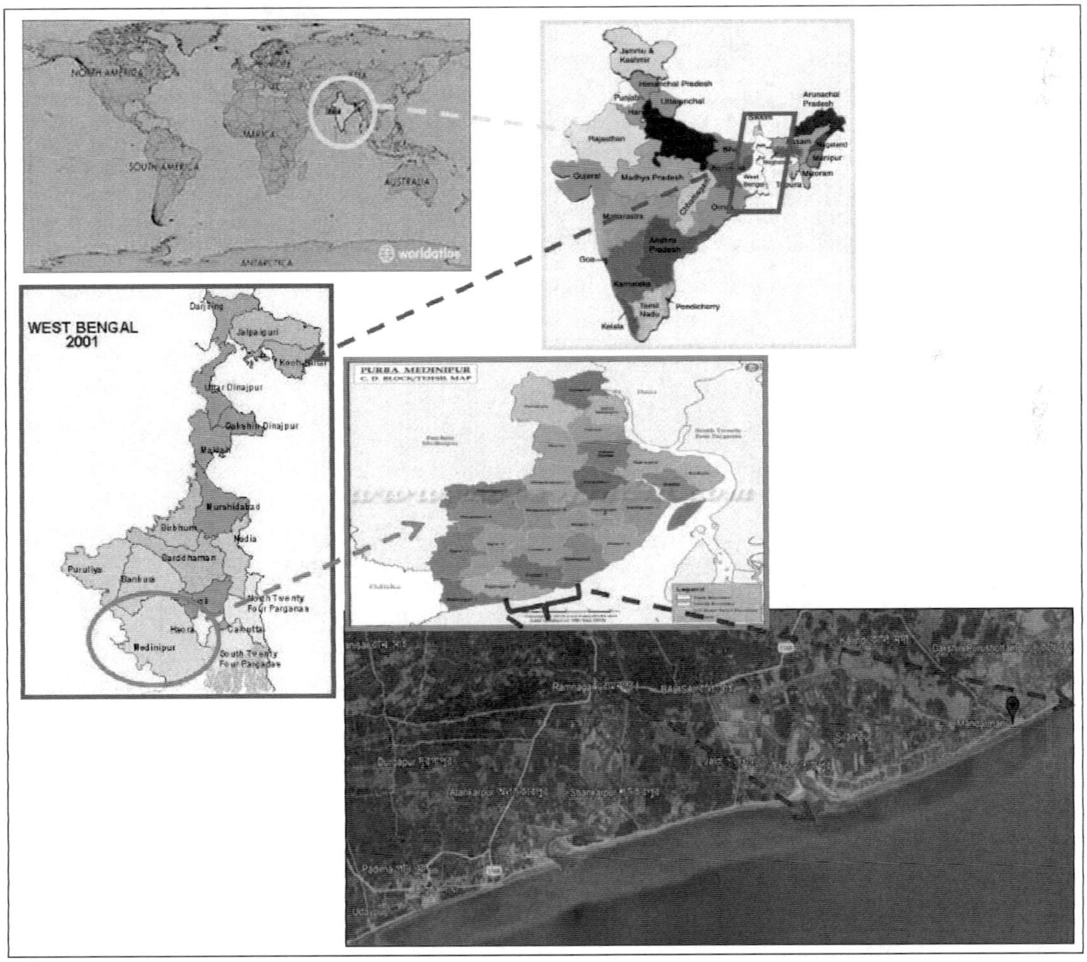


**Figure 1**. Geographical Location of the Study Area (Source: Google Earth Images)

## 2. Study Area

The coastal area of Midnapore District extended along the west bank of Hooghly estuary [9]. It is stretched from the Orissa border on the west via Udaipur, Digha, Tajpur, and Shankarpur then curving around Dadanpatrabar, Sona Muhi, Mandarmani, Junput and Haldia on the east and extends up to Tamluk [10] (figure 1). It is situated at the northern end of the Bay of Bengal and is located in the SOI Toposheets No. 73-O/6 and 73-O/10 and has a Mean Seas Level of 6m or 20ft [11]. Mandarmani falls under Jalda Mouza which is a part of the Contai coastal area of East-Medinipur district in West Bengal state [12]. The latitudinal and longitudinal position of Mandarmani is 21°40'N to 21°41'N latitude and 87°39'E to 87°45'E latitude as indicated in figure 2 [13].

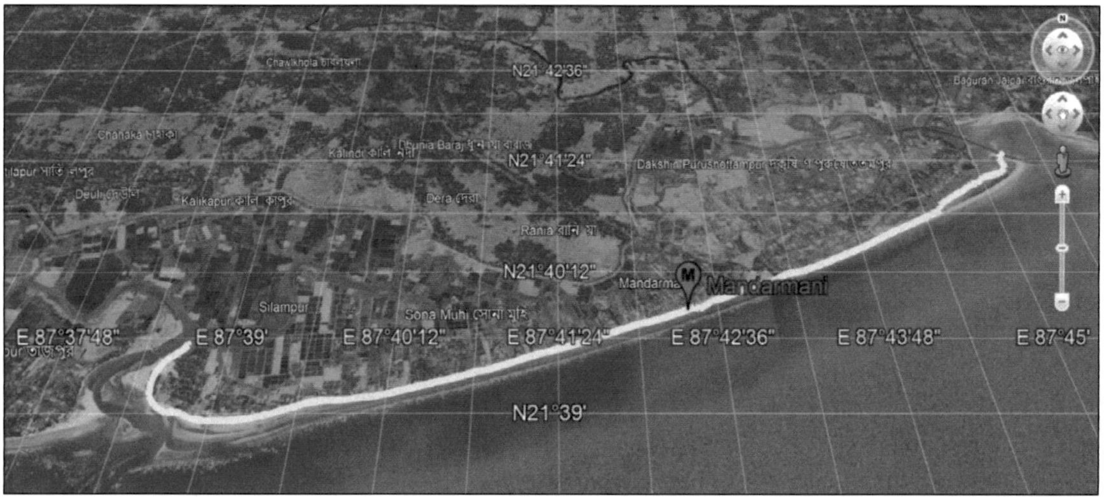


**Figure 2.** Latitudinal & Longitudinal Location of the Study Area (Source: Google Earth Images)

## 3. Materials and Methods

### *3.1. Data*

Satellite images (Landsat 5) of 2000, 2005, 2010, and 2015, from Earth Explorer (USGS). Through layer stack and subset, the images were made usable for further study. Satellite images of 2017 from Google Earth were used to identify the study area. The detailed methodology is presented in figure 3 and described as follows.

### *3.2. Digitization of Shoreline*

The coastline from Udaipur to Mandarmani was digitized from Google Earth to accurately identify the coastline. The coastlines for the years 2000, 2005, 2010, and 2015 from Udaipur to the Mandarmani area were digitized from satellite images. Locations were chosen along the coastal tract according to the level of tourism activity for focussed analysis. These locations were Sona Muhi, and Mandarmani (figure 4a to 4d and figure 5a to 5d)

### *3.3. Analysis of Shoreline Change*

All such 4 digitized coastlines of 2000, 2005, 2010, and 2015 were overlaid and analysed in high spatial resolution to identify and compare the erosion-accretion zones and to produce maps of shoreline changes for all the selected locations in a GIS platform.

### *3.4. Estimation of Shoreline Changes*

In the present study, the Fishnet application of Arc GIS extension was used. The multiple numbers of transacts according to necessity were drawn and coastlines were projected on it, according to the requirement of the model. Altogether 72 transacts were drawn to accurately estimate erosion and accretion rate over coastlines of 4 selected years. Finally, the erosion and accretion rates were measured in the Arc GIS platform for the time windows of years 2000-2005; 2005-2010; 2010-2015 for short-term analysis and from 2000-2015 for long-term analysis to determine the average retreat or advancement of shoreline.

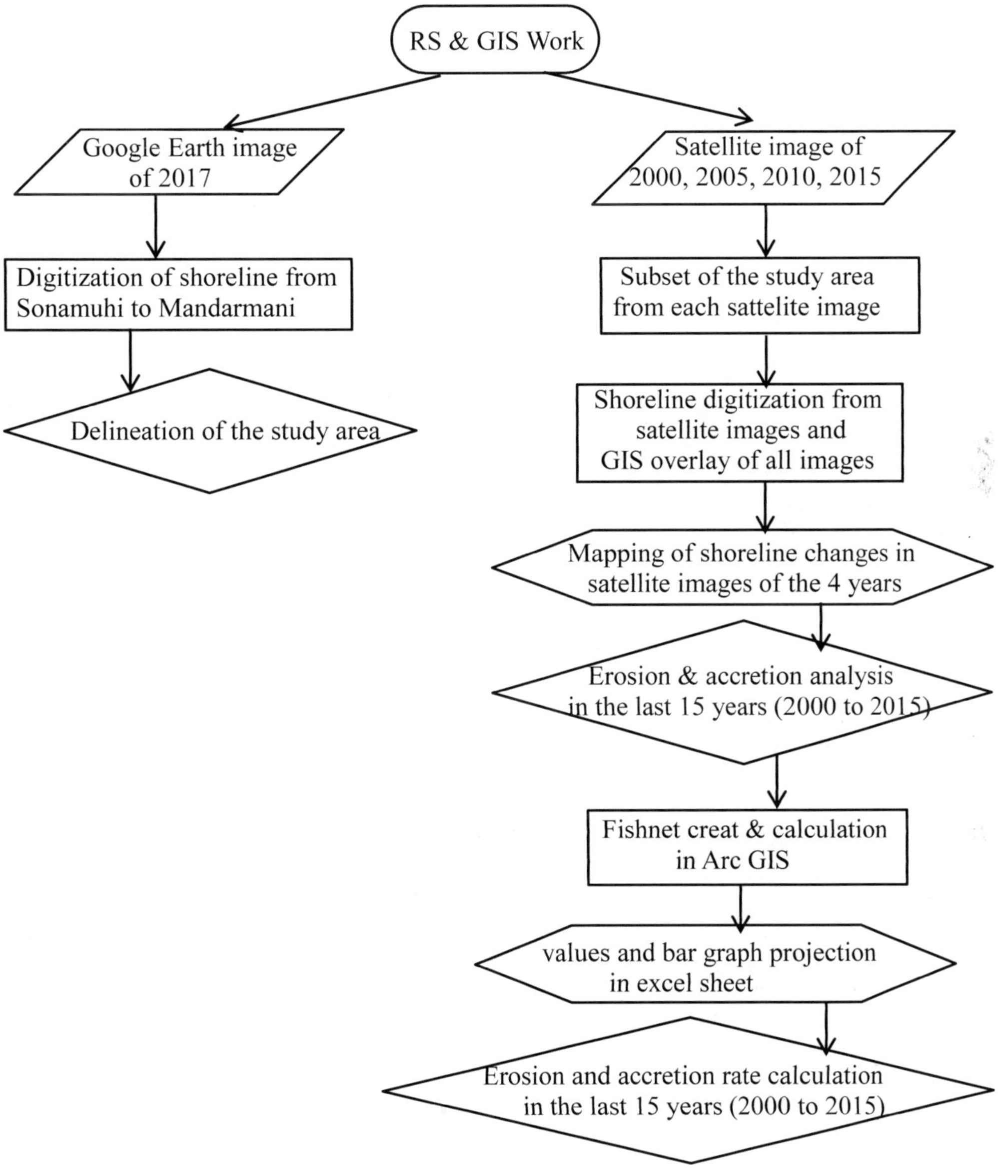


**Figure 3.** Research Methodology Snap Shot.

*3.5. Future Prediction of Shoreline Movement*

By calculating the average erosion and accretion rate from 2000 to 2015, an average yearly changing rate was estimated along each transact and linearly projected for 2050 along the same transact. The projected shoreline of 2050 is thus created and overlaid along with other shorelines of selected years on a satellite image of 2019 to compare the impact of erosion–accretion on the coast in future in the study area. Predicted GIS maps of 2050 for Mandarmani were generated.

### *3.6. Field Survey*
Field verification or ground truthing of the present coastline is a very important part of this study. The field survey was conducted at the end of the study in 2017. There is fierce evidence of severe erosion and very less accretions.

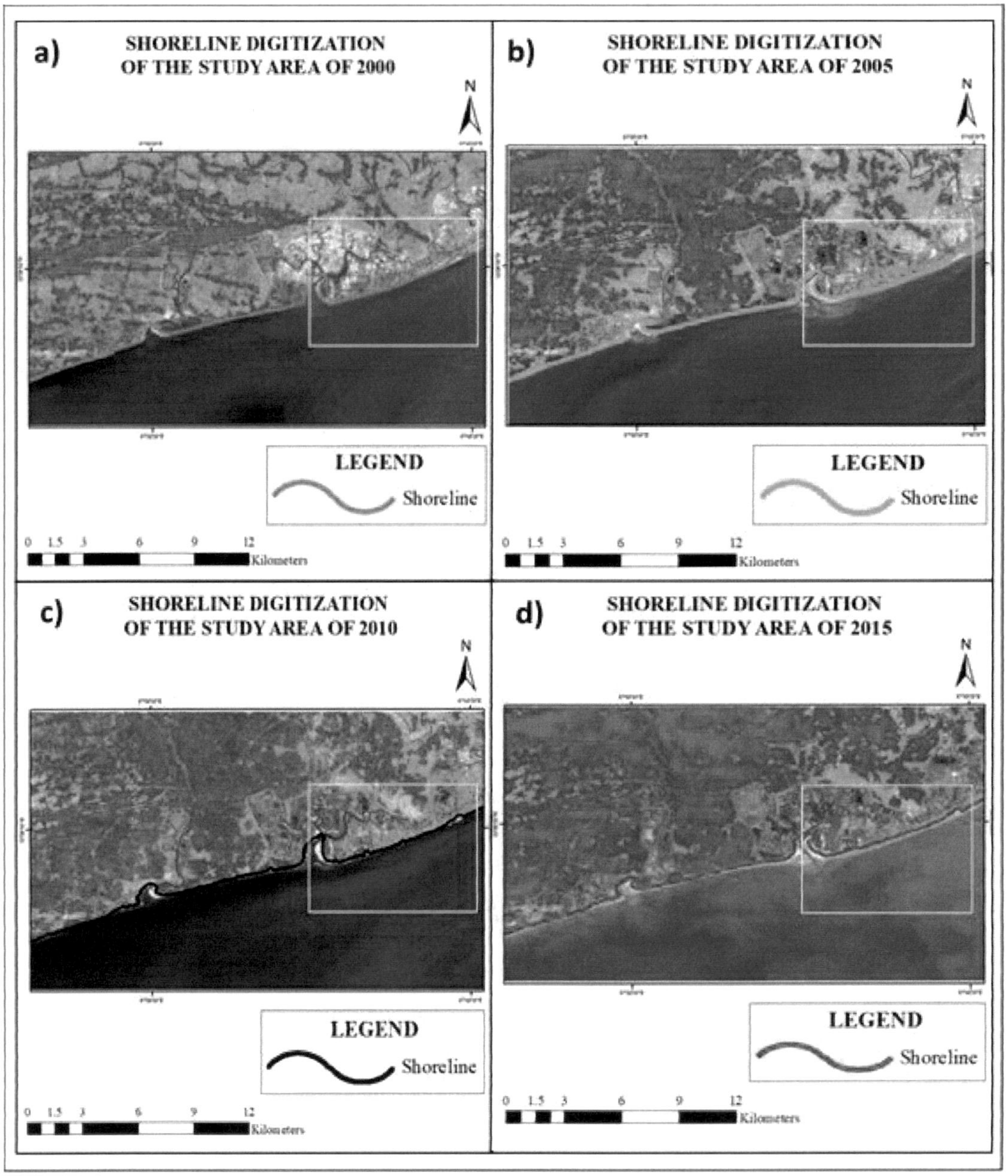


**Figure 4.** Temporal changes in shoreline from Udaipur to Mandarmani of a) 2000, b) 2005, c) 2010, and d) 2015. The yellow marked area shows the Mandarmani coast.

## 4. Result & Discussion

### *4.1. Shoreline Delineation*

The present analysis is concerned with the coastal width of 5km from the High Tide Line, where the major tourism establishments are located and which bears the brunt of major coastal hazards like erosion/accretion, cyclones and storm surges.

The coastlines of different years were digitized from multitemporal satellite data of winter months of similar tidal phases using the tonal and colour differences between the land and the sea from Udaipur to Mandarmani. For the study, the two segments that are Sona Muhi and Mandarmani were (highlighted in yellow marked line) taken as it is an upcoming tourist activity zone (figure 4a to 4d). The shorelines of 2000, 2005, 2010, and 2015 were overlaid to detect and compare the erosion and accretion rate at different time windows in the study area.

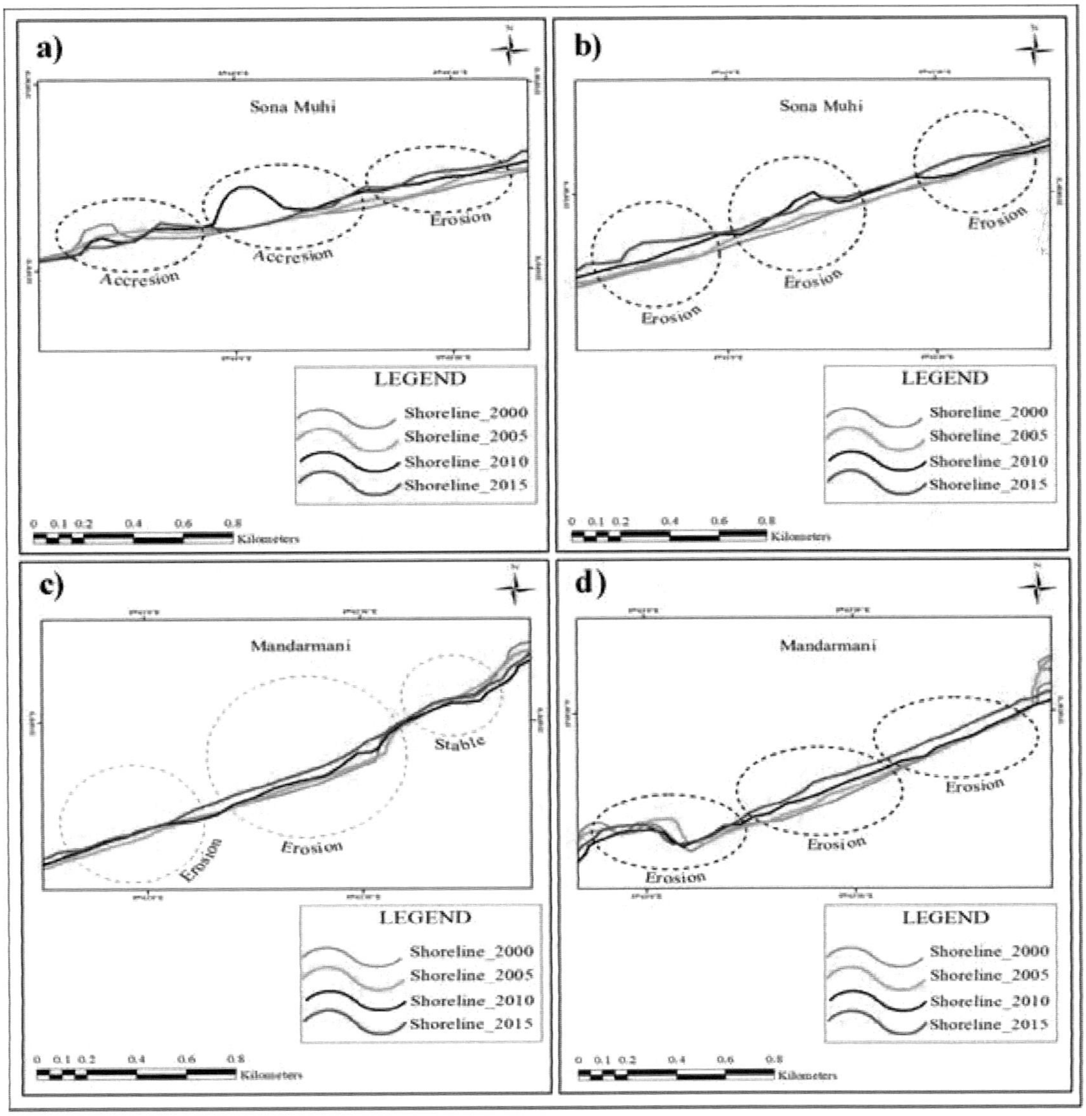


**Figure 5.** Erosion and accretion zone identified by overlaying shoreline of 2000, 2005, 2010 and 2015 in (a-b) Sona Muhi, and (c-d) Mandarmoni

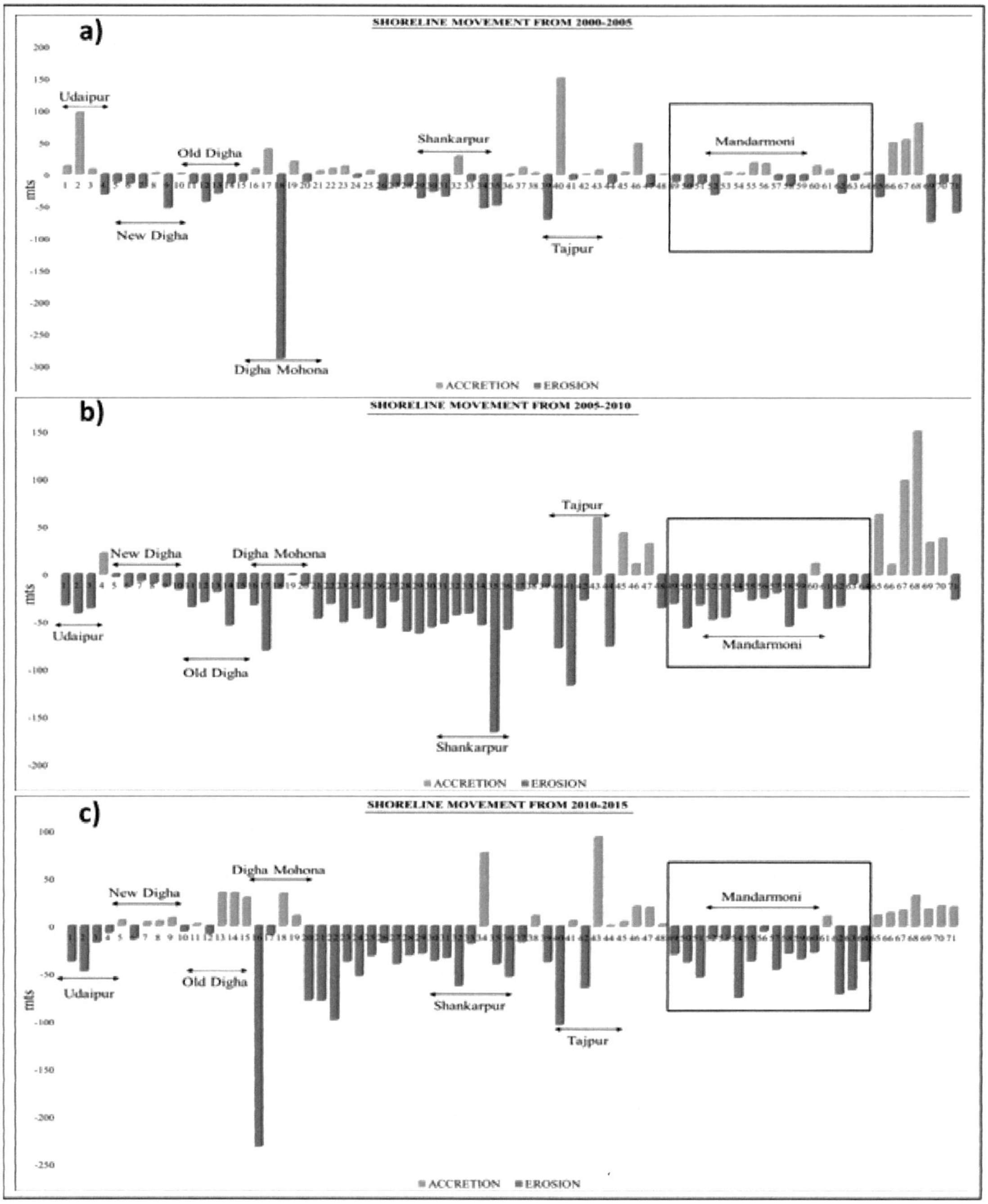


**Figure 6**. Short-term erosion and accretion rate estimated along the shoreline from Udaypur to Mandarmani for the period: a) 2000-2005, b) 2005-2010, and c) 2010-2015

### *4.2. Identification of Erosion & Accretion Zones*

The results obtained are presented in figure 5 (a to d) and discussed hereafter. The shoreline change (retreating or advancing) was identified for already developed and developing tourism establishments of Mandarmani. From 2000 to 2005 Mandarmani were observed to have undergone medium-scale erosion. Limited accretion was observed only at Sona Muhi. During the period 2005 to 2010

Mandarmani suffered severe erosion. The already developed tourism establishments along the Mandarmani coast from 2010-2015 were found to be facing severe threat as 90% of the coastline was found under erosion.

*4.3. Short Term Shoreline Movement Estimation*

The results of the DSAS analysis using the ARC GIS platform brought out the changes in the rate of erosion along the coast during 5 years intervals of 2000-2005; 2005-2010 & 2010-2015 (figure 6a to 6c) and during the longer time frame of 2000-2015 (figure 7). The 23 km long coastal stretch between Udaipur to Mandarmani was evaluated along 72 transect lines at equal spacing.

The erosion and accretion rate per year according to the transact number and tourist spot in short-term observation is as follows in table 1 with qualitative and quantitative estimates from 2000 to 2005 Mandarmani faced both medium rated erosion and accretion but from 2005 to 2010 and from 2010 to 2015 Mandarmani faced only high rated erosion shown in figure 6a, 6b, 6c and table 1.

**Table 1.** Short-Term Shoreline Movement Rate (Erosion & Accretion Rate) Estimation.

| **Coastal Segment (Map No)** | **Transact No** | **2000-2005** | | **2005-2010** | | **2010-2015** | |
|---|---|---|---|---|---|---|---|
| | | Accretion meters/ year | Erosion meters/ year | Accretion meters/ year | Erosion meters/ year | Accretion meters/ year | Erosion meters/ year |
| Mandarmani & Sona Muhi (14-17) | 52-60 | 10.97 | 12.39 | 2.2 | 53.33 | 0 | 54.56 |

*4.4. Long-Term Shoreline Movement Estimation*

The long-term analysis of shoreline change excludes short-lived impacts of storms and tidal surges but indicates impacts of littoral drifts, sediment supply and sea level change. Along the coast under study, impacts of a number of manmade structures like sea walls and groins could produce some anomalous patterns by altering the sediment dynamics of the region. The Sea wall caused a deprivation of sediment supply from the beach and induced beach lowering as indicated by continuing erosion despite the presence of the sea wall. Thus, the erosion and accretion of the shorelines in this region in a long run were observed to be caused more by the process of littoral drift than the effect of tide, wave and sea level rise. The long-term trend of shoreline change was further applied to understand the future shoreline dynamics in this region, particularly for the stability and sustainability of the tourism establishment on the Mandarmani Coast.

The erosion and accretion rate of per year measured through these transacts of the selected tourist spots from the West to East in long-term observation were given in table 2 with qualitative and quantitative estimates from 2000 to 2015. Mandarmani faced only erosion at a high rate shown in figure 7 and table 2.

**Table 2.** Long-Term Shoreline Movement Rate (Erosion & Accretion Rate) Estimation.

| **Tourist Spot** | **Transact No** | **2000-2015** | |
|---|---|---|---|
| | | Accretion m / year | Erosion m / year |
| Mandarmani & Sona Muhi | 52-60 | Nil (0) | Very High (29.78) |

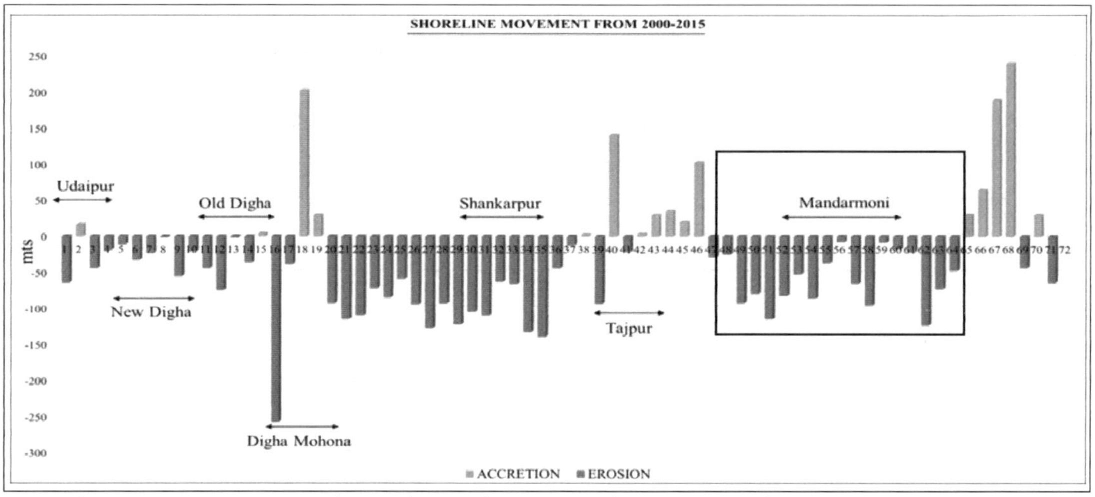


**Figure 7.** Long-term erosion and accretion rates estimated along the shoreline (Udaypur to Mandarmani) from 2000 to 2015.

### *4.5. Future Prediction of Shoreline Movement*

Using the Fishnet model the shoreline of Mandarmani has been assessed to decipher the trend of shoreline change along the preselected transacts around tourist establishments from 2000 to 2015, and the future position of 2050 of the shoreline was estimated (figure 8 and 9). In this estimation, the shoreline change rate has been computed from historical observations, and short-term impacts like tidal and storm surges were not considered. Thus, the cumulative effect of various processes like sediment transport, littoral process, and sea level rise are relevant to predict future shoreline change. The predicted total Erosion and Accretion along the 23 km long coast by 2050 along the selected transact from 1-72 between Udaipur to Mandarmani (figure 8) is given in table 3.

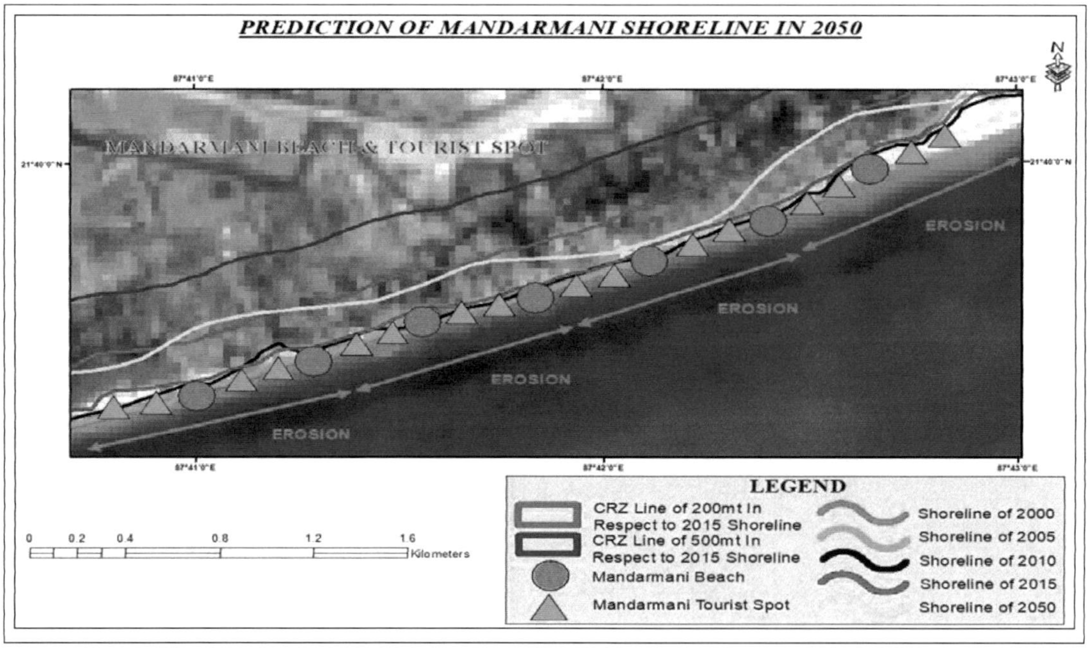


**Figure 8.** Predicted shoreline of 2050 and associated erosion and accretion zone in Mandarmoni

**Table 3.** Erosion and Accretion Prediction along the shoreline by 2050.

| **Tourist Spots** | **Transact No** | **Shoreline Length** | **Estimated Shoreline Movement In 15 Years (2000-2015)** | | **Predicted Shoreline Movement In 35 Years (2015-2050)** | |
|---|---|---|---|---|---|---|
| | | | Accretion (m) | Erosion (m) | Accretion (m) | Erosion (m) |
| Udaipur to Mandarmani | 1 - 72 | 23 km | 1134 | 3615 | 2645 | 8434 |

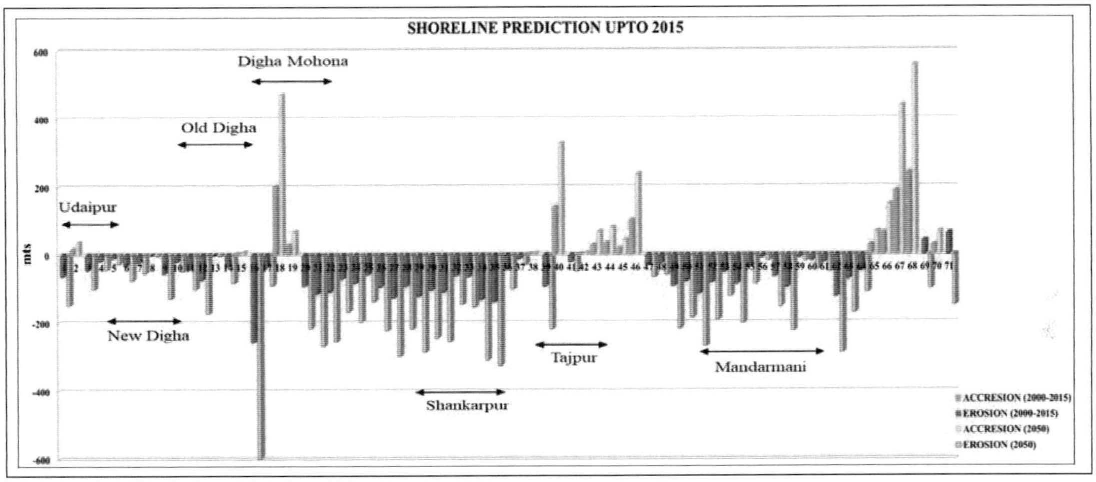


**Figure 9.** Showing Predicted Erosion & Accretion in Shoreline of 2050 on the Beach & Tourist Spot from Udaipur to Mandarmani.

According to the prediction for 2050 the most severe impact would likely to be felt at the newly developed unprotected tourist resorts of Mandarmani where the beach and almost all the tourist establishments might be washed out by 2050. It is observed from the analysis that the entire coastal stretch would be dominantly under erosion in the future. The net line meter lost from the coast during this period and the difference between the accretion and erosion was estimated to be 5789 meters by 2050 (figure 9, table 3).

## 4. Conclusion

The main purpose of the study was to understand the behaviour of shoreline change (erosion and accretion) along the Mandarmani coast, the most favoured coastal tourism destination of West Bengal, by the application of Remote Sensing and GIS technique. The area covered seven tourist spots between Udaipur and Mandarmani. The study observed both the strongly retreating coastline with temporal and special variability with minor localised accretion between the periods of 2000 to 2015. However, the most alarming situation of the coastal retreat was found along the beach of Mandarmani, the new yet unprotected coastal tourism destination of West Bengal. The analysis revealed that only about 20% of the coastlines were under stable to accreting conditions while the remaining 80% region experienced moderate to strong magnitude of coastal retreat. The Erosion Rate of 34m/year in such areas highlights the alarming trend of the shoreline retreat which is of serious concern, especially from the point of view of the sustainability of the tourism industry at Mandarmani. Also, urbanization due to tourism leads to immense pressure on this coastal area more than other coastal processes. Historic rates of shoreline change provided valuable input to the erosion process which helped to forecast the shoreline movement in future shoreline positions up to the year 2050. Five of the seven tourist locations along the coast under study may be non-existent unless some coastal protection measures are taken. Particularly in Mandarmani, 90% of hospitality establishments might be washed away by 2050.

As human interventions are much more immense than anthropogenic impacts on this coastline that greatly influences the coastal processes, the construction of unscientific infrastructure along the coast along with the cumulative action of waves and tides are responsible for the erosion/accretion trends, where wave run-up is 2.16m and the tidal range varies from 2.3m to 3m [14]. On the other hand, a large number of sea walls and breakwaters are influencing and altering the shoreline by disturbing the sediment dynamics of the region through these seawalls and groins were made for the protection of the beach. Moreover, this sea shore is also most sensitive to sea level change which has a rising trend in the last 20 years. The Mean Sea Level is rising 3 millimetres/year [14]. Thus, the retreating and advancing activity of this region in the natural condition, influenced mainly by human interference along with the combined effect of sea level fluctuation, sediment transport and coastal processes which are beyond the ambit of the present study.

Based on the present study, it can be concluded that accurate prediction and a better understanding of shoreline movement can be done cost-effectively by using high-resolution satellite data at smaller and longer intervals according to necessity and selecting short-spaced km for fishnet modelling accompanied by a field survey. The accurate prediction and better understanding of shoreline movement will enable the decision makers to identify the present and future tourism zones at high risk and accordingly suggest appropriate management measures. The present research on the shoreline movement might be helpful to devise a framework for the conservation and restorations of this coastline under severe sea erosion.

**References**

[1] Mukhopadhyay, A. Mukherjee, S. Mukherjee, S. Ghosh, S. Hazra, S. & Mitra, D. 2012 ‘Automatic shoreline detection and future prediction: A case study on Puri Coast, Bay of Bengal’, India European Journal of Remote Sensing, 45(1), pp 717-729. [online]. Available at: https://www.tandfonline.com/action/journalInformation?journalCode=tejr20. (Accessed: 25 December 2019)

[2] Maiti, S. & Bhattacharya, A.K. 2009 ‘Shoreline change analysis and its application to prediction: A remote sensing and statistics based approach’, International Journal of Marine Geology, 257(1-4), pp 11-23 [online]. Available at: www.elsevier.com/locate/ margeo. (Accessed: 25 December 2019)

[3] Kankara, R.S. Selvan, S.C. Rajan, B. & Arockiaraj, S. July 2014 ‘An adaptive approach to monitor the Shoreline changes in ICZM framework: A case study of Chennai coast’, Indian Journal of Geo-Marine Sciences, 43(7) [online]. Available at: www.researchgate.net/publication /305248177. (Accessed: 25 December 2019)

[4] Misra, A. & Balaji, R. 2015 ‘A study on the shoreline changes and land-use / land-cover along the South Gujarat coastline’, International Journal of Procedia Engineering, 116(1), pp 381-389. [online] Available at: http://creativecommons.org/licenses/by-nc-nd/4.0/. (Accessed: 25 December 2019)

[5] Nandi, S. Ghosh, M. Kundu, A. Dutta, D. & Bakshi, M. 2015 ‘Shoreline shifting and its prediction using remote sensing and GIS techniques: a case study of Sagar Island, West Bengal (India)’, Journal of Coastal Conservation, 20(1), pp 61-80. [online]. Available at: https://link.springer.com/article/10.1007/s11852-015-0418-4. (Accessed: 25 December 2019)

[6] Selvan, S.C. Kankara, R.S. & Rajan, B. July 2014 ‘Assessment of shoreline changes along Karnataka coast, India using GIS & Remote sensing techniques’, Indian Journal of Marine Sciences, 43(7) [online]. Available at: http://nopr.niscair.res.in/handle/123456789/34441. (Accessed: 25 December 2019)

[7] Chand, P. & Acharya, P. 2010 ‘Shoreline change and sea level rise along coast of Bhitarkanika wildlife sanctuary, Orissa: An analytical approach of remote sensing and statistical techniques’,

International Journal of Geomatics and Geosciences, 1(3), pp 436-455. [online]. Available at: www.researchgate.net/publication/301772465. (Accessed: 25 December 2019)

[8] Joevivek, V. Saravanan, S. & Chandrasekar, N. July 2013 'Coastal Vulnerability and Shoreline Changes for Southern Tip of India – Remote Sensing and GIS approach', The International Journal of Earth Science & Climatic Change, 4(4) [online]. Available at: http://dxdoi.org/10.4172/2157-7617.1000144. (Accessed: 25 December 2019)

[9] Mandal, M. Dandapath, P.K. & Bhushan, S. April, 2013 'Digha Shankarpur Littoral Tract A Geographical Case Study', International Journal of Humanities and Social Science Invention, 2(4), pp 46-54. [online]. Available at: www.ijhssi.org. (Accessed: 25 December 2019)

[10] Das, R. 2012 'An Analytical Study on the Phytoresources and Vegetation Ecology of Coastal Medinipur of West Bengal in India', International Journal of Science and Research, 3(10), pp 240-249. [online]. Available at: www.ijsr.net. (Accessed: 25 December 2019)

[11] https://en.wikipedia.org/wiki/Mandarmani [Dated: 16/12/2019]

[12] Baitalik, A. & Majumdar, S. 2015 'Coastal Tourism Destination in West Bengal: Historical Background and Development', International Journal of Social Sciences and Management, 2(3), pp 267-272. [online]. Available at: http://nepjol.info/index.php/IJSSM/issue/archive. (Accessed: 25 December 2019)

[13] Google Earth [Dated: 16/12/2019]

[14] Hazra, S. Sen, G.K. & Bhandary, G. 2006. 'Integrated Coastal Zone Management of Digha – Shankarpur Area, West Bengal'. 1st edn Kolkata: School of Oceanographic Studies, Jadavpur University.

# Implication of Applying IPCC AR4 and AR5 Framework for Drought-based Vulnerability and Risk Assessment in Bankura and Purulia Districts, West Bengal

**A Singha[1,2], N Pramanick[2 *], R Acharyya[3]**

[1]University of Florida, Department of Geological Sciences, 241 Williamson Hall, Gainesville, FL 32611, USA
[2]School of Oceanographic Studies, Jadavpur University, Kolkata 700 032, W.B., India
[3]Department of Geography, School of Earth Sciences, Central University of Karnataka, Kadaganchi, 585367, India

[*]Corresponding Author Email: niloy.pramanick@yahoo.com

**Abstract:** In the districts of Purulia and Bankura, this study offers a methodology for spatial assessment of vulnerable and risk-prone areas. These districts are adjacent in space and have identical geographic characteristics (other than the eastern portion of Bankura district). Vulnerability and risk assessment could be used to measure the interactions between individuals and their surroundings. This research aims to pinpoint the areas in these two districts that are particularly susceptible to natural, social, and meteorological disasters. The natural and climate-induced factors considered are rainfall distribution and vegetation conditions. The social factors are agricultural dependence, percentage of farmers, female population, labor dependence on agriculture, and literacy rate. The potential impacts of developmental and environmental degradation processes can be examined and assessed by classifying regions according to their vulnerability and risk levels. The fundamental factors impacting susceptibility and risk, which are recognized, and the associated thematic-based outputs are produced in this study based on the persistent phenomenon of drought within these two districts. The elements of vulnerability selected for this study are exposure, sensitivity, and adaptive capacity (IPCC AR4) and risk, which is the combined outcome of hazard, exposure, and vulnerability (IPCC AR5). The aim of this research is to create a simplified, scalable assessment model for evaluating both vulnerabilities and threats, which can help with drought mitigation. It has been observed from the results that the western portion of the study area (Arsha, Purulia-I, Baghmundi blocks of Purulia district) with relatively higher risk and vulnerability needs more attention for reducing the vulnerability and risk than the eastern part. As a result, this research can serve as a platform for district-level prioritizing efforts, emergency response protocols, and policy interventions aimed at reducing disaster susceptibility (mostly drought) in Bankura and Purulia districts.

**Keywords:** Bankura, Drought, IPCC AR4, IPCC AR5, Purulia, Vulnerability Assessment.

## 1. Introduction

The two districts, Purulia and Bankura of West Bengal, India, are prone to recurrent droughts, often continuing for years. These two districts also rank quite low compared to other districts of West Bengal in terms of the composite human development index [1]. Purulia district is the most vulnerable to drought risks across the state, West Bengal. As a result, the district is known as "Ahalya Bhumi," or "the region with a stone heart." (Hindustan Times, Kolkata, 2021). While in Bankura, the land is undulating, and lateritic porous, resulting in inadequate subsoil moisture, which becomes a potent threat to the crop [1]. The Fourth Assessment Report (AR4) aims to evaluate scientific, technological, and economical knowledge regarding climate change, its probable consequences, and viable solutions adaptation and mitigation options [2]. In light of this (IPCC AR4 report), a conceptual model has been developed to identify the vulnerable areas in these two regions. The Fifth Assessment Report (AR5) supersedes AR4 in 2014. The outline of the AR5 was established through a painstaking process that included climate change experts from all related fields as well as users of IPCC reports, as it had been in the past. As per AR5, vulnerability has further broadened to encompass the threat of climate change effects [3]. While there is no commonly agreed definition of vulnerability [4, 5] and it is unclear how it could be employed as an environmental assessment mechanism [6, 7] has become an increasingly significant part of understanding the man-environment interaction [8]. So, with the help of this (IPCC AR5 report), we have also deciphered the risk-prone areas in these two districts. Among other natural calamities, Purulia and Bankura districts are remarkably affected by drought. Short-range variations in terrain, soil, and water conditions result from undulating topography and severely dissected environments [9]. Droughts have wreaked havoc on human health, the climate, and the economy around the world in recent decades [10]. Droughts are the most common natural disaster on the planet, affecting more people than any other natural threat [11]. In view of that, drought-related vulnerability and risk assessments are conducted. Droughts are characterized as extended periods of unusually low rainfall that led to a water shortage. Large temporal datasets with complex interrelationships within climatological and meteorological results are needed for drought research [12]. A detailed evaluation of social vulnerability is essential for understanding natural hazard threats and developing effective response capabilities [13]. Drought tracking, early warning, and risk management are all supported by advances in Geographic Information Systems (GIS) and remote sensing (RS) techniques [14]. This project used RS and GIS to create a vulnerability and risk management model.

## 2. Study area

Purulia and Bankura districts, which are part of the eastern plateau zone of India, are located in West Bengal's westernmost region. The Bankura district is situated between 22° 38'-23° 38'N latitude and between 86° 36'- 87° 46' E longitude, covering 6788 sq.km area and, the Purulia district extends between 22° 42' 35" N - 23° 42' 00" N latitudes and 85° 49' 25" E - 86° 54' 37" E longitudes covering 6259 sq.km area (figure 1).

## 3. Physiographic Characteristics of the Study Area

Drought is a protracted climatic phenomenon that is difficult to forecast in terms of occurrence, severity, intensity, and duration. Drought forecasting is critical in the dry portions of West Bengal for water planning and efficient use of irrigated water, as agriculture is the main economic sector and is heavily reliant on rainfall [15]. As the rate of plant growth(vegetation) and development are influenced by soil and air temperature [16], [17] it is important to consider these variables in drought-affected areas. The climate in Puruliya and Bankura is warm and temperate, which is as per the Koppen-Geiger climatic class of Cwa ("Monsoon-influenced humid subtropical climate"). Another key Physiographic characteristic would be vegetation. The vegetation's geographical distribution and quality can have an impact on a variety of environmental and biological factors [18]. The research area's vegetation state varies throughout time. The variability of vegetation status in the irrigation-based region is quite high, and decreased in the forest region due to minor differences in quality of vegetation, and it can go very low for the remaining area [19]. Purulia's geology is characterised by Proterozoic hard granitic gneiss rocks, including mica schist and soft phyllite from the Singhbhum group, which is portion of the Chotanagpur plateau [20]. Similarly, the Bankura district is situated on the eastern side of the Precambrian Chota Nagpur plateau, which is characterized by granite gneiss and Gondwana deposits

[21]. The different soil texture types are mainly Fine Loamy, Fine loamy Sandy, Coarse Loamy, and Gravelly loam. The majority of the study area is dominated by Fine loamy type of soil [22].

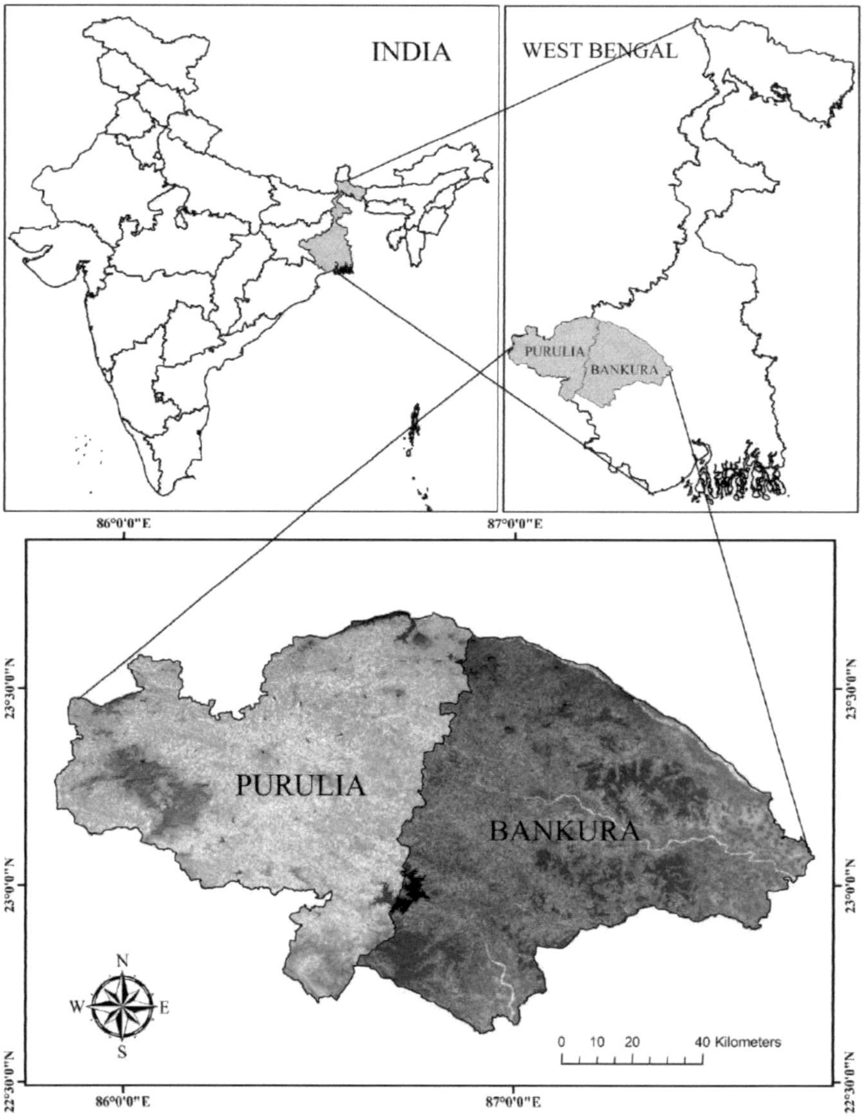


**Figure 1**: Study area: Purulia and Bankura Districts, West Bengal, India

## 4. Rainfall Deficiency

Rainfall is the most important predictor of drought, and most drought warning systems are based on it [23]. The unpredictability and uneven distribution of rain is a serious issue in the research area's rainfall-dependent agriculture. Crop failures as a result of dry periods and occasional droughts remain common as a result of this unpredictability [24]. Although the yearly rainfall data of the area under study, show that there is enough rain to prevent drought, the monthly rainfall data of the same region indicates that rain is not evenly distributed throughout the year. These two districts receive the most rainfall, especially during the monsoon season.

IOP Conf. Series: Earth and Environmental Science 1164 (2023) 012009 doi:10.1088/1755-1315/1164/1/012009

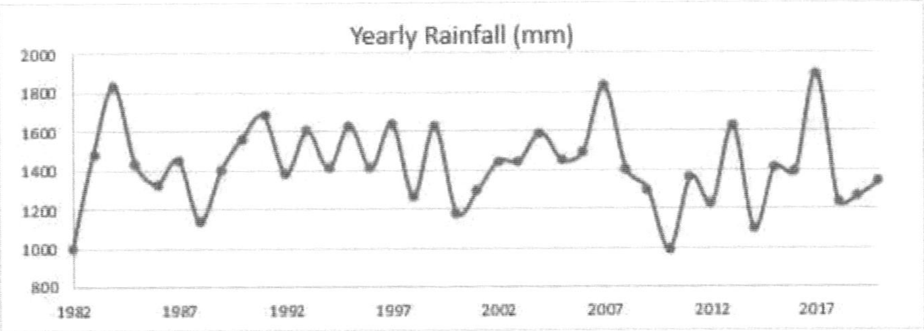


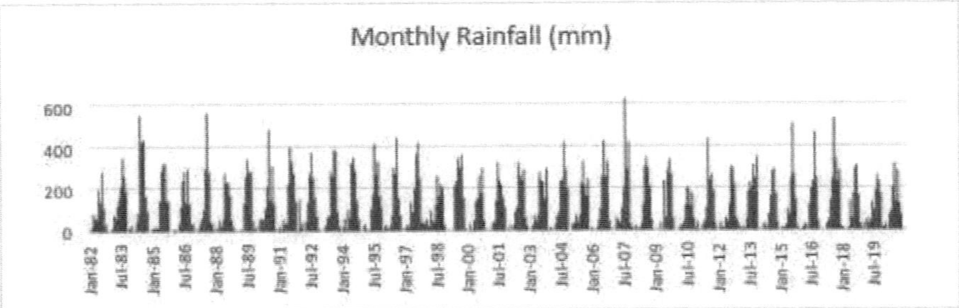


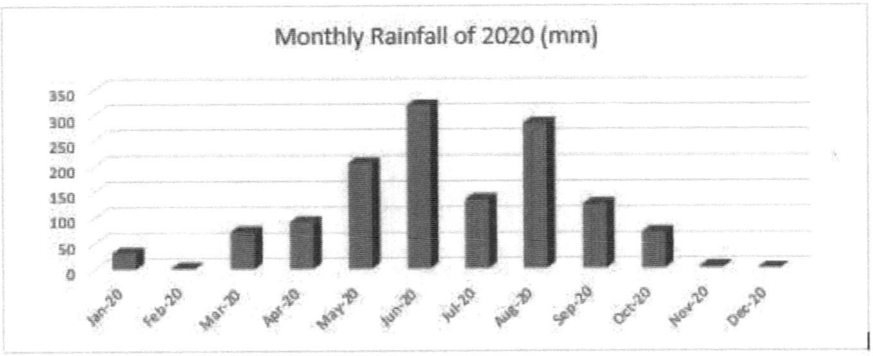


**Figure 2**: Annual rainfall (up), monthly rainfall (middle) and Bar diagrams showing monthly rainfall of Bankura in 2020

Figures 2 and 3 show that the amount of rainfall, these two districts received varied markedly. e.g., in 2007, both these districts received around 1800mm rainfall, while in 2010 it was around 900mm. about 50% reduction of rainfall in just three years leads to imbalance, which indicates a possibility of drought. Considering the same two years (2007 and 2010), from the monthly rainfall graphs (figures 2 and 3), it is observable that during the monsoon season (mainly July 2007), the rainfall amount for both of these districts was above 600 mm, while in 2010, both of these districts received less than 200 mm of rainfall during the monsoon (mainly May).

The most recent and previous year's (2020) monthly rainfall data shows that these two districts do not receive the same amount of rainfall throughout the year. Rainfall has been heavy in both districts only during June and August.

IOP Conf. Series: Earth and Environmental Science 1164 (2023) 012009 doi:10.1088/1755-1315/1164/1/012009

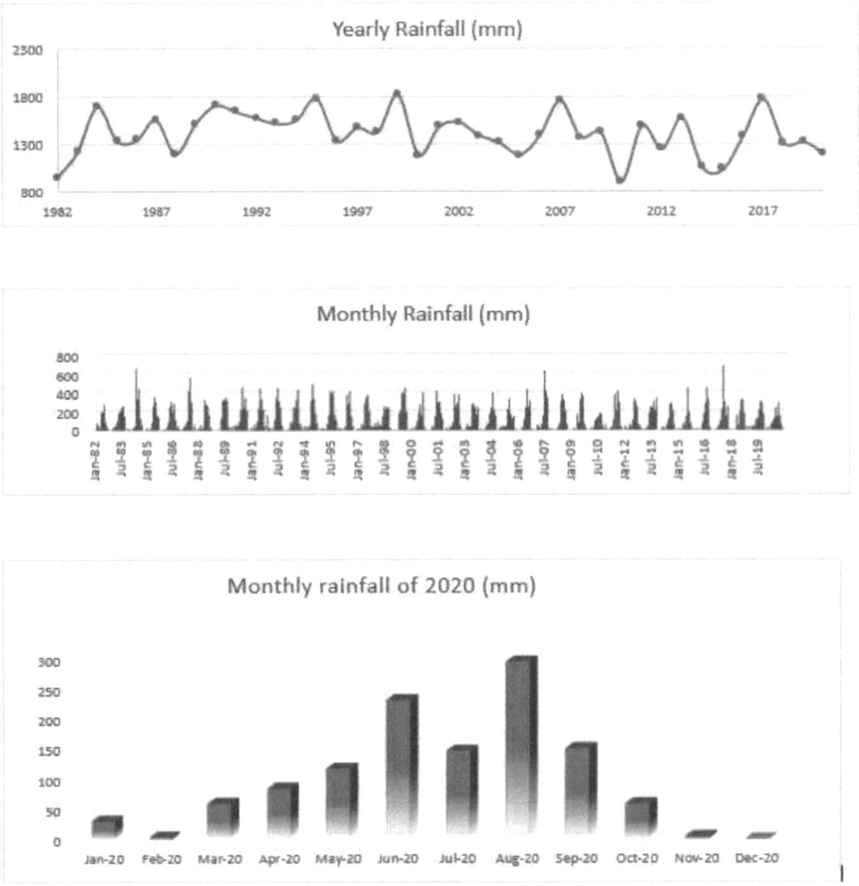


**Figure 3:** Annual rainfall (up), monthly rainfall (middle), and Bar diagrams showing monthly rainfall of Purulia in 2020

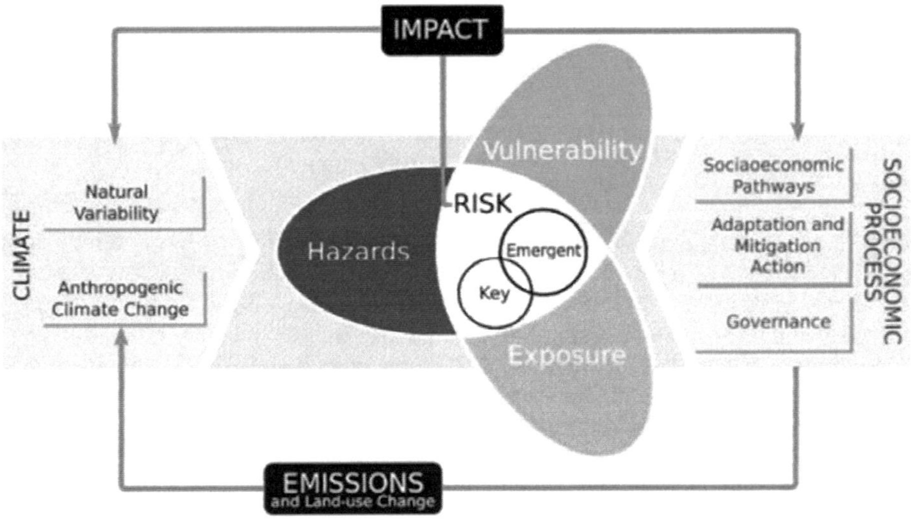


**Figure 4:** Linking Risk and Vulnerability [2]

## 5. Data and methodology

### *5.1. Data software*
The remotely sensed images from LANDSAT OLI 8 and TIRS data and MODIS ("Moderate Resolution Imaging Spectroradiometer") data have been utilized in the current research. Then, in terms of Census data, the data from the Census 2011 for Puruliya (Purulia) and Bankura Districts have been considered for the study. In terms of precipitation analysis, rainfall datasets of Global Weather Data for SWAT (1982-2020) have been incorporated. ArcGIS 10.3.1 has been used to assess remotely sensed and geospatial datasets in a GIS environment. Other quantitative data analyses and estimations have been performed using Microsoft Office Excel 2010.

### *5.2. Methodology*
Figure 4 is adapted from IPCC AR5, 2014. From this model, two distinct methods have been developed to decipher Vulnerability (figure 5) and Risk (figure 6).

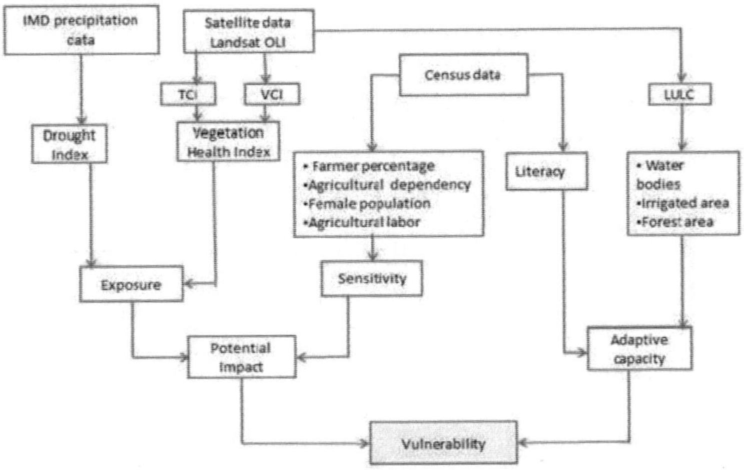


**Figure 5**: General logic to calculate Vulnerability

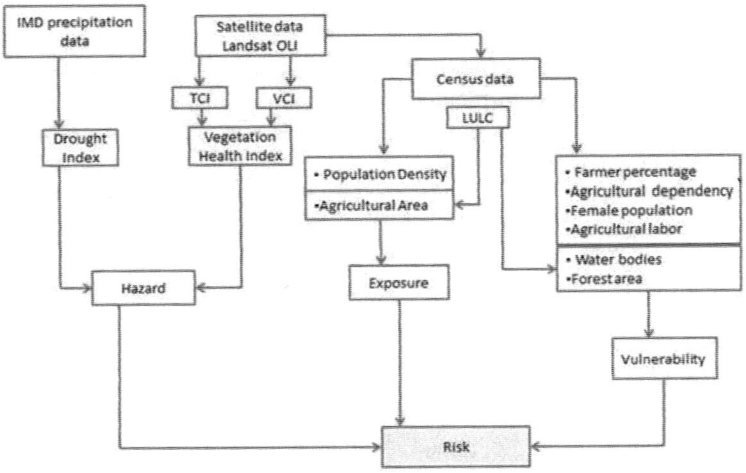


**Figure 6:** General logic to calculate risk

## 6. Assessment of Different Criteria

In the 4th Assessment Report (AR4), the IPCC ("Intergovernmental Panel on Climate Change") defines vulnerability as the degree to be able to cope and handle the adverse effects of climate change, particularly climate variability and extremes, in geophysical, biological as well as socio-economic settings. [2]. The term vulnerability in AR4 states that the vulnerable structure itself and the impact on this system (Schneider SH et al., 2007). As per IPCC AR4, the three key factors for measuring vulnerability are sensitivity, exposure, and adaptive capacity [25,26]. In AR4, exposure refers to the extent and length of stress resulting from the climate induced drought or variations in rainfall, while vulnerability refers to the degree to which the system is impacted by climate-related stress or severe events [2]. In AR4, the ability to withstand or recover from extreme events/damage is referred to as adaptive capability [25,27,28]. According to the IPCC's Fifth Assessment Report (AR5), risk is defined as "the potential for repercussions where something of importance is at risk and the result is unpredictable, taking into account the range of values." If specific occurrences or trends occur, the probability of incidences of hazardous events area increasing [29]. As AR5 is a developed version of AR4, some components are interrelated and interchangeable. The terms 'vulnerability' and 'exposure' are very common, but they are utilized contrarily in AR4 and AR5. As per AR5, 'exposure' is defined as the persistence of human population, communities, organisms or ecosystems, ecological processes, infrastructure, services, and physical, cultural, or social resources across locations and environments which could be unfavorably affected [29]. While vulnerability refers to the tendency or the predisposition to be negatively impacted [29]. In AR5, vulnerability includes the principles of adaptation (vulnerability to damage) and adaptive capacity (Das et al, 2020). According to AR5, a human- or naturally-caused physical incident, pattern, or effect that may result in fatal damages and losses to property, infrastructure, livelihoods, provision of services, ecosystems, and natural resources are also possible. [29].

Both the IPCC's AR5 and AR4 operating concepts make it clear that risk and vulnerability have an external factor, which is climate-related stress (drought), defined as 'exposure' in AR4 and 'risk' in AR5. There is also an internal component that includes 'adaptive capability' and 'sensitivity' in AR4 and 'vulnerability' and 'exposure' in AR5 [2].

The vegetation health and drought indices for vulnerability are considered to be the components of exposure. On the other side, these same components act as the constituents of hazard to compute risk-prone zones.

As per the framework suggested by Fussel [30], potential impact (PI) is composed of sensitivity (S) and exposure (E), whereas adaptive capability (AC) has been the capacity of a mechanism to address these effects, as seen in equation below (1).

$$PI = E \times S \tag{1}$$

If a system's adaptive capacity is low or inadequate and the system is exposed and has a sensitivity to the effects of climate change scenarios (e.g., drought), then the system should be considered a more vulnerable one. Vulnerability can thus be represented mathematically using equation (2).

$$V = PI - (PI \times AC) \text{ or } V = PI\,(1 - AC) \tag{2}$$

Hazard, Vulnerability, Exposure, and Adaptive Capacity can be used to quantify risk using equation (3).

$$R = H \times E \times S\,(1 - AC) \tag{3}$$

The vulnerability and risk indices have a range of 1 - 5, with higher values indicating a greater degree of risk and vulnerability. Finally, the entire range is divided into 5 equal classes with qualitative measures of risk and susceptibility that range from very low to very high. Separate maps were generated in a GIS environment to interpret and evaluate the data in a spatial context.

For vulnerability assessment with the help of IPCC AR4, vegetation indices and drought indices are considered exposure factors.

### *6.1. Vegetation Condition Index (VCI):*

In years when the weather is exceedingly unfavorable (mostly arid and hot), only a small amount of vegetation develops, suppressing growth of vegetation both directly and indirectly through a reduction in the rate of ecosystem resource usage (e.g., in years of drought spell, soil nutrient uptake is drastically reduced due to a lack of water) [31]. As a result, to identify drought condition over the years, Vegetation condition is needed to recognize; thus, VCI has been implemented [equation (4)].

$$VCI = 100 \times (NDVI - NDVI_{MIN}) \div (NDVI_{MAX} - NDVI_{MIN}) \quad (4)$$

Where, NDVI, $NDVI_{MAX}$, and $NDVI_{MIN}$ are the smoothed annual normalized difference vegetation index.

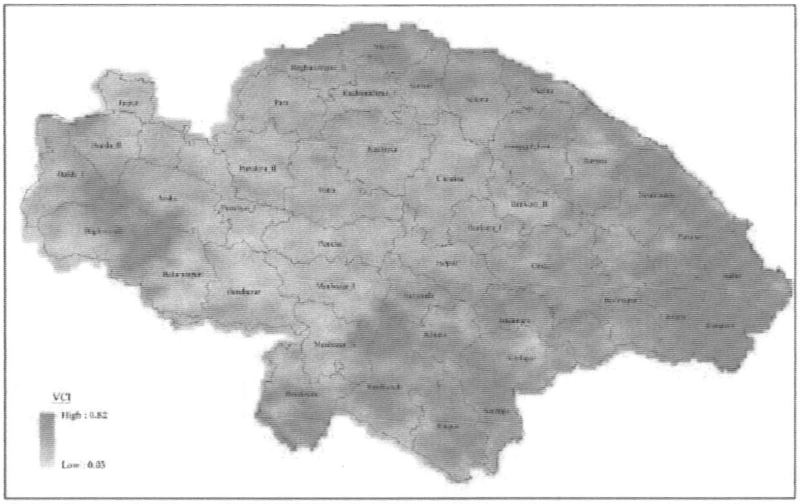


**Figure 7**: Vegetation Index: VCI

*6.1.1. NDVI ("Normalized Difference Vegetation Index").* NDVI is an effective monitoring tool, the vegetation cover of large areas and detecting the occurrence and persistence of droughts (Wattanakij et al., 2006). The NDVI is a vegetation index that indicates stable and dense vegetation. NDVI can be calculated using spectral radiance in red and near-infrared reflectance from satellite imagery. This formula is applied to calculate NDVI [equation (5)].

$$NDVI = (NIR - RED) \div (NIR + RED) \quad (5)$$

The measurements of spectral reflectance in the red (visible) and near-infrared (NIR) regions are denoted by red and NIR, respectively. This spectral reflectance ranges from 0.0 to 1.0 since they are ratios of reflected over incoming radiation in each spectral band. As a result, the NDVI itself differs between -1.0 and +1.0 by design (figure 8).

*6.1.2. SAVI ("Soil-Adjusted Vegetation Index").* The SAVI is comparable to the NDVI but includes a 'Soil brightness correction factor,' where NIR denotes reflection from the near-infrared band, RED denotes reflectance from the red band, and L is the soil brightness adjustment factor. The stability of Empirical NDVI estimates have been found to vary with soil colour, soil moisture, and saturation effects from thick vegetation. In order to improve NDVI, Huete (1988) developed a vegetation index which compensated for the variable red and near-infrared absorption throughout the plant canopy. The indicator is a method for minimizing the influence of spectral vegetation indicators with RED and NIR bands. on soil brightness. The SAVI index has been proposed by Huete (1988) and is calculated by the following equation (6).

$$SAVI = [(1 + L)(NIR - RED)] \div (NIR + RED + L) \quad (6)$$

Where L is a consideration for adjusting the canopy background. In reflectance space, A L valuation of 0.5 was discovered in reflectance space to limit soil brightness changes and eliminate the requirement for additional soil calibration. It was discovered that the transformation almost completely eliminated soil-induced disparities in vegetation indices [32] (figure 8b).

*6.1.3. EVI ("Enhanced Vegetation Index").* EVI is a variable for vegetation greenness that could be used to forecast net primary production [33]. By decoupling background signals from the canopy and reducing atmospheric influences, the enhanced vegetation index (EVI) was created for improved vegetation detection [34]. As a result, the EVI is now regarded as a reformed NDVI with improved vegetation control capabilities and increased exposure to high biomass regions. The following equation (7) is used for EVI calculation.

$$EVI = G \times \left((NIR - RED) \div (NIR + CI \times RED - C2 \times BLUE + L)\right) \quad (7)$$

Where, L=1, C1 = 6, C2 = 7.5, and G (gain factor) = 2.5

EVI is a plant greenness metric that is comparable to the NDVI. EVI, on the other hand, corrects certain ambient settings and canopy background noise and is further alert in densely vegetated areas [35]. It includes a 'L' value for canopy history, "C" values regarding atmospheric resistance coefficient and blue band (B) values (Fig. 8c).

*6.1.4.* Temperature Condition Index (TCI). Since favorable weather allows for optimum moisture levels, high VCI values indicate stable, unstressed vegetation [36]. As a result, a high VCI suggests a low level of vulnerability. Low TCI values, on the other hand, indicate vegetation stress instigated by high temperature and dryness. The TCI (equation 8) allows for the detection of subtle changes in plant health caused by the thermal effect when drought spreads and is followed by high temperatures [37] [equation (8)].

$$TCI = 100 \times (BT_{MAX} - BT) \div (BT_{MAX} - BT_{MIN}) \quad (8)$$

Where BT =Brightness Temperature, $BT_{MAX}$, and $BT_{MIN}$ are the smoothed yearly radiant temperature. VCI and TCI differ from 0 (for highly unfavorable circumstances) to 100 (optimal circumstances).

The TCI and Vegetation indices are combined to create the Vegetation Health Index. The health of the vegetation is analyzed using figure 7 and figure 8 (d).

### *6.2. Drought Indices*

Drought monitoring and forecasting are critical methods for putting in place effective mitigation strategies to mitigate negative consequences [38]. For this purpose, four indices (SPI, PNI, RAI, ZSC) are used to interpret the drought frequency over our study area. A drought index is a tool for tracking drought conditions that can be defined as a single value that expresses the dryness or wetness of a situation (Bhunia, Das, and Maiti 2020). Using these four indices from rainfall data (1982-2020), we calculated the DDI (Degree of Dryness Index) to deduce drought frequency. The different Drought Indices are as follows:

*6.2.1. SPI ("Standardized Precipitation Index").* SPI is the most frequently used drought indicator [39] and it is used to describe meteorological droughts [40, 41]. Because of its versatility and special features for monitoring drought using precipitation data, the SPI is commonly used to indicate drought and is ranked first [42]. The only needed input parameter is precipitation. Furthermore, it is equally useful in evaluating rainy cycles as it is in analysing dry cycles. The output values ranged from -2 to 2. [43], The used equation for SPI is in equation (9). where $\Gamma(\alpha)$ is the gamma factor; x (mm) is the quantity of precipitation ($x > 0$); $\alpha$ is the shape function ($\alpha > 0$), and scale parameter is $\beta$ ($\beta > 0$) (figure 9 a).

$$g(x) = \frac{1}{\beta^{\alpha} r(\alpha)} x^{\alpha-1e-\frac{x}{\beta}(x>0)} \quad (9)$$

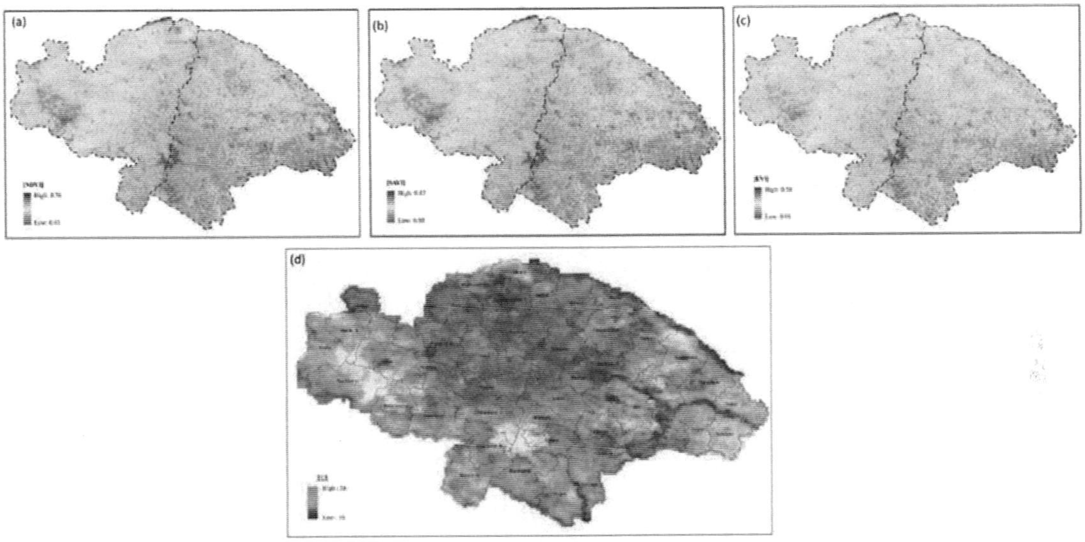

**Figure 8:** (a) NDVI, (b) SAVI, (c) EVI, (d) Temperature condition composite map.

*6.2.2. PNI ("Percent of Normal Index").* Willeke et al. [44] defined PNI as a proportion of normal rainfall. It could be computed on a variety of different timeframes (monthly, seasonally, and annually). For defining drought in a single region or season, PNI has been found to be appropriate [45]. PNI is calculated following the equation (10).

$$PNI - {}^{P_i}/_{P} \times 100 \quad (10)$$

Where, P is the normal rainfall for the study period, and Pi is the rainfall in time step *i*(mm) (Figure 9 b).

*6.2.3. ZSC("Z-Score Index").* ZSI is sometimes misunderstood with SPI [43]. It is more akin to CZI, though, in that it does'nt entail fitting rainfall data to a distribution function or a Pearson type III distribution. The coefficient of variation for the precipitation anomaly is known as the Z-score, which does not require the data to be adjusted by fitting them to any particular theoretical distribution. [46] (figure 13c). ZSI can be calculated by the following equation (11).

$$ZSI = (P_i - \bar{P}) \quad (11)$$

where P denotes the average monthly rainfall (mm), Pi denotes rainfall in a single month (mm), and SD denotes the standard deviation for particular time frame (mm) (figure 9c).

*6.2.4. RAI ("Rainfall Anomaly Index").* RAI can analyze agriculture, water resources, and other segments that have been affected by droughts. As a hydro-climatic index, the rainfall anomaly index (RAI) can be used to estimate the wetness and dryness conditions of climatic variations [47].

IOP Conf. Series: Earth and Environmental Science 1164 (2023) 012009 doi:10.1088/1755-1315/1164/1/012009

The RAI considers two types of anomalies: negative and positive anomalies. The precipitation data are presented in descending order. A positive anomaly threshold is formed by combining the ten highest values, and a negative anomaly threshold is generated by averaging the ten lowest values. The following equation (12) calculates the threshold value.

$$RAI = 3 \times [(\rho - \bar{\rho}) \div (\bar{m} - \bar{\rho})] \ and \ RAI = -3 \times [(\rho - \bar{\rho}) \div (\bar{m} - \bar{\rho})] \qquad (12)$$

Where p represents the actual rainfall for every year (mm), p represents the long-term mean rainfall (mm), and m represents the average of the 10 greatest p values for the positive anomalies and the average of the ten lowest p values for the negative anomalies ( (figure 9d).

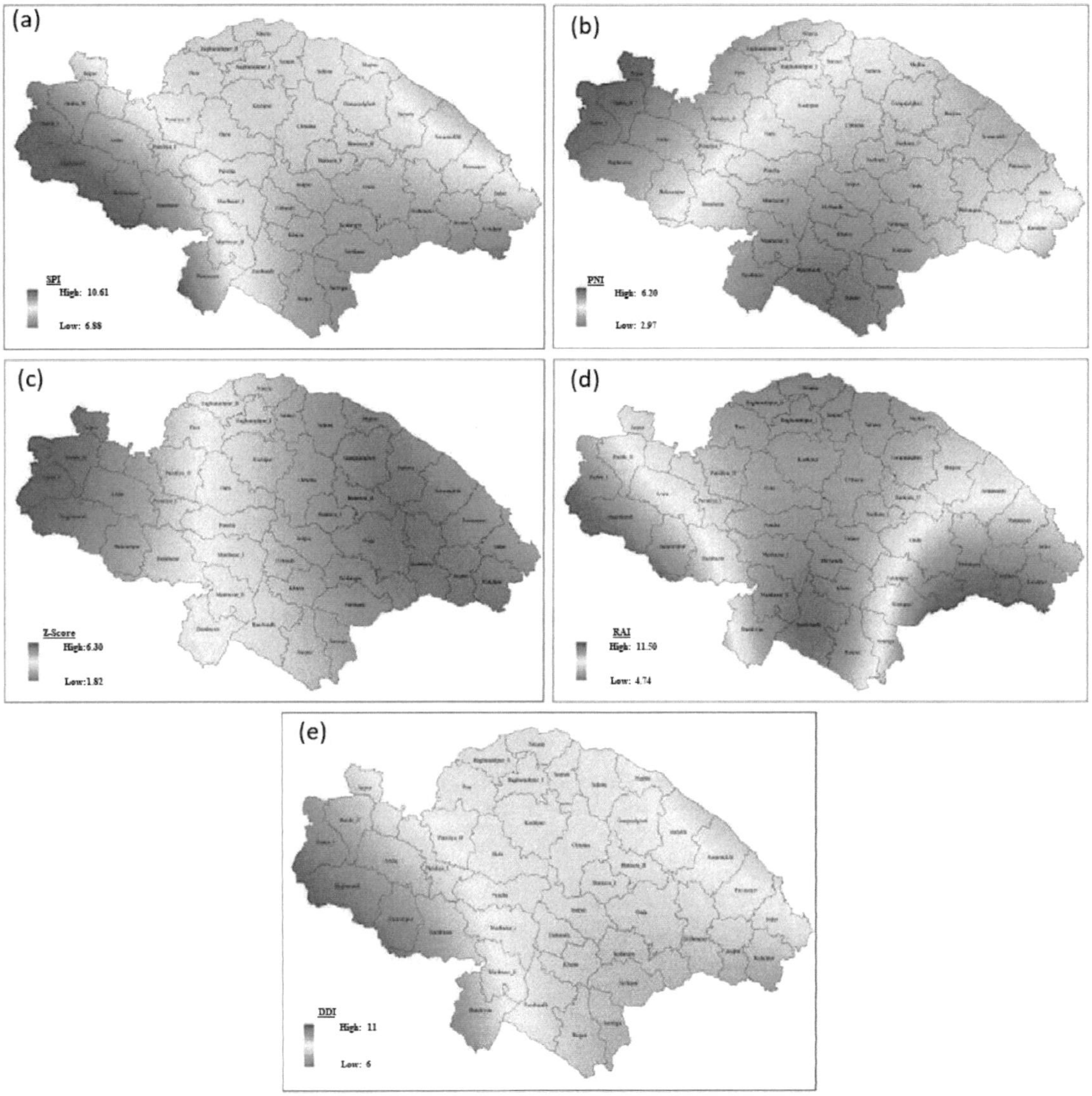


**Figure 9: (a)** SPI (b)PNI (c) ZSC (d) RAI (e) DDI

*6.2.5.* Degree of Dryness Index (DDI). 38 years of rainfall data (from 1982 to 2020) were used for this. Severe, moderate, and low are the three classifications for areas with relatively low rainfall. 21 position points (in and around our study area) were calculated using excel sheets. Afterward, four different drought indices were processed upon this. For example, in the SPI index from 1982 to 2020, the study area had a severe drought in 1983, Moderate drought in 1989 and 1992, with 2016 being the extremely dry year. Following that, multipliers of three for extreme drought months, 2 for medium drought months, and one for moderate drought months were employed to calculate a yearly degree of dryness index (DDI) (equation 13). of 8 for SPI at the study site. [43] (figure 9e).

$$DDI = (1 \times 3) + (2 \times 2) + (1 \times 1) = 8 \quad (13)$$

*6.3. Sensitivity*

The degree in which the mechanism is influenced by climate-related stresses or severe occurrences is referred to as sensitivity [2]. Basically, to operationalize a system's sensitivity, a 'cause-effect' or 'dose-response' mechanism is used **[48].** The following factors are considered components of sensitivity:

*6.3.1. Agricultural dependency.* In one family, it is more stable to have two agricultural labors who share their income than in a family with one agricultural labor and one dependent. So, if the number of dependents increases, the stability of a family decreases. In drought-prone areas, where agricultural labors have nominal income, a growing number of dependents can generate more vulnerability (figure 10a).

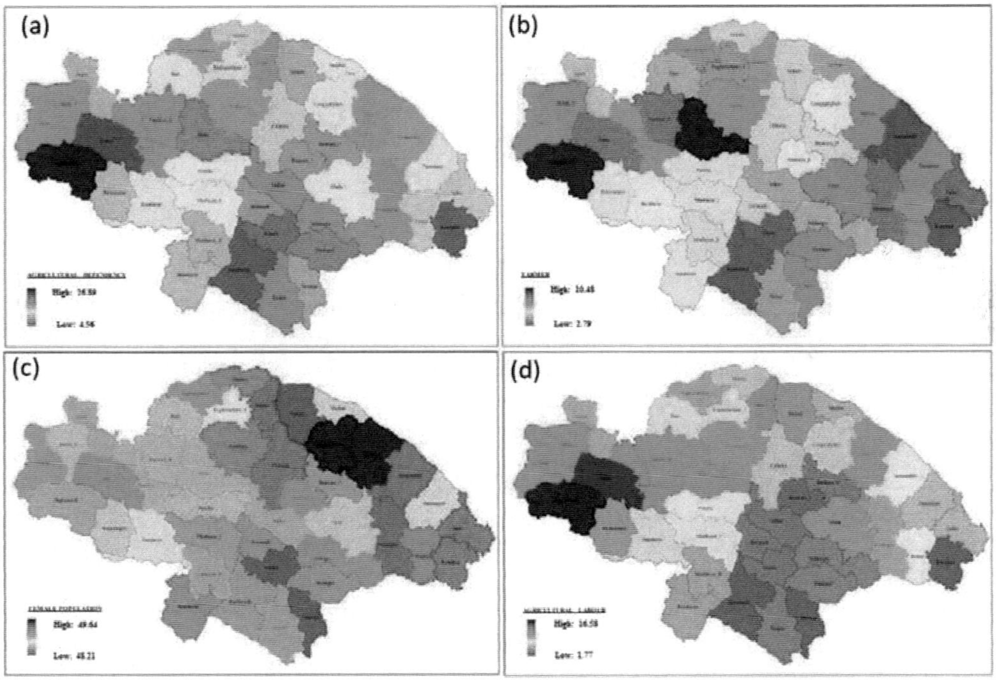


Figure 10 (a): Agricultural dependency; (b) Farmer percentage; (c) Female population; (d) Agricultural Labour

*6.3.2. Farmer percentage.* This factor represents the percentage of a primary cultivator in any particular location. Drought, as a normal and gradual occurrence, causes a slew of problems for agricultural societies [49]. A higher farmer percentage implies more vulnerability in that area. As water resource is the prime requirement for cultivation, in drought-prone areas, a rise in farmers can mark any particular

IOP Conf. Series: Earth and Environmental Science 1164 (2023) 012009 doi:10.1088/1755-1315/1164/1/012009

region vulnerable (Figure 10b).

*6.3.3. Female Population.* Women's contribution is limited and not well acknowledged in various developing countries [50]. Since their family lives are affected, women are more likely to share knowledge about community health, which is vital for resilience, and they are more able to respond to environmental changes. In addition, women are often the first responders in communities affected by natural disasters. In order to mitigate the effects of climate change, women's involvement in decision-making is critical. Therefore, it can be considered that a more female population means a less vulnerable zone (figure 10c).

*6.3.4. Agricultural Labour.* A person, who works on another person's land for wage money, can be named agricultural labor (Census 1981). Adverse climatic condition (e.g., drought) has a damaging effect on plants which declines their productivity. With that, the income of agricultural labors plummets. Therefore, areas with high drought frequency will have high vulnerability when the number of agricultural labors increases (figure 10d).

### *6.4. Adaptive capacity*

The global climate mechanism and contemporary society are both dynamic, ever-changing systems. [51]. They occasionally evolve in reaction to external influences, and sometimes they evolve on their own [52]. The capability of a framework to respond to climatic change in order to survive is referred to as adaptive capacity to minimize future damages, seize opportunities, or cope with the consequences (IPCC 2001, also referenced in the 2009 California Climate Adaptation Strategy). Adaptive capability is a key component of long-term climate change adaptation, and it is the subject of an increasingly increasing body of research [53]. The following factors are considered as features of adaptive capacity:

*6.4.1. Surface Water Bodies.* Streams, rivers, lakes, reservoirs, and oceans are examples of surface-water sources (USGS). Small water bodies are becoming more recognized for their abundance, importance for freshwater biodiversity, sensitivity and vulnerability, as well as contribution to ecosystem services [54-58]. Natural and human-made processes that interfere with the hydrological cycle, altering the natural water volume regime in aquatic bodies, may make water supplies vulnerable [59]. The higher availability of surface water bodies indicates healthy irrigation conditions. Apart from agricultural purposes, the water body is needed for everyday purposes. Consequently, any zone with a greater number of surface water bodies is less pressurized to use groundwater. So, the adverse effect can be mitigated by increasing the supply of surface water bodies over an area for a brief time span. For this reason, the surface water body is inversely proportionate to vulnerability. The area with Surface water bodies in these two districts has been projected from the land use land cover map (figure 11a).

*6.4.2. Literacy rate.* The overall number of literate persons in a certain age group is indicated as a percentage of the entire population within that age group. The adult literacy rate is a measurement of literacy among people aged 15 and above [60]. Vulnerability is reduced by having a high literacy rate. By enhancing people's capacities and having access to information on far-reaching changing climate, people might well be able to devise any necessary solutions to deal with the adversities. (figure 11b).

*6.4.3. Forest Area.* Forests can adapt to changes in mean climate variables and increase variability with a greater risk of extreme weather events, such as prolonged drought [61]. Forests are a source of livelihood for many people. Forests are the most important terrestrial habitats, providing food for both wildlife and agricultural populations [62]. Currently, forest conversation is a pressing matter because biodiversity-rich forests are less vulnerable due to the varying temperature tolerance of plant species. Hence area with high forest availability represents lesser vulnerability. From the land use land cover map, the forest area of these two districts has been generated (figure 11c).

*6.4.4. Irrigated area.* Human survival and social stability are dependent on food security [63]. Irrigation is, applying the controlled amount of water for plantation at needed intervals. Irrigation helps to grow crops, and maintain landscapes and has a positive effect on re-vegetation in disturbed soils in dry areas and during periods of less than average rainfall. Groundwater provides about 75% of irrigation water, with surface water providing the remaining 25%. So, when drought frequency is relatively high in any area, it can be estimated that higher availability of irrigated land will generate more vulnerable areas than zones with low availability of irrigated land (figure 11d). For risk assessment with the help of IPCC AR5, Literacy rate, Forest area, and Water bodies are considered adaptive capacity factors.

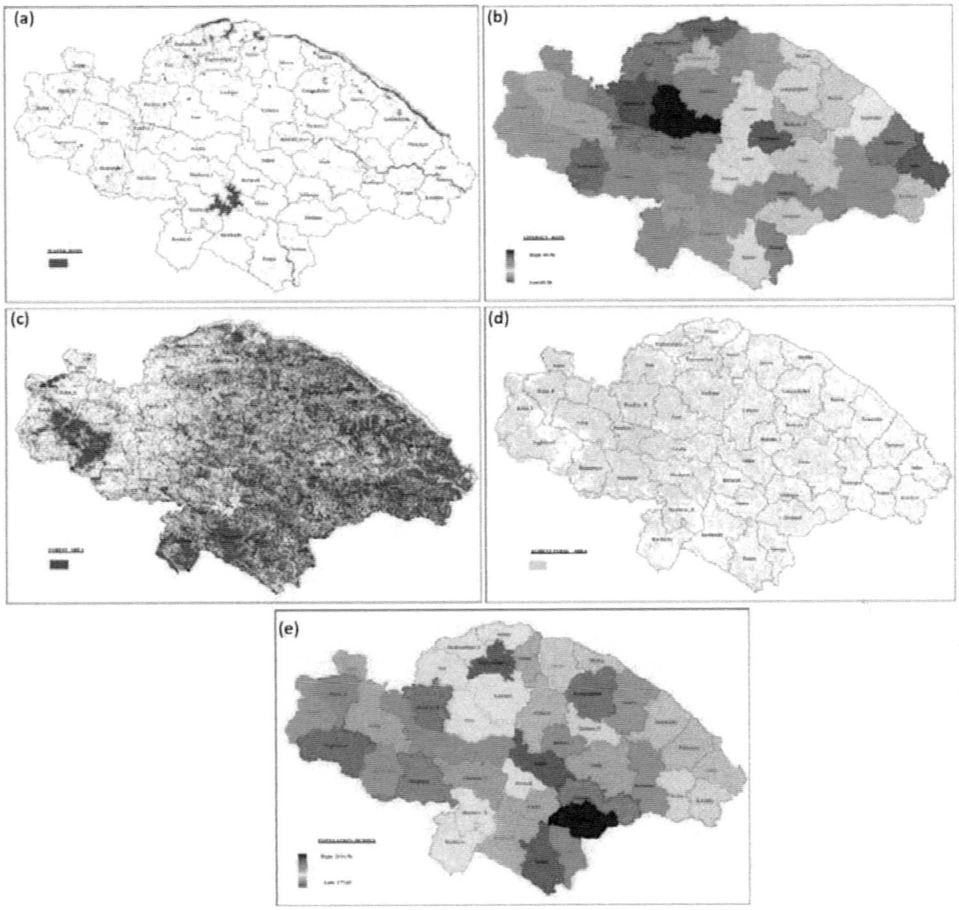


Figure 11: (a) Surface Water Bodies; (b) Literacy rate; (c)Forest Area; (d) Agricultural Area; (e) Population Density

## 6.5. Exposure

*6.5.1. Population density.* People per sq.km are measured by population density. The population density of the impacted region will be related to the severity of the consequences of a disaster (such as a drought) [64]. Consequently, the region hit by the drought that has a larger population density becomes more vulnerable than the area with a low population density. People's vulnerability is acknowledged in the social dimension of vulnerability, and the focus is on coping efficiency [65] (figure 11e).

IOP Conf. Series: Earth and Environmental Science 1164 (2023) 012009 doi:10.1088/1755-1315/1164/1/012009

### *6.5.2. Irrigated Area*

As discussed above, for risk assessment with the help of IPCC AR5, irrigated area and population density are considered exposure factors.

## 7. Result and Discussion

On different timelines, different system characteristics can adapt to drought [66]. As a result, drought indicators aren't usually closely linked, and they can differ widely [67, 68]. But with the above-mentioned factors and comparatively recent data, two resulting maps were produced.

figures 12 and 13 are the resulting maps. Considering all the constituents mentioned above and executing the thematic maps, it can be concluded from the vulnerability map that Arsha, Purulia I, Purulia II, and Jaipur are the most susceptible blocks in the study area. Additionally, it is evident that Arsha and Purulia I are the most risk-prone blocks.

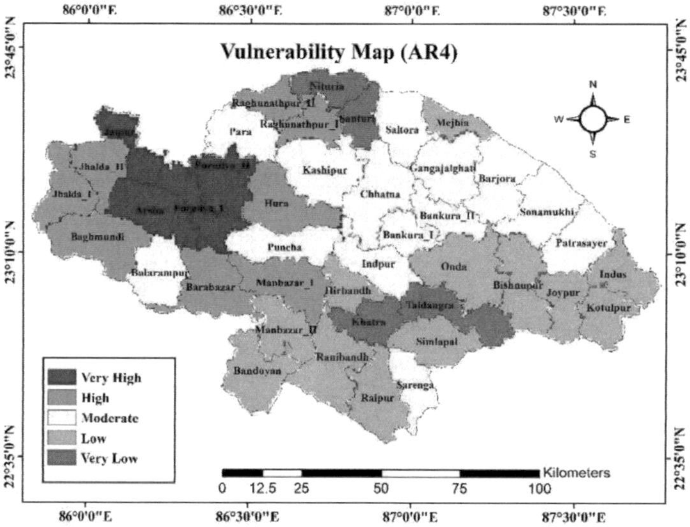


**Figure 12**: Resulting Vulnerability Map

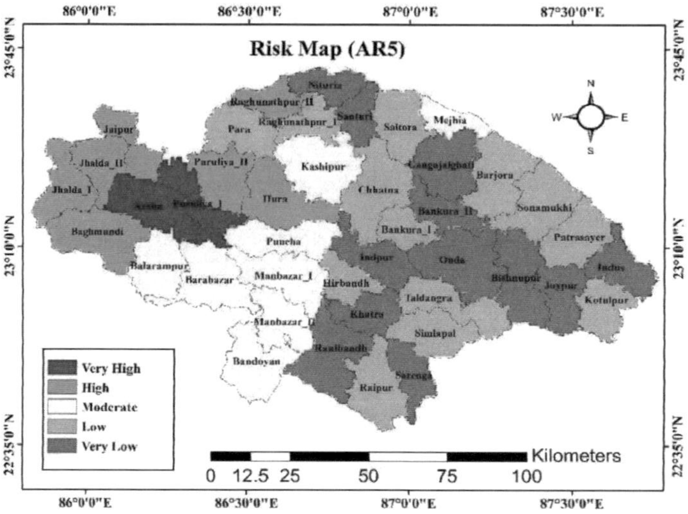


**Figure 13:** Resulting Risk Map

In both these figures 14 and 15 it is evident that due to low adaptive capacity, these particular blocks [Arsha, Purulia-I (common blocks in both cases), Purulia – II, Jaipur] are more vulnerable and risk-prone.

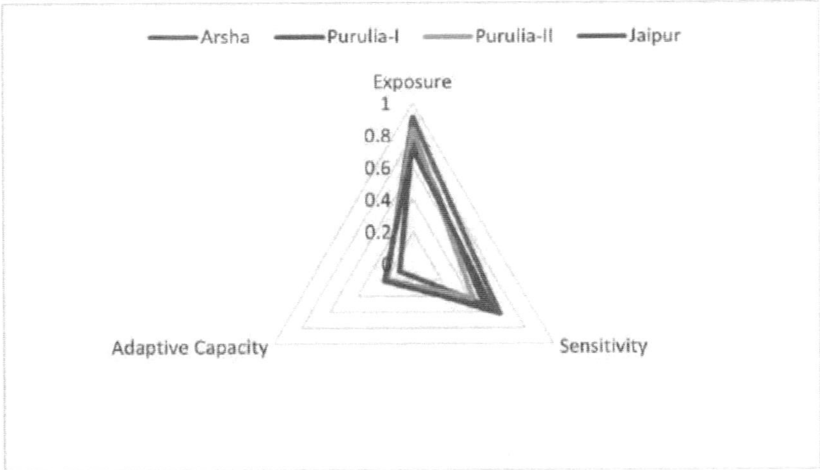


**Figure 14**: Top four vulnerable blocks

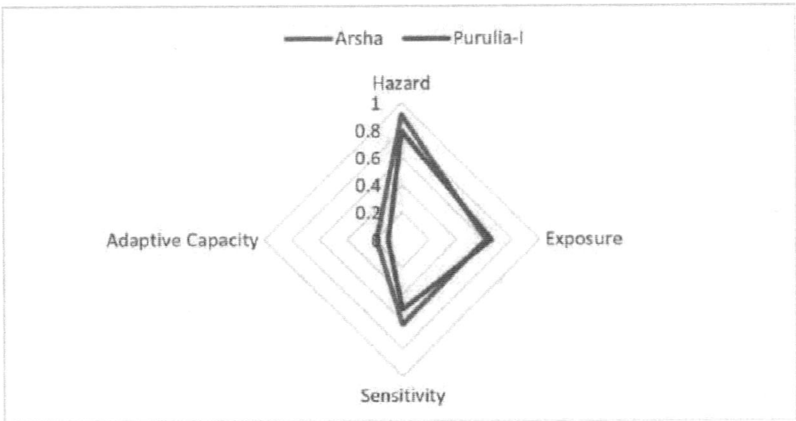


**Figure 15:** Top two risk-prone blocks

## 8. Conclusion and Recommendation

This project was carried out in the West Bengal districts of Purulia and Bankura to determine the blocks that are most vulnerable and dangerous. In view of the drought, vulnerability and risk assessments have been made. From several thematic maps, it can be inferred that based on drought vulnerability and risk, the western section of the study area (Purulia) is in a more vulnerable and riskier situation than the eastern part (Bankura). In this study, we manifested how all the factors (rainfall, vegetation, agricultural labor, farmer percentage, female population, agricultural dependency labor, literacy rate, forest area, irrigated area, water bodies, percentage of population, and population density) are affected (directly or indirectly) by drought. Blocks in the Purulia and Bankura districts can be prioritised for rapid improvement using the IPCC's AR4 and AR5 reports. Since agriculture is the primary source of income in these areas, improving irrigation alone will help to mitigate vulnerability and risk.

## References

[1] Hazra S, Roy S, Mitra S. 2017 Enhancing adaptive capacity and increasing resilience of small and marginal farmers of Purulia and Bankura districts, West Bengal to climate change *DRCSC Report* **1114.**

[2] Das S, Ghosh A, Hazra S, Ghosh T, de Campos RS, Samanta S. 2020 Linking IPCC AR4 & AR5 frameworks for assessing vulnerability and risk to climate change in the Indian Bengal Delta. *Progress in Disaster Science* **7** 100110.

[3] Vergage F, Cramer L, Thornton PK, Campbell BM. 2018 Climate risk assessment and agricultural value chain prioritisation for Malawi and Zambia. CCAFS Working Paper no. 228

[4] Cutter SL. 1996 Vulnerability to environmental hazards. *Progress in human geography* **20**(4) 529- 39.

[5] O'Brien K, Eriksen SE, Schjolden A, Nygaard LP. 2004 What's in a word? Conflicting interpretations of vulnerability in climate change research. *CICERO Working Paper.*

[6] Gallop´ın GC. 2006 Linkages between vulnerability, resilience, and adaptive capacity. *Global environmental change* **16**(3) 293-303.

[7] Fu¨ssel HM. 2007 Vulnerability: A generally applicable conceptual framework for climate change research. *Global environmental change* **17**(2) 155-67.

[8] Mukhopadhyay A, Dasgupta R, Hazra S, Mitra D. 2012 Coastal hazards and vulnerability: a review. *International journal of geology, earth and environmental sciences* **2**(1) 57-69.

[9] Wagesho N, Bekele S. 2004 GIS based irrigation suitability analysis (A case study of Abaya-chamo basin, southern rift valley of Ethiopia). *In: In Lake Abaya Res Symp Proc.* **4** 79-89.

[10] Hagenlocher M, Meza I, Anderson CC, Min A, Renaud FG, Walz Y, et al. 2019 Drought vulnerability and risk assessments: state of the art, persistent gaps, and research agenda. *Environmental Research Letters* **14**(8) 083002.

[11] Wilhite DA, Vanyarkho OV. 2000 Drought: Pervasive impacts of a creeping phenomenon.

[12] Muthumanickam D, Kannan P, Kumaraperumal R, Natarajan S, Sivasamy R, Poongodi C. 2011 Drought assessment and monitoring through remote sensing and GIS in western tracts of Tamil Nadu, India *International journal of remote sensing* **32**(18) 5157-76.

[13] Tate E 2012 Social vulnerability indices: a comparative assessment using uncertainty and sensitivity analysis *Natural Hazards* ***63***(2) 325-47.

[14] Jeyaseelan A. 2003 Droughts & floods assessment and monitoring using remote sensing and GIS. *Satellite remote sensing and GIS applications in agricultural meteorology* **291**.

[15] Bhunia P, Das P, Maiti R 2020 Meteorological drought study through SP Iinthree drought prone districts of West Bengal, India *Earth Systems and Environment* **4**(1) 43-55.

[16] Wheeler TR, Craufurd PQ, Ellis RH, Porter JR, Prasad PV 2000 Temperature variability and the yield of annual crops *Agriculture, Ecosystems & Environment* **82**(1-3) 159-67.

[17] Tubiello F, Rosenzweig C, Goldberg R, Jagtap S, Jones J 2002 Effects of climate change on US crop production: simulation results using two different GCM scenarios. Part I: wheat, potato, maize, and citrus *Climate research* **20**(3) 259-70.

[18] Naidoo R, Gerkey D, Hole D, Pfaff A, Ellis A, Golden C, et al. 2019 Evaluating the impacts of protected areas on human well-being across the developing world *Science Advances* **5**(4) eaav3006.

[19] Malik S, Pal SC, Das B, Chakrabortty R 2020 Intra-annual variations of vegetation status in a subtropical deciduous forest-dominated area using geospatial approach: a case study of Sali watershed, Bankura, West Bengal, India *Geology, Ecology, and Landscapes* **4**(4) 257-68.

[20] Dolui G, Chatterjee S, Das Chatterjee N 2016 Geophysical and geochemical alteration of rocks in granitic profiles during intense weathering in southern Purulia district, West Bengal, India *Modeling Earth Systems and Environment* **2**(3) 1-22.

[21] Mahala A 2021 Delineating the Status of Groundwater in a Plateau Fringe Region Using Multi-influencing Factor (MIF) and GIS: A Study of Bankura District, West Bengal, India. In: Geostatistics and Geospatial Technologies for Groundwater Resources in India *Springer;* p. 215-

37.
[22] Das A, Mondal M, Das B, Ghosh AR.2012 Analysis of drainage morphometry and watershed prioritization in Bandu Watershed, Purulia, West Bengal through Remote Sensing and GIS technology- A case study *International journal of Geomatics and Geosciences* **2**(4) 995-1013.
[23] Pandey RP, Pandey A, Galkate RV, Byun HR, Mal BC.2010 Integrating hydro-meteorological and physiographic factors for assessment of vulnerability to drought *Water resources management* **24**(15) 4199-217.
[24] Banik S, Nandi R 2004 Effect of supplementation of rice straw with biogas residual slurry manure on the yield, protein and mineral contents of oyster mushroom *Industrial crops and products* **20**(3) 311-9.
[25] Hahn MB, Riederer AM, Foster SO. 2009 The Livelihood Vulnerability Index: A pragmatic approach to assessing risks from climate variability and change—A case study in Mozambique *Global environmental change* **19**(1) 74-88.
[26] Dewan A 2013 Floods in a megacity: geospatial techniques in assessing hazards, risk and vulnerability *Springer.*
[27] Patwardhan A, Semenov S, Schnieder S, Burton I, Magadza C, Oppenheimer M, et al. 2007 Assessing key vulnerabilities and the risk from climate change *Climate change* 779-810.
[28] Ebi KL, Kovats RS, Menne B 2006 An approach for assessing human health vulnerability and public health interventions to adapt to climate change *Environmental health perspectives* **114**(12) 1930-4.
[29] Pachauri RK, Allen MR, Barros VR, Broome J, Cramer W, Christ R, et al. 2014 Climate change 2014: synthesis report. Contribution of Working Groups I, II and III to the fifth assessment report of the *Intergovernmental Panel on Climate Change. IPCC.*
*[30]*Fussel HF, Klein RJT 2006 Climate change vulnerability assessments: An evolution of conceptual thinking. *Springer*
[31] Domenikiotis C, Spiliotopoulos M, Tsiros E, Dalezios N 2004 Early cotton yield assessment by the use of the NOAA/AVHRR derived Vegetation Condition Index (VCI) in Greece *International Journal of Remote Sensing* **25**(14) 2807-19.
[32] Huete AR 1988 A soil-adjusted vegetation index (SAVI) *Remote sensing of environment* **25**(3) 295309.
[33] Gurung RB, Breidt FJ, Dutin A, Ogle SM 2009 Predicting Enhanced Vegetation Index(EVI)curves for ecosystem modeling applications *Remote Sensing of Environment* **113**(10) 2186-93.
[34] Huete AR, Miura T, Gao X 2003 Land cover conversion and degradation analyses through coupled soil- plant biophysical parameters derived from hyperspectral EO-1 Hyperion *IEEE Transactions on Geoscience and Remote Sensing* **41**(6) 1268-76.
[35] Liu HQ, Huete A 1995 A feedback-based modification of the NDVI to minimize canopy background and atmospheric noise *IEEE transactions on geoscience and remote sensing* **33**(2) 457-65.
[36] Bhuiyan C 2008 Desert vegetation during droughts: response and sensitivity *Int Arch Photogramm Remote Sens Spat Inf Sci* **37** 907-12.
[37] Dabrowska-Zielinska K, Kogan F, Ciolkosz A, Gruszczynska M, Kowalik W 2002 Modelling of crop growth conditions and crop yield in Poland using AVHRR-based indices *International Journal of Remote Sensing* **23**(6) 1109-23.
[38] Cancelliere A, Mauro GD, Bonaccorso B, Rossi G 2007 Drought forecasting using the standardized precipitation index *Water resources management* **21**(5) 801-19.
[39] Karabulut M 2015 Drought analysis in Antakya-Kahramanmaras Graben, Turkey *Journal of Arid Land* **7**(6) 741-54.
[40] Hayes MJ, Svoboda MD, Wiihite DA, Vanyarkho OV 1999 Monitoring the 1996 drought using the standardized precipitation index *Bulletin of the American meteorological society* **80**(3) 42938.

[41] Deo RC, Byun HR, Adamowski JF, Begum K 2017 Application of effective drought index for quantification of meteorological drought events: a case study in Australia *Theoretical and Applied Climatology* **128**(1) 359-79.

[42] Adnan S, Ullah K, Gao S, Khosa AH, Wang Z 2017 Shifting of agro-climatic zones, their drought vulnerability, and precipitation and temperature trends in Pakistan *International Journal of Climatology* **37** 529-43.

[43] Salehnia N, Alizadeh A, Sanaeinejad H, Bannayan M, Zarrin A, Hoogenboom G 2017 Estimation of meteorological drought indices based on AgMERRA precipitation data and station-observed precipitation data *Journal of arid land* **9**(6) 797-809.

[44] Willeke G, Hosking J, Wallis JR, Guttman NB 1994 The national drought atlas. Institute for water resources report 94.

[45] Hayes MJ 2002 Drought indices. National drought mitigation center, University of Nebraska.

[46] LoukasA, Vasiliades L, Dalezios N 2003 Intercomparison of meteorological drought indices for drought assessment and monitoring in Greece. In: Proceedings of the International Conference on Environmental Science and Technology. vol. 2, p.484-91.

[47] Sadiq A, Suleman M, Mohammed U 2020 An estimation of rainfall anomaly index and its impact on crop production in Yola and environs *African Journal of Environment and Natural Science Research* **3**(4) 35-53.

[48] Sharma J, Ravindranath NH 2019 Applying IPCC 2014 framework for hazard-specific vulnerability assessment under climate change *Environmental Research Communications* **1**(5) 051004.

[49] Zarafshani K, Sharafi L, Azadi H, Hosseininia G, De Maeyer P, Witlox F 2012 Drought vulnerability assessment: The case of wheat farmers in Western Iran *Global and Planetary Change* **98** 12230.

[50] Hameed WU 2019 Is Vulnerability Factors Discourages Women-Empowerment? The moderating role of Environmental, Health, Economic and Political Vulnerability between Microfinance Institutions and Women-Empowerment *Academic Press*.

[51] Gupta J, Termeer C, Klostermann J, Meijerink S, van den Brink M, Jong P, et al. 2010 The adaptive capacity wheel: a method to assess the inherent characteristics of institutions to enable the adaptive capacity of society *Environmental Science & Policy* **13**(6) 459-71.

[52] Gilbert AJ 2006 Coevolution in complex networks. An analysis of socio-natural interactions for wetlands management VU Vrije Universiteit, Print Partners Ipskamp BV, Enschede.

[53] Siders A. Adaptive capacity to climate change: A synthesis of concepts, methods, and findings in a fragmented field *WIREs Climate Change* **10**(3) e573. doi org/101002/wcc.2019;573.

[54] Williams P, Whitfield M, Biggs J, Bray S, Fox G, Nicolet P, et al. 2004 Comparative biodiversity of rivers, streams, ditches and ponds in an agricultural landscape in Southern England *Biological conservation* **115**(2) 329-41.

[55] Downing JA, Prairie Y, Cole J, Duarte C, Tranvik L, Striegl RG, et al. 2006 The global abundance and size distribution of lakes, ponds, and impoundments *Limnology and Oceanography* **51**(5) 2388-97.

[56] Verdonschot RC, Keizer-vlek HE, Verdonschot PF 2011 Biodiversity value of agricultural drainage ditches: a comparative analysis of the aquatic invertebrate fauna of ditches and small lakes *Aquatic Conservation: Marine and Freshwater Ecosystems* **21**(7) 715-27.

[57] EPAU 2015 Connectivity of Streams and Wetlands to Downstream Waters: A Review and Synthesis of the Scientific Evidence (Final Report). US Environmental Protection Agency, Washington, DC. EPA/600/R-14.

[58] Bartout P, Touchart L, Terasmaa J, Choffel Q, Marzecova A, Koff T, et al. 2015 A new approach to inventorying bodies of water, from local to global scale *DIE ERDE–Journal of the Geographical Society of Berlin* **146**(4) 245-58.

[59] Minea I 2020 The vulnerability of water resources from Eastern Romania to anthropic impact and climate change. In: Water resources management in Romania. Springer, p.229-50.

[60] Wagner DA 2011 What happened to literacy? Historical and conceptual perspectives on literacy in UNESCO *International Journal of Educational Development* **31**(3) 319-23.

[61] Lindner M, Maroschek M, Netherer S, Kremer A, Barbati A, Garcia-Gonzalo J, et al. 2010 Climate change impacts, adaptive capacity, and vulnerability of European forest ecosystems *Forest ecology and management* **259**(4) 698-709.

[62] Pokhriyal P, Rehman S, Areendran G, Raj K, Pandey R, Kumar M, et al. 2020 Assessing forest cover vulnerability in Uttarakhand, India using analytical hierarchy process *Modeling Earth Systems and Environment* **6**(2) 821-31.

[63] Wang Z, He F, Fang W, Liao Y 2013 Assessment of physical vulnerability to agricultural drought in China *Natural hazards* **67**(2) 645-57.

[64] Shahid S, Behrawan H 2008 Drought risk assessment in the western part of Bangladesh *Natural hazards* **46**(3) 391-413.

[65] Kumpulainen S 2006 Vulnerability concepts in hazard and risk assessment. Special paper-Geological Survey of Finland, 42:65.

[66] Sall I, Jarchow CJ, Sigafus BH, Eby LA, Forzley MJ, Hossack BR 2021Estimating inundation of small waterbodies with sub-pixel analysis of Landsat imagery: long-term trends in surface water area and evaluation of common drought indices *Remote Sensing in Ecology and Conservation* **7**(1) 10924.

[67] Vicente-Serrano SM, Beguer´ıa S, L´opez-Moreno JI 2010 A multiscalar drought index sensitive to global warming: the standardized precipitation evapotranspiration index *Journal of climate* **23**(7) 1696-718.

[68] Bachmair S, Stahl K, Collins K, Hannaford J, Acreman M, Svoboda M, et al. 2016 Drought indicators revisited: the need for a wider consideration of environment and society *Wiley Interdisciplinary Reviews: Water* **3**(4) 516-36.

# Disparities in exposure and accessibility to health: A Spatial Assessment of Covid 19 Pandemic Situation of South Bengal

**S Chatterjee[1*]**

[1]Department of Geography, Adamas University, Kolkata, India.

*Corresponding Author Email: suman.chatterjee27@yahoo.in

**Abstract.** Inequality and disparities of health is prevalent in a developing nation like India. The emergence of the Novel Corona virus has further aggravated this existing problem. The major issue that has been observed in dealing with the pandemic situation is the lack of facilities for testing and inadequate health infrastructure to support the huge population of this nation. This infrastructural deficiency has evidently become severe as we move away from the city areas. In this study, we have tried to assess this rural-urban disparity of health infrastructure in the settings of the COVID 19 pandemic situation. The study has been conducted to measure rural-urban disparities of eight districts i.e. Howrah, Hooghly, Kolkata, North 24 Parganas and South 24 Parganas concerning risk factors and access to health care infrastructure. To achieve the desired objectives, we have used a multidisciplinary technique incorporating statistical, GIS techniques, composite index with several aspects e.g., geographical coverage and sufficiency of COVID care and testing facilities (CCF, CTF), economic status of people, medical costs, and susceptibility to covid risk factors i.e., number of infected populations, and population density. A significant correlation between access to CCF and CTF and the proportion of urban inhabitants has been found, i.e., urban populations mostly enjoy better access to treatment and testing with dense CTFs and CCFs over space. The density and serviceability of care facilities over rural regions decrease with less urbanized areas. Urban areas are found to be at higher risk in terms of the number of active cases, population density. The study helps us to geo visualize the current COVID scenario of South Bengal in terms of regional disparities. This will help us identify the nearest testing and care facilities to any location within the study area and will be useful in mitigating infrastructural lacuna.

**Keywords**: Covid 19, Health Risk, Accessibility, Disparity.

## 1. Introduction

Geographical disparity e.g. between Rural and Urban areas, is a common concern in a developing nation like India when it comes to health inequalities, notably concerning access to good medical care services or even just for medical facilities. As a direct result of the socio-economic condition, proximity to a dense network of medical facilities, good transportation networks, and well-developed infrastructure, people who live in urban regions always have greater access to health care services. People living in rural or semi urban areas, conversely are forced to commute a long distance in order to receive adequate health care, even though the travel costs are higher than the actual medical charges. Additionally, their economic condition, educational attainment, and the scarcity of medical

IOP Conf. Series: Earth and Environmental Science 1164 (2023) 012010 doi:10.1088/1755-1315/1164/1/012010

care facilities in their immediate surroundings have all contributed to this inequality. Rural and suburban health care facilities are not only scarce but also crippling in terms of infrastructure, such as a shortage of personnel, physicians, specialists, beds, and other critical resources[1,2].Private practitioners, personal health care, and clinics are not exempt from this lack of health care providers and resources. In order to access the services of city hospitals, people must travel greater distances. While we are referring to exposure to health risk variables, numerous research on disease burdens show that a an urban resident is at higher risk from environmental, psychological, and behavioural risk factors to a certain level.

Corona virus is the greatest threat the world has faced in recent times (COVID-19). China's Wuhan city is the first to get infected. The illness is rapidly expanding throughout the world, with 6.54 million individuals affected in 181 countries by the year 2020. COVID-19 was first discovered in India on January 30th 2020 in Kerala. A severe global public health issue, the COVID-19 pandemic continues to pose a challenge to healthcare services and their accessibility[3]. In order to test and treat COVID-19 cases, a large number of resources and staff are diverted from their routine activities. Healthcare providers are in short supply, and people are afraid to seek them out. Combined with a fear of infection, access to health care, surgeries, and other hospital services has been significantly reduced.

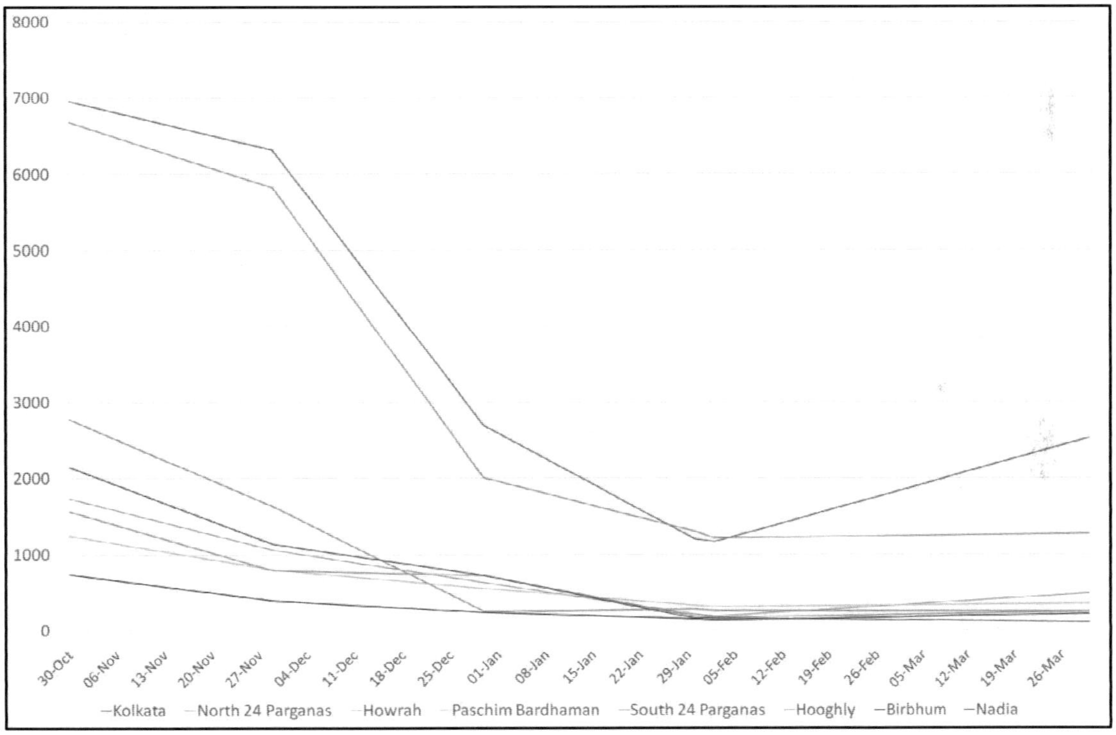


**Figure 1**.District-wise trend of COVID cases in West Bengal.

It was considerably later in West Bengal than the other parts of India and the world for COVID-19 to be triggered and disseminated. West Bengal has 564,032 COVID-19 patients as of 15th January 2021, which is 0.618 percent of the state's total population. As a result, the COVID-19 scenario for India and the globe is respectively 0.843 percent and 1.341 percent of the total population of each country. Our already overburdened healthcare system has been strained by the COVID 19 pandemic. Some researchers have looked at the geographical pattern of COVID 19 in West Bengal and found that it is unequal. Kolkata has the most overall instances, followed by North 24 Parganas, South 24 Parganas, Haora, and Hugli. The first COVID-19 case was reported in Kolkata, with the highest number of patients and fatalities documented till March 2021. In the huge metropolis of Kolkata, we may state that the burden of COVID-19 illnesses is significantly higher. The nature and growth of COVID-19 cases and deaths are influenced by factors such as urbanization, population density, concrete roofs, and distance from a large metropolis (Kolkata) [4]. District wise growth trends indicate a similar pattern for Kolkata and North 24 Parganas with the highest number of cases but the growth rate is a little higher for North 24 PGS. The South 24 PGS exhibits the third highest number of cases which has surpassed Howrah after May 2020 (figure 1).

This research examines three aspects: 1) the differences between rural and urban regions concerning of COVID care accessibility (ACCF); 2) the discrepancy between rural and urban regions regarding COVID transmission risk factor (CTRF) and socioeconomic status (SS); and 3) the ability to identify high-risk areas by comparing Risk-Access ratios. This research examines the differences between the availability of medical facilities and susceptibility to covid risk factors in rural and urban regions. The research adopted inter-disciplinary methodologies, such as the composite indicator, Geospatial tools, to interpret a group of parameters, including geographical coverage and sufficiency of medical facilities (CCF), socio-economic status, medical costs, and Susceptibility for illness [5].

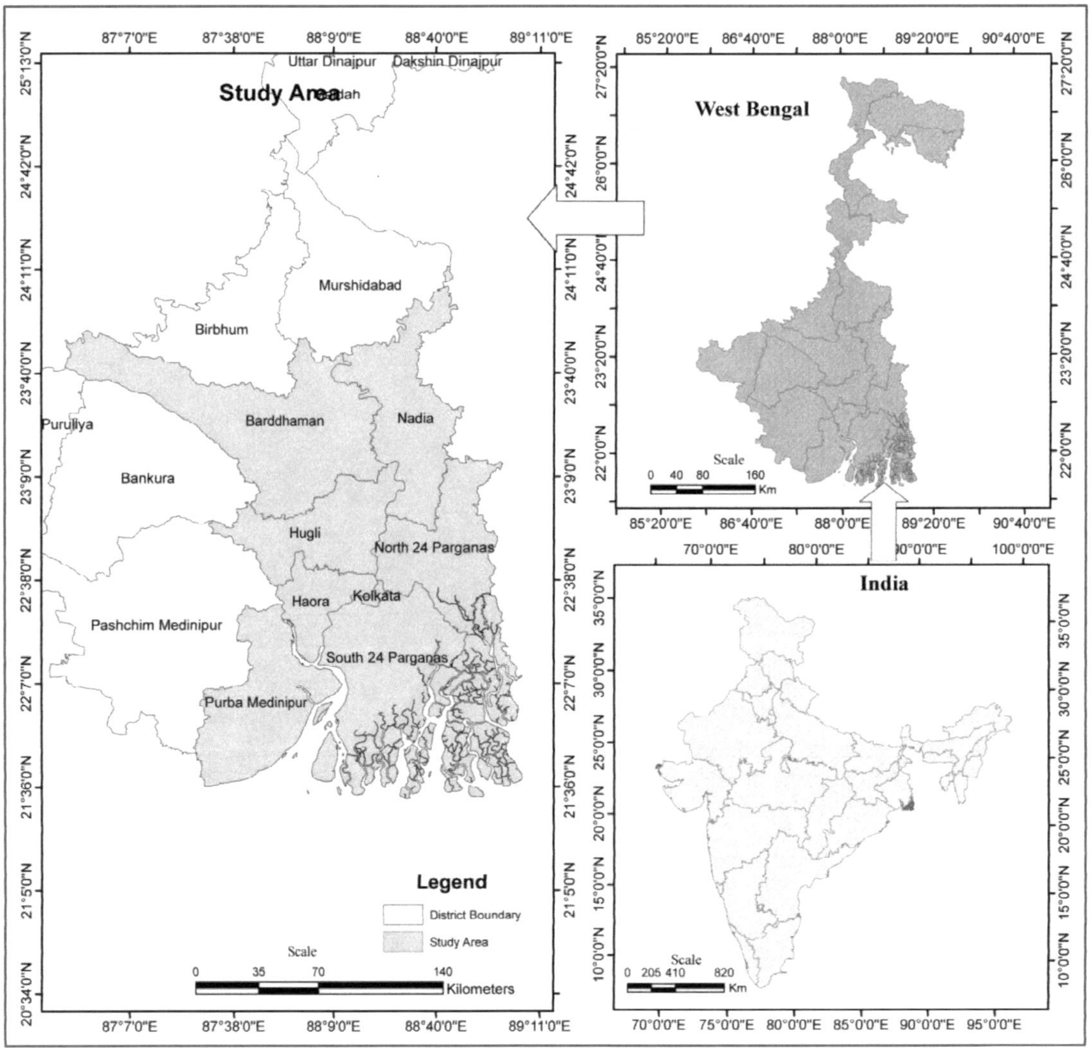


**Figure 2.** Location Map of the study area.

IOP Conf. Series: Earth and Environmental Science 1164 (2023) 012010 doi:10.1088/1755-1315/1164/1/012010

## 2.Study Area

Eight districts of South Bengal i.e., South 24 PGS, North 24 PGS, East Midnapore, Burdwan, Hooghly, Howrah, Nadia, and Kolkata have been selected as the study area (figure 2). Although initially block wise study was intended to be performed but due to lack of block wise data, it has been decided to conduct district wise study. However, all the care units within the study area have been mapped and coverage for each unit has been estimated. Location information of the COVID care facilities and testing centres have been extracted and prepared the maps in the GIS environment.

## 3.Data Collection

The ICMR, the West Bengal Department of Health and Family Welfare, Census 2011, the District Statistical Handbook, and other sources provided the majority of the secondary data that was used in this research. Table 1 displays the information used in this investigation and where it came from.

**Table 1.** Data Acquisition from USGS Earth Explorer

| Data | Source |
|---|---|
| COVID testing lab | ICMR: icmr.gov.in/arcctestlab.html |
| Treatment facilities and Bed availability, isolation beds, ICU, ventilators | Department of Health & Family Welfare Govt. of West Bengal: wbhealth.gov.in/pages/corona/bed_availability https://www.google.com/maps/ |
| Standard of living | Source: Census 2011 |
| Percentage of households having permanent houses in a block | |
| % of homes using tap water as their primary source of drinking water | |
| % of homes using electricity as their primary illumination source | |
| % of homes with a bathroom on the property | |
| The percentage of homes with closed or open sewerage connections for waste water outflows | [6], [7],[8–10] |
| The percentage of homes having kitchen facility | |
| % of homes having television, pcs/laptop | |
| Percentage of households availing banking service | |
| Net Income from Small Savings ('000 Rs.) | |
| Active recovered cases of COVID_19 | Bing COVID 19 tracker: https://bing.com/COVID/local/haora_westbengal_india |

## 4.Methodology

The study employs techniques like the composite index, creating Voronoi diagram and Extracting Built-up area in GIS environment, disparity and inequality measuring tools such as the Quintile ratio and Sopher's Index, and statistical measures such as the test of significance, Pearson Correlation

IOP Conf. Series: Earth and Environmental Science 1164 (2023) 012010 doi:10.1088/1755-1315/1164/1/012010

coefficient, and linear regression [5]. Initially, we discussed the presumptions, variables, and dimensions, as well as how to calculate them (table 2). Figure 3 depicts the entire structure of this research.

**Table 2:** Assumptions of the study

| Sl. No. | Assumptions | Explanations |
|---|---|---|
| 1 | QCCF α ACCF | People will choose to use a COVID care facility (QCCF) if it provides improved services; as a result, there will be more people who can get COVID care (ACCF) |
| 2 | LS α AHC | Access to COVID care (ACCF) will increase if a person's economic wellbeing (LS) rises |
| 3 | AC = k/%PGOD | The more a district's inhabitants is compelled to migrate (PGOD) for access to medical services, the lesser its index's value will be. |
| 4 | CHS = k/AHC | If a facility's COVID care service (CCS) costs are higher, people will not opt to use it unless the higher costs are justified by the COVID care service's suitability. As a result, if CCS is higher, ACCF would be lower. |

**Source:** Author

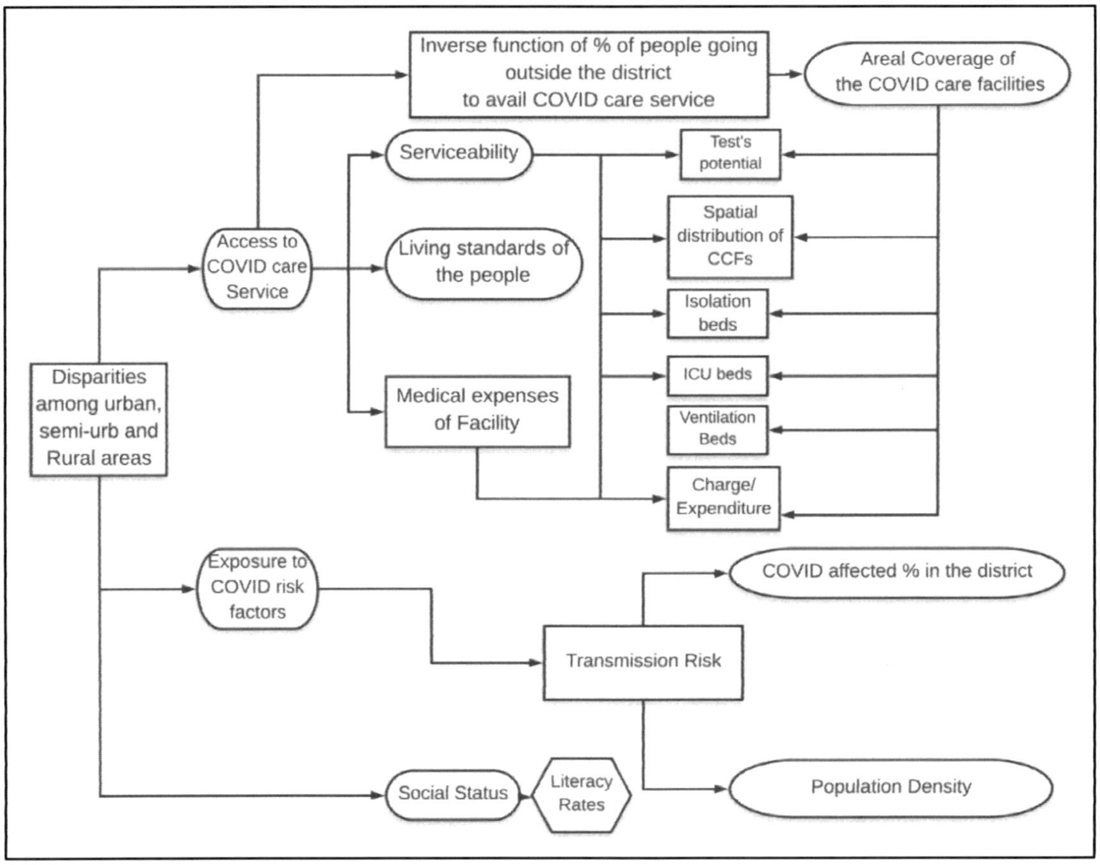


**Figure 3.** Structure of the study, Source: Author.

IOP Conf. Series: Earth and Environmental Science 1164 (2023) 012010 doi:10.1088/1755-1315/1164/1/012010

### *4.1. CCF and its serviceability*

Two types of COVID care facilities have been considered here i.e. COVID testing facilities and COVID treatment facilities where, the first one functioning only in terms of testing operations and the second one accounting for both testing and providing care facilities. The serviceability of COVID testing has been decided with existing test facilities of testing centre i.e. RT-PCR, RT-PCR and TrueNat/CBNAAT, only TrueNAT or only CB-NAAT (figure 4). Cost score is calculated as per the nature of the testing centre i.e. score of 1 for govt. and 0.0006 for private facilities. Serviceability of the COVID care facilities has been decided with Number of isolation beds, ICU bedsand Number of ventilation beds (figure 3).

### *4.2. Access to medical facilities*

This metric indicates how easily a group of people may access quality COVID care facilities. The following dimensions were used to create a composite index for access to COVID care facilities (ACCF).

- The percentage of population that commute outside of the district to get CCF (PGOD)
- The district's serviceability of CCFs in terms of Test potential, number of Isolation beds, ICU beds, Ventilation beds under the district
- The district's economic status
- Expenses (healthcare costs) and the district's CCF serviceability.

### *4.3. Proportion of inhabitants going outside the district to avail CCF (PGOD)*

The dimension of ACCF is the percentage of the population who travelling outside the district to get CCF. This sub-indicator has been employed as an inverse function, which means that the higher the PGOD, the lower the ACCF for that district. PGOD is the percentage of a district's total population who travels outside the district to receive COVID care. If the COVID care facilities in a district are insufficient or inadequate, people are forced to go further to obtain better care outside the district.

To get the PGOD, we first calculated the geographical coverage of each CCF and retrieved the built-up area from the Copernicus Global Land Cover Image of 2019 (figure 3). Following that, we calculated the total populated area covered by individual CCF in each administrative district, as well as the overall populated area covered by the CCFs in individual district.

### *4.5 CCF's spatial coverage*

Thiessen polygons containing weighted factors for CCF prices and serviceability have been used to compute and map CCF geographical coverage. The number of isolation beds, ICU beds, and ventilation beds have all been considered surrogates of CCF serviceability (figure 3). Thiessen polygons are polygons created inside a specific space surrounding a group of points by allowing all of the space within the nearest set of points. The nearest point within the Thiessen polygon is closer to any other representative of the point range than any other point outside of it. Voronoi diagrams are known as spatial tessellations [11]. Points that stand for a certain characteristic they possess might also be given particular weights. For instance, this study's definition of CCFs as sites with their prices and serviceability as weights indicates that CCFs with lower costs and higher serviceability would posses more geographical coverage than other CCFs (sites). Following the computation of area coverage, we retrieved the habitation area of each district as well as the habitation area inside the CCF's coverage area below each district. To determine the share of population going to CCFs out of the district, we subtracted CCF service area inside the district from the district's overall settlement area. This amount was then converted as a percentage of the entire settlement.

**Figure 4**.COVID diagnostic centres, Spatial service areas, and Percentage of Percentage of people accessing CTF outside of their districts

IOP Conf. Series: Earth and Environmental Science 1164 (2023) 012010 doi:10.1088/1755-1315/1164/1/012010

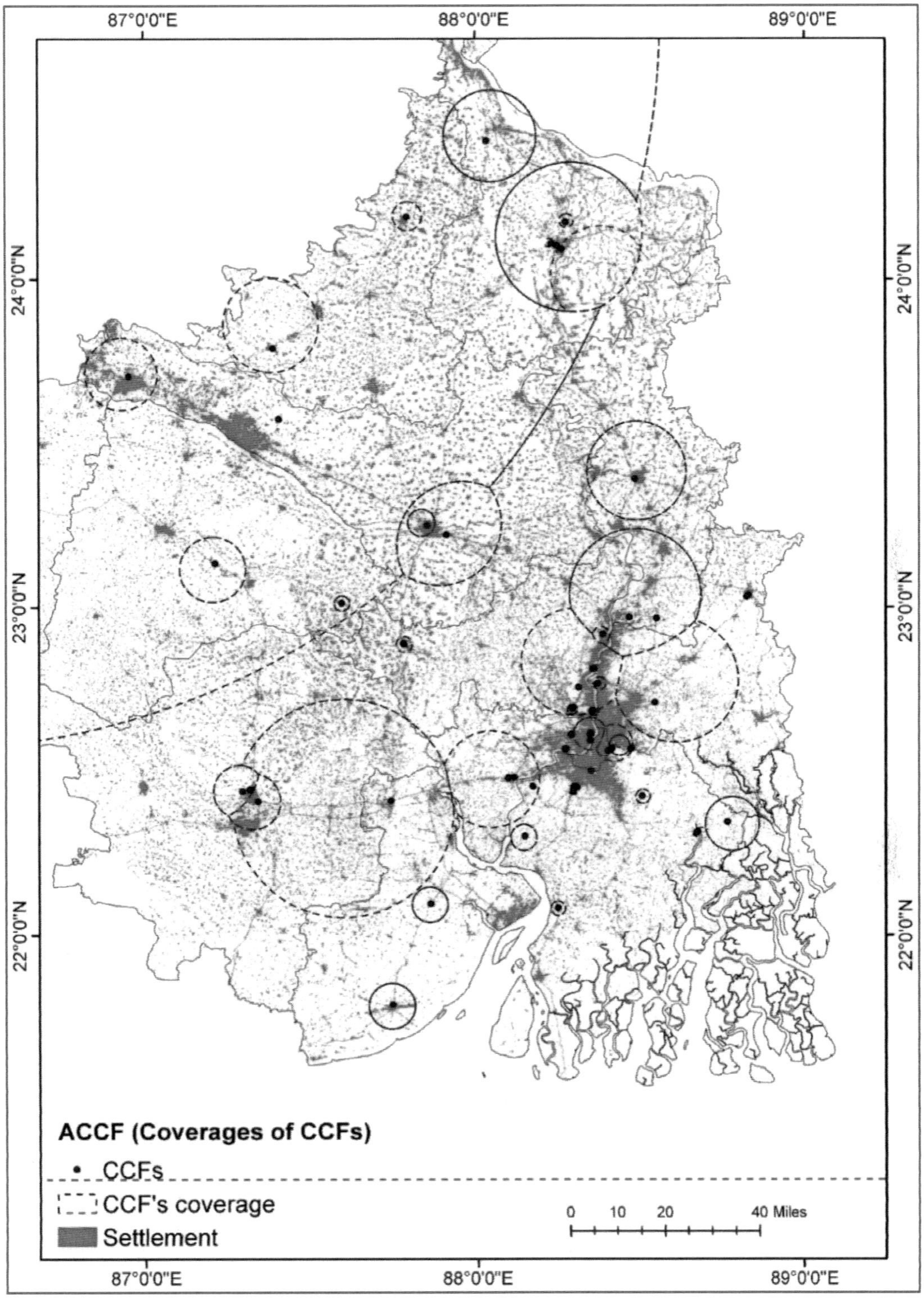


**Figure 5**. COVIDcare facilities, Spatial service areas, and Percentage of Percentage of people accessing CCF outside of their districts

### *4.6. Living standards*

While GDP is a valuable tool for gauging market production, it does not provide a complete picture of economic status. The European Statistical System Committee (ESSC) has well-documented quality of life with an indicator that consists of numerous variables that have been updated by researchers on periodically [12]. In this study, comparable indicators were applied to indicate the quality of life with the following aspects (table 4).

**Table 3:** Component of Living Standards

| Indicators | Components | Source |
|---|---|---|
| Material Living Condition Index | Percentage of households having permanent houses<br>Percentage of homes using purified water from the tap for consumption<br>percentage of homes using electricity as their primary source of illumination<br>Percentage of homes with a toilet on the property<br>Percentage of homes with waste water outlets connected to the sewage system<br>Percentage of families using certain electronic components, such as a PC or TV | District census handbook and district statistical handbook |
| Economic | Percentage of homes with banking facilities<br>Gross Domestic Product (GDP) of the districts | |

**Source:** Adapted from Kundu. A [12]

### *4.7. Cost (medical bills) and CCFs' Serviceability.*

In India, the expense of receiving medical treatment differs depending on the sort of service provider. Broadly speaking, costs are greater for privately held institutions and significantly cheaper for those owned by the government or the public sector. For each of these two categories of service providers, we compared their fees and set an adjusted number value. According to state government order memo no. 258 HFW/Secy, diagnosis and treatment costs for COVID19 are nil to insignificant in government institutions but they are estimated to cost 1,500 in private care facilities..

### *4.8.COVID transmission risks factor*

The level of exposure has been determined using transmission risk factors. Two dimensions i.e., population density of the administrative units and percentage of total active cases. The population density and Active COVID cases of these two dimensions have been normalised to 0-1 range and aggregated [equation (1) and (2)].

### *4.9.Aggregating sub-indices*

We aggregated various variables to create one value using composite index because we employed many variables. We utilised two sorts of normalisation algorithms to normalise the sub-indices: more is better and less is better [equation (1) and (2)]. Depending on the properties of the data, we employed arithmetic and geometric means to aggregate the sub-indices.

$$More\ is\ better: Score = (Value - Minimum)/(Maximum - Minimum) \quad (1)$$

$$Less\ is\ better{:}Score = 1 - [\frac{Value\ - Minimum}{Maximum\ - Minimum}] \quad (2)$$

### 4.10. Measures of Disparities

This study used multiple approaches to quantify disparities, including the Sopher Index, sample t-test, Test of significance Correlation coefficient, and Quintile ratio. The data has been divided into two groups for the Sopher Index and t-test: districts with less than 39% of the urban population and districts with more than 39% of the urban population. Disparities are measured by either absolute or relative quantities. Absolute values are simply the actual difference between the values for the classes. Relative values are expressed as the ratio of two or as a change of the values in accordance with the chosen reference [13]. The Sopher index is a popular technique for assessing disparities that was introduced by Sopher [14], and modifeid by Kundu and Rao [15]. The Sopher index estimates the difference in logarithmic odds ratios between the two categories that have a certain element [equation (3)].

$$Ds = log\frac{X_2}{X_1} + log\frac{(200 - X_1)}{(200 - X_2)} \quad (3)$$

Where, $X_2 \geq X_1$ and $X_1$ and $X_2$ are various attributes.

It is common to use the Quintile Share Ratio as a measure of inequality (QSR). In the set of social cohesion indicators maintained by the European Union serves as the primary indicator of income inequality. Twenty percent of the maximum value (top quintile) divided by twenty percent of the minimum value is known as the quintile ratio (lowest quintile). [16].

To investigate the link between proportion of urban inhabitants and the factors i.e., ACCF, Exposure, and Socioeconomic status we have performed Karl Pearson correlation coefficient with the significance level of 0.05.

### 4.11. Stressed areas (disparity between Exposure factors and ACCF)

We calculated the ratio [equation (4), between transmission risk value (numerator) and ACCF] to determine the region in stress, i.e., districts with lower accessibility to COVID care facilities but greater exposure to transmission risk (denominator).

$$\frac{Normalised\ \ AQI\ value}{Normalised\ \ AHCF} \quad (4)$$

## 5. Result and Discussion

In terms of access to COVID-care facilities and living standards, the highest scores have been estimated for districts with high percentages of the urban population in the study area i.e. Kolkata, Howrah, North 24 Parganas. East Midnapore, 24-Parganas (S), and Hooghly have been recognized as the districts that have the lowest access to health care services and standards of life. These districts have low percentages of the urban population (table 4 and 5). Despite the least urban population percentage, East Medinipur has observed better social status in terms of literacy rates followed by Kolkata with an urban population percentage of 8.29 and 100 respectively. The highest ratio between exposure (CRF) and access (ACCF) has been observed for North 24 Parganas, South 24 Parganas, Kolkata, and Howrah. These districts have a higher urban population percentage and higher active cases (table 6). In addition, district south 24 Parganas possess the lowest socio-economic status. Numerous studies have examined the role that private organizations play in ensuring healthcare access in the Indian setting due to their improved infrastructure [17]. Instead of having higher access to COVID care facilities in Kolkata, North 24 Parganas, and Howrah, much higher exposure risk was identified in these districts due to the higher number of active cases and a population density which is eventually surplus the access factor and making them areas under stress. Contrarily the comparative higher accessibility in Burdwan district neutralized the moderate risk factor.

Due to better health care accessibility and socioeconomic conditions, Kolkata, Burdwan, North 24 Parganas districts are less stressed. This has been attained by the predominance of various private CCFs, resulting in a closely packed distribution of CCFs (Table 6). These CCFs collectively offer improved services by delivering services at all health care levels (primary, secondary, and tertiary).. On the other hand, the districts under stress e.g. Kolkata, North 24 Parganas, South 24 Parganas have limited access to CCF due to the absence of CCFs with improved infrastructure. Fewer COVID-care facilities have been observed in the South 24 Parganas, and besides, the number of COVID cases has increased rapidly, which is making the district of concern. In other words, if the cases continue to increase at the same rate, the infrastructure will be insufficient to deal with the situation. Although the density of CCF in the Burdwan district is sparse, the district has enough access to health care facilities due to the greater serviceability of the care facilities.

**Table 4. Living conditions of the blocks**

| Districts | Living Condition Index | Economic condition Index | Living Standard | Urban Population % |
|---|---|---|---|---|
| 24-Parganas (N) | 0.56 | 0.315575 | 0.438132 | 57.27 |
| 24-Parganas (S) | 0.19 | 0 | 0.097185 | 25.58 |
| Hooghly | 0.38 | 0.336073 | 0.360289 | 38.57 |
| Nadia | 0.18 | 0.005962 | 0.095405 | 27.84 |
| Midnapore East | 0.22 | 0.649385 | 0.432353 | 8.29 |
| Howrah | 0.49 | 0.300142 | 0.394 | 63.38 |
| Kolkata | 1.00 | 1 | 1.000004 | 100 |
| Burdwan | 0.40 | 0.468489 | 0.434245 | 39.89 |

**Source:** District census Handbook (2011)

**Table 5.Comparing ACCF, Exposure and social status of different districts of South Bengal**

| Districts | Urban Population % | ACCF | Risk Factors | Social status | Standard of living | Risk/Access |
|---|---|---|---|---|---|---|
| Burdwan | 39.89 | 0.434245 | 0.083572 | 0.880105 | 0.434245 | 0.434245 |
| East Medinipur | 8.29 | 0.194996 | 0.038426 | 1 | 0.432353 | 0.432353 |
| Howrah | 63.38 | 0.186387 | 0.133261 | 0.956537 | 0.394 | 0.394 |
| Hooghly | 38.57 | 0.135907 | 0.074145 | 0.941707 | 0.360289 | 0.360289 |
| Kolkata | 100 | 0.673032 | 1 | 0.994068 | 1.000004 | 1.000004 |
| N24PGS | 57.27 | 0.197905 | 0.302411 | 0.969085 | 0.438132 | 0.438132 |
| Nadia | 27.84 | 0.133255 | 0.051797 | 0.862195 | 0.095405 | 0.095405 |
| S24PGS | 25.58 | 0.102542 | 0.076259 | 0.792266 | 0.097185 | 0.097185 |

U.P: Urban Population

**Source:** Refer to Table 2

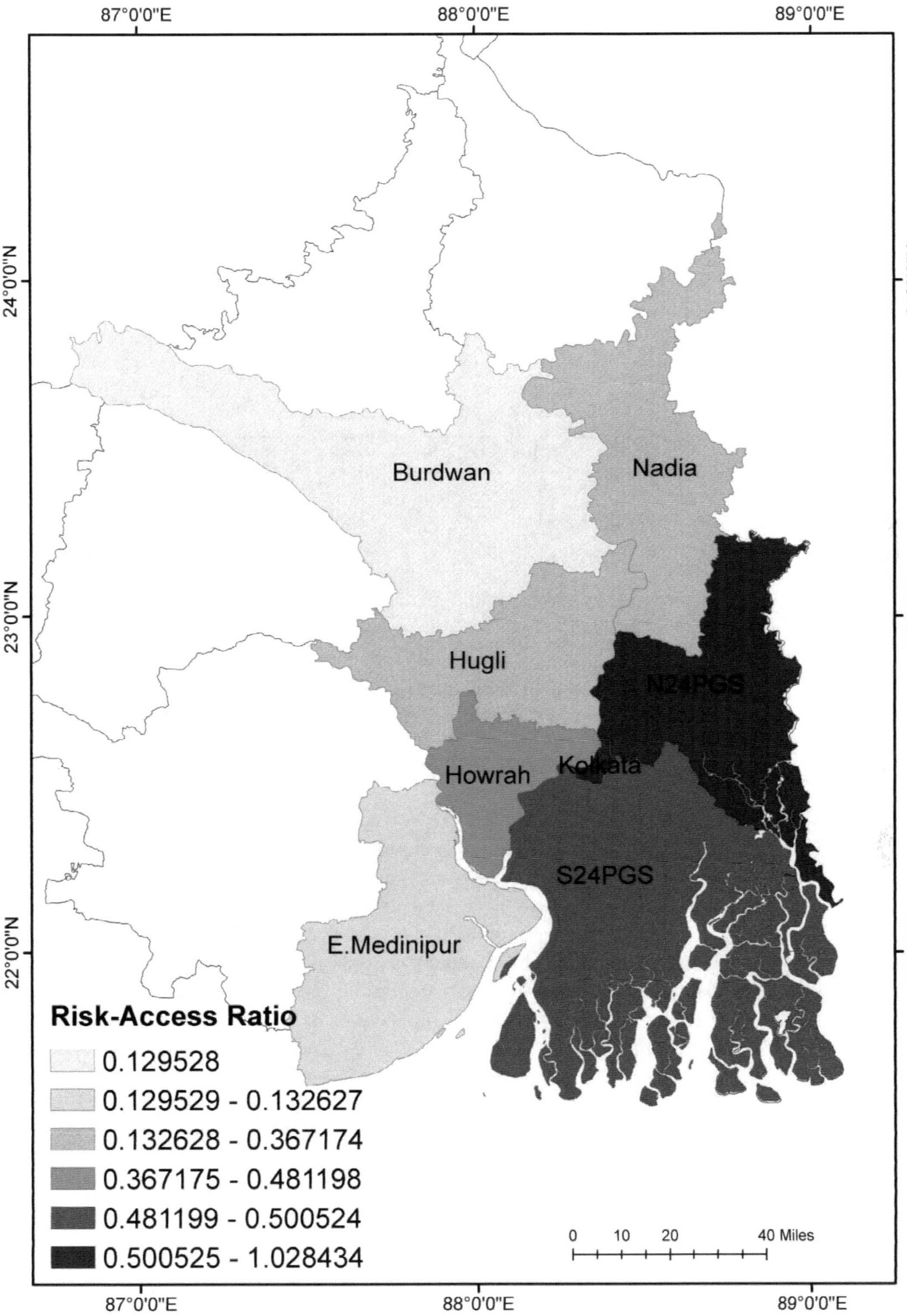


**Figure 6.** Areas under stress (Risk-Access Ratio).

**Table 6.Rural-Urban Disparities: Pearson Correlations**

| Pearson Correlation Coefficient | Urban Population % | (ACCTF) | Transmission Risk | Social status | Economic status | Urban Population % |
|---|---|---|---|---|---|---|
| Urban Population % | 1 | | | | | 1 |
| Access to COVID care and treatment facilities (ACCTF) | **0.722569673** | 1 | | | | **0.722569673** |
| Risk Factors | **0.871845996** | 0.832073492 | 1 | | | **0.871845996** |
| Social status | 0.411880449 | 0.37424319 | 0.447628 | 1 | | 0.411880449 |
| Economic status | **0.776281279** | **0.897896523** | **0.86857** | 0.704617 | 1 | **0.776281279** |

**Statistically significant correlation

**Table 7.Sopher Index of the factors**

| **Factors** | **Sopher's index AHCF** |
|---|---|
| Parameters | Sopher's index AHCF |
| ACCTF | 0.**420792021** |
| Risk Factor | **0.79958996** |
| Social status | 0.024031332 |
| Economic Status | 0.369023065 |

Table 7 shows the significant amount of rural-urban disparity that exists in the state in terms of COVID care infrastructure. Sopher index value is indicating the degree of disparity which is exhibiting much higher risk factors and access to COVID care and treatment facility.

## 6. Conclusion

According to the study, districts with a greater urban population have ready access to COVID care centres, a higher quality of living, and higher socio-economic conditions. Because of greater transmission rates, vulnerability to health risk factors is also high in those districts. In addition, the urban areas are more closer to the pollution source, which might aggravate the symptoms. Some rural areas, such as districts in South 24 Parganas, are under a critical state in terms of exposure and ACCF. N24PGS, Kolkata, and S24PGS have been noted as districts with the maximum risk of transmission and the least access to health care infrastructure. These districts have been classified as high-risk areas with limited access to COVID care facilities but a greater risk of transmission aspects. Burdwan, East Medinipur, Hooghly, and Nadia have been noted as districts with lower distinctions in exposure and ACCF (Table 6 and 7). As per general observation, most of the locations experiencing stress have a lower proportion of urban inhabitants and vice versa. It is concerning that districts with lower urban populations, such as south 24 Parganas, are stressed in all facets, including lesser ACCF and socioeconomic condition, as well as greater exposure to health risk. With time, urban and industrial pollution has altered the surrounding environment and exacerbated health risk factors. This study will help us not only to identify the nearest testing and care facilities to any location within the study area but also a similar approach can be adopted for demarcating the coverage of the vaccination centers and will be useful in mitigating infrastructural lacuna. The study aids in visualizing the geographical variations of health and exposure discrepancy, allowing us to pinpoint regions of concern. This will assist policymakers in recognising and prioritising political ramifications in these areas. Where infrastructure will be constructed, where infrastructure will be decentralised, and where environmental conservation actions will be implemented. This disparity assessment process correctly portrayed not only the geographical distribution and service areas of CCFs, but also their accessibility, because it integrated other important aspects such as living conditions, social standing, air quality, and ailment rate all at the same time, enabling it to interpret and understand.

**References:**

[1] Patil A V, Somasundaram K V. and Goyal R C 2002 Current health scenario in rural India *The Australian journal of rural health*
[2] Bhalla A S 1990 Rural-urban disparities in India and China *World Development* **18** 1097–110
[3] Chanda A 2020 COVID-19 in India: transmission dynamics, epidemiological characteristics, testing, recovery and effect of weather *Epidemiology and Infection* **148** e182
[4] Biswas B, Roy R, Roy T, Chowdhury S, Dhara A and Mistry K 2021 Geographical Appraisal of COVID-19 in West Bengal, India *GeoJournal*
[5] Chatterjee S and Sarkar K 2021 Appraisal of urban–rural disparities in access to health care facilities and exposure to health risk factors: A case study of Durgapur Industrial region, India *GeoJournal*
[6] Office of the Registrar General and Census Commissioner 2011 *Tables on Houses, Household Amenities and Assets* (New Delhi)
[7] Office of the Registrar General and Census Commissioner 2011 *DISTRICT CENSUS HANDBOOK BARDDHAMAN* (New Delhi: DIRECTORATE OF CENSUS OPERATIONS WEST BENGAL)
[8] Bureau of Applied Economics & Statistics Burdwan 2011 *District statistical handbook Burdwan* (Department of Statistics & Programme Implementation, Government of West Bengal)
[9] Bureau of Applied Economics & Statistics 2011 *District Statistical Handbook Bankura* (Kolkata: Department of Statistics & Programme Implementation, Government of West Bengal)
[10] Bureau of Applied Economics & Statistics 2011 *District Statistical Handbook Birbhum* (Kolkata: Department of Statistics & Programme Implementation, Government of West Bengal)
[11] Yamada I 2016 Thiessen Polygons *International Encyclopedia of Geography: People, the Earth, Environment and Technology*
[12] Kundu A 2016 *Quality of population and resource utilization in the districts of West Bengal adjoining Chotanagpur plateau* (University of Calcutta)
[13] Krishan G and Singh N 2019 *Researching Geography* (Routledge India)
[14] Sopher D E 1980 Sex disparity in Indian literacy *An exploration of India: Geographical perspectives on society and culture* ed D . Sopher (New York, NY: Cornell Press) pp 130–88
[15] Kundu A and Rao J M 1986 Inequity in educational development: Issues in measurement, changing structure and its socio-economic correlated with special reference to India *Educational planning: A long term perspectiv* ed M Raza (New Delhi, India.: New Delhi: NIEPA and Concept Publishing Company) pp 435–66
[16] Drezner T, Drezner Z and Hulliger B 2014 The Quintile Share Ratio in location analysis *European Journal of Operational Research* **238** 166–74
[17] Berendes S, Heywood P, Oliver S and Garner P 2011 Quality of private and public ambulatory health care in low and middle income countries: Systematic review of comparative studies *PLoS Medicine*

IOP Conf. Series: Earth and Environmental Science 1164 (2023) 012011 doi:10.1088/1755-1315/1164/1/012011

# Monitoring Land Use & Land Cover Change in an Urban Landscape: A Case Study of Siliguri, India

**R Sarkar[1*], M Baral[1]**

[1]Department of Geography, Adamas University, Kolkata, West Bengal- 700126, India.

*Corresponding Author Email: rajib.sarkar@adamasuniversity.ac.in

**Abstract.** Information related to Land use and Land Cover (LULC) change is important for conservation of natural resources and sustainable development of an area. Siliguri is one of the populated metropolitan of West Bengal, India that has witnessed rapid urbanization. The major challenge is to identify the reason and pattern of urbanization due to inadequate and insufficient data in small cities like Siliguri. This paper has mainly assessed the LU & LC change of Siliguri municipality for the period of 1991-2001, 2001-2011 and 2011-2021.The main aim of the study is to determine the pattern and direction of future expansion. Satellite images from Landsat 4-5(1991-2011) and Landsat 8(2021) were used to extract LU & LC map of the municipality. Supervised classification using maximum likelihood classification was used to developed LULC maps. The supervised classification produced good results with overall accuracy of more than 80%.The change of the LULC pattern was obtained using change detection method. Four major classes viz. Vegetation, water body, sand and built-up have been identified. The analysis shows that maximum transformation has happened in the buildup environment from other land surface features.

**Keywords:** Remote sensing, geographic information system, urban environment, landsat, maximum likelihood classification

## 1. Introduction

Land use and land cover (LULC) change have become perineal issue across the globe [1]. LULC change are driven by anthropogenic acts with different spatial and temporal resolutions. LULC change lead to different social, economic, and environmental externalities. Some of these factors are positive for human well-being and many of them also pose a serious threat to human life. Urbanisation is a process by which the population of the urban areas rises with time [1, 2]. The rise of the population transforms the city morphology with varying intensity. LULC transformation are not uniform across different countries of the world. Global data shows that countries of the south are experiencing a larger change in land use in comparison to the northern hemisphere countries. At the country level, similar differences are also observed in the urbanisation level as well as land use transformation. Some states enjoy supremacy in urbanisation where land use transformation is higher than the other states. Many scholars like highlighted that urban land use transformations have negative impacts on the urban landscape. Through the land use transformation different externalities like the development of urban heat islands, increased runoff in shorter duration, sedimentation influx in lakes and streams, loss in biodiversity, etc. [3, 4, 5]. So identification of LULC change has remained a crucial factor. Land-use change is also contributing to climate change by infusing temperature through different human activities [6, 7] . LULC change havealong-term impact on the overall environment of urban space. Thus, LULC change measurement and mapping in an urban environment are very important considering all the above factors.

The urban landscape is a complex system where different componentsofthe urban system interact with one another to form a pattern [8] . The components of urban systems are interwoven by myriad factors such as socio-economic, political, technological, natural, as well as cultural factors. All these components can act independently but in most cases, they are highly interlinked. So urban landscape is a multifaceted element. LULC mapping in the urban landscape also can be controlled by different factors[5] .

The present study aims at mapping and measureofLULC change in an urban frame. The main objective of this study is to the identification of LULC change of the urban environment. Apart from that percent change in land use categories are also mapped to understand the relative transformation of the land surface feature.

## 2. Study Area

As a study area, Siliguri Municipal Corporation (SMC) is located in the northern part of the state of West Bengal, India. It is also popularly recognised as the "Gateway of Northeast India" or Tarai region of West Bengal. Siliguri is situated on the riverbank of Mahananda and Teesta and at the foothill of the eastern Himalayan range. Siliguri is the third-largest urban agglomeration preceded by Kolkata and Asansol in West Bengal. It shares its international border with Bangladesh, China, Bhutan, and Nepal. The latitudinal and longitudinal location of Siliguri is 26.71°N 88.43°E, respectively (figure 1). This study area covers an area of 83.557 sq. Km.

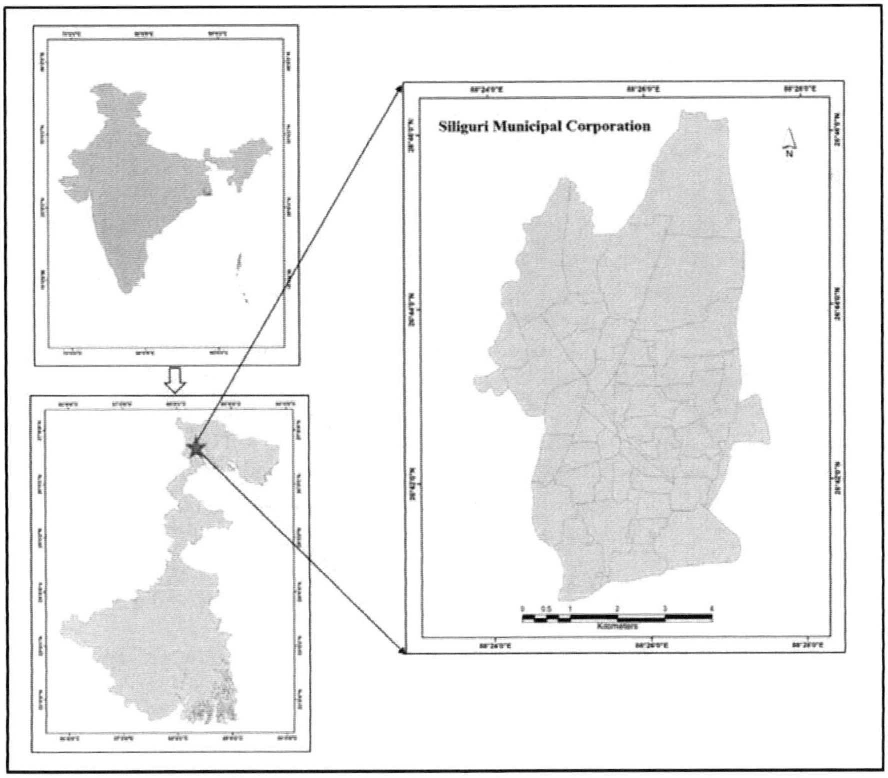


**Figure 1**. Study area map of Siliguri Municaipal Corporation

The average elevations of the study area is between 300m to 350m above mean sea level and the slope is twards south-west from north east direction. The climate of Siliguri Municipal Corporation (SMC) is mainly humid subtropical and features warm and humid summer with cool and dry winter. Annual mean temperature of the area is 23.7 °C. August is the most hottest month of the year i.e. 28.5°C and while coldest month, January experiences 16.1°C on an average.The average rainfall of Siliguri is about 3340 mm per year. More than 80 percent of the total rainfall occurs between June to September. The type of soil is mostly sandy.

## 3. Data and Methodology

Satellite imagery was used to determine LULC change, while images were selected based on clear atmospheric conditions, data quality, and availability. Four different Landsat imageries were collected for the years 1991, 2001, 2011, and 2021 from the USGS Earth Explorer. Landsat 4-5 are used for the years 1991, 2001, and 2011 and Landsat 8 is used for 2021. The path and row of the images are 139/41 for 1991, 2001, and 139/42 for the year 2011, and 2021 (table 1). These mentioned data were used for the identification of LULC change study in Siliguri Municipal Corporation. The spatial resolution of all the images is 30 meters.Details information related to satellite images used for the study is given here below.

**Table 1.**Characteristics of the collected Landsat data

| Satellite type | Spatial Resolution | WRS Path/raw | Sensor Type | Acquisition Date | No. of Bands |
|---|---|---|---|---|---|
| Landsat-V | 30m | 139/41 | TM+ | 30/04/1991 | 7 |
| Landsat-V | 30m | 139/41 | TM+ | 09/04/2001 | 7 |
| Landsat-V | 30m | 139/42 | TM+ | 21/04/2011 | 7 |
| Landsat-VIII | 30m | 139/42 | OLI & TIRS | 15/03/2021 | 11 |

## 4. Data Preparation

First, Landsat images were retrived from United States Geological Survey (USGS) Earth Explorer web portal. Before making the LULC map need to do some Image pre-processing. Image pre-processing is necessary for the correction of some geometric and radiometric distortion this may take place by remote sensing technology in combination with earth rotation, atmospheric conditions and platform instability. Pre-processing processes include geometric transformations, atmospheric correction, georeferencing, and image enhancement. In this study, image pre-processing was performed by using QGIS 3.16 software package. Radiometric correction and atmospheric correction were performed for this study.For atmospheric corrections Dark Object Subtraction (DOS) based method is preferred over other methods because it requires minimum data in comparison to other methods. After getting the atmospherically corrected data, images were further processed to develop false colour composite to measure different LULC types. Based on the image, five major land-use types were identified. Four LULC maps were prepared by using the supervised classification method with maximum likelihood algorithm after completing the pre-processing of the images. Landsat 4-5 (TM) and Landsat 8 (OLI &TIRS) satellite images have 7 and 11 bands respectively. For preparing the LULC classification, individual bands were transformed into a single image by using the composite tool. After that four composite images (the year 1991, 2001, 2011, 2021) were clipped to the study area. There are various methods for making LULC map, however, the most frequent methodologies for mapping are supervised and unsupervised procedures. In this study, the supervised method was used because it gives more accurate results than the unsupervised one. The five main classes of LULCtypeswere used to prepare the maps. Thirty training samples for each class have been taken. Total 150 samples were collected from each image to construct the LULC map and after that maximum likelihood algorithm was applied by using remote sensing and GIS environment. From these supervised maps, changes in LULC were detected over the 30 years at the 10 years interval.

Lastly, the accuracy assessment was carried out in the study. Accuracy assessment is important for individual classification if the classification map is used in a change detection study. For the accuracy assessment, the prepared LULC are compared with available reference datasets, which are assumed to be correct for defining a classification. Accuracy assessment was measured using user accuracy, producer accuracy, overall accuracy and Kappa Coefficient.

## 5. Result

Generally, different LULC classes are selected and fixed as per the demand of a particular study and its application. In this study, we have classified the study area into five LULC classes. The five LULC classifications used in this study are Waterbodies, Sand, Vacant land, Vegetation, and Built-up area. Supervised classification with a maximum likelihood algorithm is used for LULC classification. The final classification of LULC provides an overview of the major LULC features of Siliguri municipality of West Bengal, India from 1991 to 2021 (figure 2).

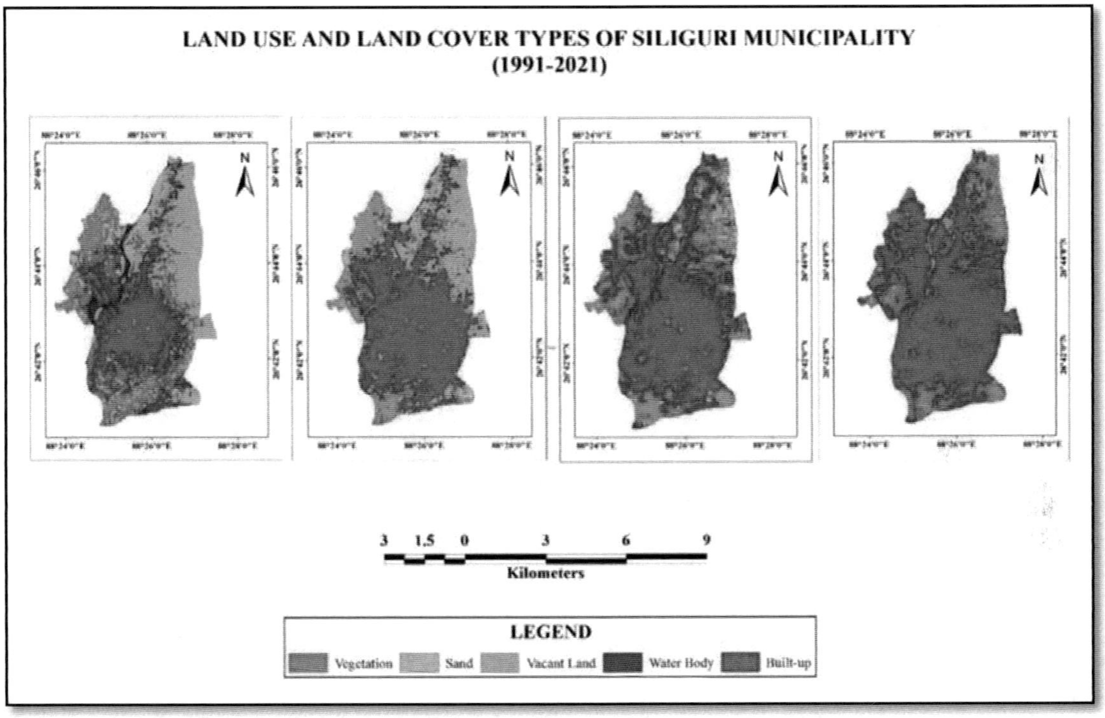


**Figure 2**. LULC maps of the study area at a different times (1991-2021).

The accuracy of the LULC classification was calculated using a confusion matrix. The confusion matrix represents the overall accuracy, producer and user accuracy, and kappa statistics with mathematical precision. The overall accuracy for the classified images of 1991, 2001, 2011, and 2021 were 90.67, 93.33, 90.67, and 85.71percent respectively (table 3). The kappa coefficient for the year of 1991, 2001, 2011 and 2021. LULC maps were 0.87, 0.89, .83 and 0.80, respectively, showing strong accuracy. This technique of image processing was found to be effective in producing appropriate data on LULC changes.

LULC classification map is presented in figure 2 which indicates that in the year 1991 vegetation cover is accounted for 34 percentof the total area of the municipality. It was the largest proportion of land cover in the year 1991. The built-up area occupies the second place with 31percent while the vacant land comprises about 24 percent. The sand area consisted 8 percent of the total area followed by the water body which account for a proportion of about 3 percent.

The LULC map fortheyear 2001 are presented in figure 3. It revealed that the proportion of the vegetation cover decreased to 18 percent against the proportion of 34 percent in the year 1991. Similarly, the proportion of settlement decreased to 10.87 percent against the observed proportion of 24 percent in the year 1991.Water body decreased from 2.76 percent in the year 1991 to 0.34 percent in 2001. On contrary, sand cover and built-up areas increased from 8 percent in 1991 to 14 percent in 2001 and from 31 percent in 1991 to 56.7 percent in 2001, respectively.

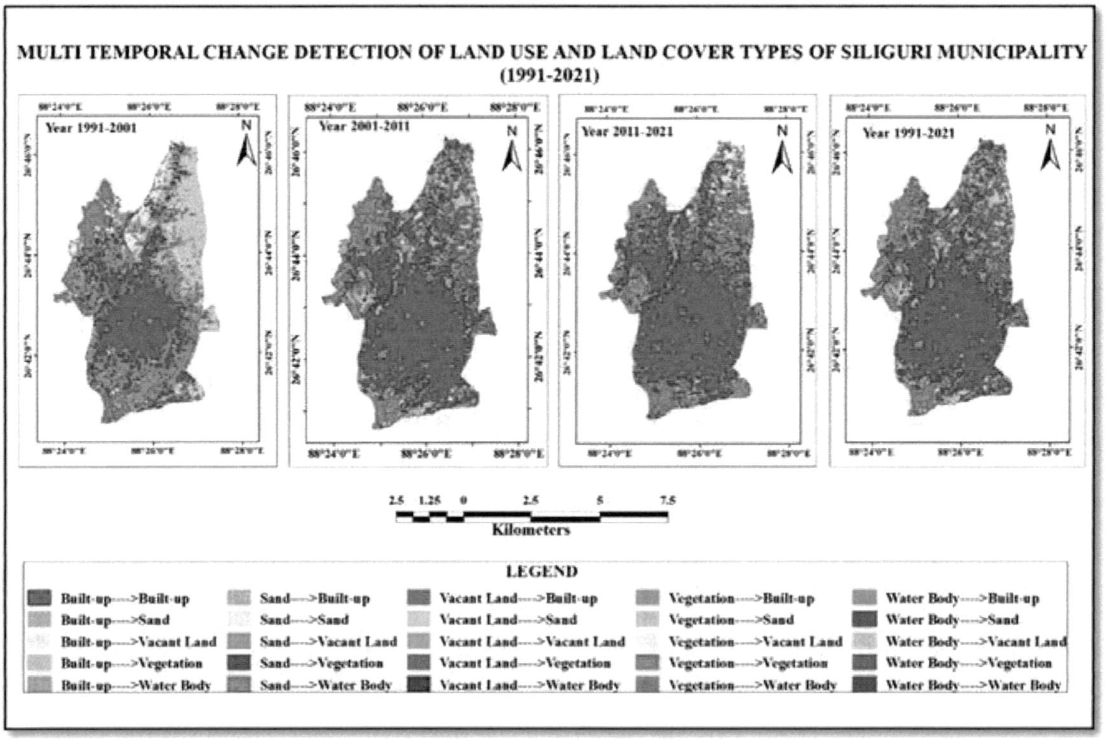


**Figure 3**. LULC change detection map of the Siliguri Municipal corporation.

**Table 2.**Temporal change of LULC in percentage

| CLASS NAME | YEAR | | | | | | | |
|---|---|---|---|---|---|---|---|---|
| | 19991 | | 2001 | | 2011 | | 2021 | |
| | SQ. KM | % | SQ.KM | % | SQ.KM | % | SQ.KM | % |
| **Vegetation** | 12.93 | 33.67 | 6.82 | 17.77 | 13.15 | 34.37 | 9.56 | 24.77 |
| **Sand** | 3.16 | 8.23 | 5.5 | 14.32 | 1.01 | 2.64 | 0.25 | 0.62 |
| **Vacant Land** | 9.32 | 24.28 | 4.17 | 10.87 | 0.74 | 1.93 | 1.68 | 4.32 |
| **Water Body** | 1.06 | 2.76 | 0.13 | 0.35 | 0.24 | 0.64 | 0.34 | 0.88 |
| **Built-up** | 11.92 | 31.05 | 21.76 | 56.7 | 23.13 | 60.42 | 26.46 | 69 |

In the year 2011, the proportion of the vegetation coverage was increased to 34 percent while the proportion of sand cover reduced to 2 percent. The proportion of the vacant land declined to 1.9 percent in 2011. The share of a water body and built-up area increased by the proportion of 0.6 percent and 60 percent in the year 2011 respectively.

In the year 2021, the proportion of the vegetation coverage was decreased to 24 percent and the proportion of sand cover was reduced to .62 percent. The share of vacant land declined to 4.39 percent in 2021.Built-up areas or settlements accounted for more than 69 percent of the total study area in the present year (2021) while the proportion of water bodies declined by almost 1 sq.km. in this year 2021.

In this study, we discover the pattern of LULC change from the year 1991 to 2021.The rate of LULC change is not uniform over the years. For example, conversion from forest to build-up area does not follow a similar pattern of change in different periods.

## 6. Discussion

The result shows the overall change in the urban landuse pattern. Siliguri Municipal corporation areas lie at the edge of the Himalayan region. So major LULC change in the area have a significant impact on the biophysical environment of the region as a whole. Land use transformation shows that built-up areas are rising faster than the other land-use types. This transformation can lead to urban flooding, the development of urban heat islands, and a significant increase in the runoff and siltation in the drainage system of the city. This phenomenon is already observed in many parts of the world where natural land use is transformed with artificial surfaces.

The result of the study closely collaborates with the other studies of LULC change. Most of the major cities of the world have experienced a rapid rise in built-up areas at the cost of other landform types like vegetation, waterbodies, agricultural land, etc. Most of the LULC change studies [5, 6, 8, 13] show that the LULC change are very much prominent across all the globe. And in all cases, built-up areas are rising at the cost of either vegetation, waterbody agricultural land, or vacant spaces. So the present study of LULC change monitoring is depicting the same picture for the Siliguri Municipal Corporation.

The biophysical changes in the natural landscape with artificial structure cause alteration in the ambient environment [14, 15]. Habitats fragmentation, loss of biodiversity, urban floods, and heat island effects are already been observed in the study area due to the LULC changes. So continuous monitoring of these biophysical alterations is of significant importance for the sustainable maintenance of the environment [16, 17].

## 7. Conclusion

The study of LULC change mapping in the Siliguri Municipal Corporation area has aimed to highlight the major changes in temporal scale of thirty years starting from 1991 to 2021. Based on the classification of these change has been monitored. Maximum Likelihood classification is applied to all images to get the classified map. Overall accuracy shows that the classification is worthy to be analysed to change mapping. The land types which show major positive change are built-up areas and vegetation has the highest negative growth. Identifying and monitoring the land-use dynamics are very pertinent and important to understanding different macro and microscaleevents of local hazardous events like landslides bank erosion to climate change. Such type of study also helps in creating the much-needed base information for future urban planning. Mapping LULC change are also part of Sustainable Goals 2030. So considering all those aspects, the study portraysthat Siliguri Municipal Corporation needs appropriate land management forthefuture.

## Reference

[1] Bocquier, P. (2005) World urbanization prospects: An alternative to the UN model of projection compatible with the mobility transition theory, *Demographic Research*. New Yorks. doi: 10.4054/DemRes.2005.12.9.

[2] Sudhira, H. S., Ramachandra, T. V. and Jagadish, K. S. (2004) 'Urban sprawl: Metrics, dynamics and modelling using GIS', *International Journal of Applied Earth Observation and Geoinformation*, 5(1), pp. 29–39. doi: 10.1016/j.jag.2003.08.002.
[3] Dwivedi, A. and Mohan, B. K. (2018) 'Impact of green roof on micro climate to reduce Urban Heat Island', *Remote Sensing Applications: Society and Environment*, 10, pp. 56–69. doi: 10.1016/j.rsase.2018.01.003.
[4] Sharma, R., Ghosh, A. and Joshi, P. K. (2013) 'Analysing spatio-temporal footprints of urbanization on environment of Surat city using satellite-derived bio-physical parameters', *Geocarto International*, 28(5), pp. 420–438. doi: 10.1080/10106049.2012.715208.
[5] Subasinghe, S. *et al.* (2016) 'Spatiotemporal Analysis of Urban Growth Using GIS and Remote Sensing: A Case Study of the Colombo Metropolitan Area, Sri Lanka', *ISPRS International Journal of Geo-Information 2016, Vol. 5, Page 197*, 5(11), p. 197. doi: 10.3390/IJGI5110197.
[6] Du, H. *et al.* (2016) 'Influences of land cover types, meteorological conditions, anthropogenic heat and urban area on surface urban heat island in the Yangtze River Delta Urban Agglomeration', *Science of the Total Environment*, 571. doi: 10.1016/j.scitotenv.2016.07.012.
[7] Suriya, S. and Mudgal, B. V. (2012) 'Impact of urbanization on flooding: The Thirusoolam sub watershed - A case study', *Journal of Hydrology*, 412–413, pp. 210–219. doi: 10.1016/J.JHYDROL.2011.05.008.
[8] Ramachandra, T. V., Aithal, B. H. and Sowmyashree, M. V. (2014) 'Urban structure in Kolkata: metrics and modelling through geo-informatics', *Applied Geomatics*, 6(4), pp. 229–244. doi: 10.1007/s12518-014-0135-y.
[9] Das, S. (2019) 'Geospatial mapping of flood susceptibility and hydro-geomorphic response to the floods in Ulhas basin, India', *Remote Sensing Applications: Society and Environment*, 14, pp. 60–74. doi: 10.1016/J.RSASE.2019.02.006.
[10] Kim, H. H. (1992) 'Urban heat island', *International Journal of Remote Sensing*. doi: 10.1080/01431169208904271.
[11] Moussa, R., Voltz, M. and Andrieux, P. (2002) 'Effects of the spatial organization of agricultural management on the hydrological behaviour of a farmed catchment during flood events', *Hydrological Processes*, 16(2), pp. 393–412. doi: 10.1002/HYP.333.
[12] Sharma, R., Chakraborty, A. and Joshi, P. K. (2015) 'Geospatial quantification and analysis of environmental changes in urbanizing city of Kolkata (India)', *Environmental Monitoring and Assessment*, 187(1), p. 4206. doi: 10.1007/s10661-014-4206-7.
[13] Tran, H. *et al.* (2006) 'Assessment with satellite data of the urban heat island effects in Asian mega cities', *International Journal of Applied Earth Observation and Geoinformation*, 8(1), pp. 34–48. doi: 10.1016/j.jag.2005.05.003.
[14] Jeevalakshmi D, Narayana Reddy S and Manikiam, B. (2017) Land surface temperature retrieval from LANDSAT data using emissivity estimation Int. J. Appl. Eng. Res. 12
[15] Sharma R, Chakraborty A and Joshi P K. 2015 Geospatial quantification and analysis of environmental changes in urbanizing city of Kolkata (India) Environ. Monit. Assess. 187 4206.
[16] Masoudi M and Tan P. Y., [2019] Multi-year comparison of the effects of spatial pattern of urban green spaces on urban land surface temperature *Landscape and Urban Planning.* **184** 44–58
[17] Li W, Cao Q, Lang K and Wu, J. [2017] Linking potential heat source and sink to urban heat island: Heterogeneous effects of landscape pattern on land surface temperature *Science of The Total Environ.* **586**

# Environmental Stress and Health Vulnerability Assessment around Kolaghat Thermal Power Plant, West Bengal

**S Chatterjee[1*], A Rai[1], S Hazra[1]**

[1]Department of Geography, Adamas University, Kolkata, India.

*Corresponding author Email: suman.chatterjee27@yahoo.in

**Abstract.** One of the seventeen types of severely polluting sectors in the nation is thermal power plants which exert pressure upon the environment and health aspects in several ways e.g. creating thermal, air, and water pollution, exacerbating physiological acute and chronic responses, and affecting the subjective wellbeing of an individual residing proximal to the source. Situated on the right bank of the Rupnarayan river, Kolaghat thermal power plant is one of the major power generating plants of West Bengal, which is also producing 7500-8000 metric tons of fly ash every day but having only 325 acres of land for its disposal. Hence the plant has been exerting pressure upon the surrounding environment since its establishment and growth between 1980-1985. This study is to evaluate the overall impact of thermal power plants on the local environment and public health using a composite index that incorporates environmental and health dimensions i.e. Air quality index, drinking water quality index, land surface temperature, social parameters, Non–communicable disease rate, etc. The study reveals the local-level impact of air pollution and the impact of fly ash emissions. Nandakumar, Tamluk, and Kolaghat have been identified as blocks under stress. The workers of the plant have reported undesirable thermal and ambient air quality conditions within the battery limit, with the prevalence of skin problems, allergies, and fatigue among them. Site-specific plantation and phytoremediation techniques with the use of riparian buffers have been recommended.

**Key Words**: Thermal Power Plants, Environmental Stressors, Health Risk, Stress Index.

## 1. Introduction

Some major concerns for achieving sustainability determined in the literatures are conservation of forests, preventing desertification, land degradation, and biodiversity loss, increasing attention to air quality, municipal and other waste management, minimizing the adverse environmental consequences of cities per capita, protecting public health and wellbeing against hazardous chemical release in the air, water, and soil, have received special attention. These concerns have been fortified through Sustainable Development Goals 3, 15 and 11 [1–3].

Over the years, thermal energy has grown into India's main power source and this trend seems to continue. India is the world's third-largest producer and consumer of electricity [4]. In India, more than 70% of the electrical power comes from thermal power plants [5].The environmental effects of thermal power generation have become an important topic of concern. The change of the occupied land began to have a detrimental impact on the local ecosystem and environment. With the establishment of power plants and during the operation, local surface water bodies may receive plant's wastewater, affecting water quality, and ambient air may receive greenhouse gases, other pollutants, particulate matter, ashes, and radionuclide, all of which can exert pressure on the local environment and public health, as well as the global climate[6].

As a consequence of leaching, volatilization, melting, decomposition, oxidation, hydration, and other chemical processes, certain pollutants such as CO2, SO2, NO2, particulate matter (PM), and heavy metals are released into air and water, causing significant environmental and health effects. CO2, NO2, SO2, Particulate Matter, and certain heavy metal pollution have all been linked to a variety of health issues. CO2 is a key contributor to climate change as well as certain severe illnesses such as Chronic Obstructive Pulmonary Diseases COPD, and lung cancer. SO2 is not only hazardous to organisms, and produces acid rain, but it also causes a variety of illnesses e.g. heart arrhythmia, skin cancer, asthma, cough, headache, throat, and nose irritations. Another significant pollutant emitted by coal-fired power plants is NO2, which causes hypoxic respiratory failure in neonates due to chronic pulmonary hypertension (PPHN).PM together with CO2, SO2 and NO2 do serious damage to the environment as well as to human health. Heavy metals released during coal burning can also cause skin and lung cancer, cardiovascular conditions, stomach discomfort, gene mutations, leukaemia, and deadly comas [7].

The majority of the power plants in West Bengal use thermal coal as fuel, e.g. Damodar Valley Corporation (DVC), National Thermal Power Corporation (NTPC), and West Bengal Power Development Corporation Limited (WBPDCL) are the organizations engaged in coal-based power generation. The government of India plans to shut all units of thermal power plants, which are above 25 years old and are producing 11-mw or more fly ash every year [8]. Kolaghat thermal power station was established during India's sixth five-year plan (1980-85) and emerged as the second highest power generating plant in West Bengal. The plant is operating with total six units and having six ash ponds for damping the ashes. The coal type used in the thermal power plant at Kolaghat, like bituminous charcoal, is quite poor and generates a lot of ash. According to a recent report, Kalaghat Thermal Power Plant produces 1,74,577 metric ton of ashes from 401,603 metric ton of coal (Jan, 2016). Once the fly ash is collected from the base of the boiler and using the fabric bay filters, the ash has been combined with water and transported to the six ash ponds located around 4-5 kilometres south of the thermal power plant Kolaghat. During the monsoons, the ponds flood and pollute the surrounding areas. The most impacted villages are Dakhin-rakhashachak, Andulia, and Bonmecheda [9].

Several studies have assessed the environmental stress of K.T.P.P upon the local environment. These studies mostly focus on fly ash emissions affecting, surrounding natural vegetation, the wellbeing of the nearby inhabitants, degrading air quality, altering physical and chemical properties of water and soil [9–12]. Studies are conducted on the adverse effects of the continuous discharge of wastewater and deposition of fly ash on the Rupnarayan river and found deteriorating effects on water quality and aquatic lives [13]. Fly ash is gradually accumulating at the bottom of the river and rapidly reducing the depth of the river. Jena (1993) explored that the fly ash from Thermal power plants is changing soil upper layer structure, PH level, soil fertility [14]. According to studies, the amount of metals like zinc, copper, magnesium, and iron in fly ash from thermal power plants increases with the proximity of the source and decreases with distance [15]. By studying the literature, the research gaps have been identified as the following: Numerous pieces of research have mostly focused on the effects of only one environmental stressor, such as fly ash or radionuclide, with one environmental media of concern, such as groundwater, ambient air quality, soil, or river sediments. It is also necessary to compare the contribution of many environmental stressors i.e. heat, air pollutants, water contaminants, etc. overall environmental and health stress in order to pinpoint the stressor of concern.

The goal of this study is to evaluate the overall impact of thermal power plants on the local environment and public health using a composite index that incorporates environmental and health dimensions. The study incorporates the parameters of environment i.e. Air quality index, drinking water quality index, land surface temperature, non–communicable disease rate etc. The study furthermore compares the contributions of the above mentioned environmental stressors. On this basis we have estimated environmental and health vulnerability using a composite index. Moreover, subjective assessment of inhabitant's environmental and health perceptions and occupational health has been conducted within the study. The study's findings will aid in the accurate depiction of the existing situation in the region, assisting in the identification of areas and stressor of concern and suggesting mitigating actions based on the current evaluation.

## 2. Study Area

K.T.P.P is one of West Bengal's leading thermal power plants. It is situated at Mecheda, approx. 55 km from Kolkata (Latitude 22°25'52' 'W and Longitude 87°52'30' E) in the district of Purba Medinipur. The river Rupnarayana flows to the right side of this plant. It is connected with the south Eastern railway [9]. Firstly, a 15 km buffer area has been created around Kalaghat Thermal Power Point. Then all the administrative blocks that fall under that buffer area are taken as study areas. The Study area is located on the right bank of Rupnarayan River. Left bank of the river is an extended part of the urbanised area of Howrah. The river also works as the physical barrier. Hence the blocks on the left bank have not been incorporated as the study area. The study area consists of eight blocks on the right bank of the river i.e. Moyna, Panskura I, Daspur II, Tamluk, Kolaghat, Daspur I, Sahid Matangini and Nanda Kumar (figure 1).

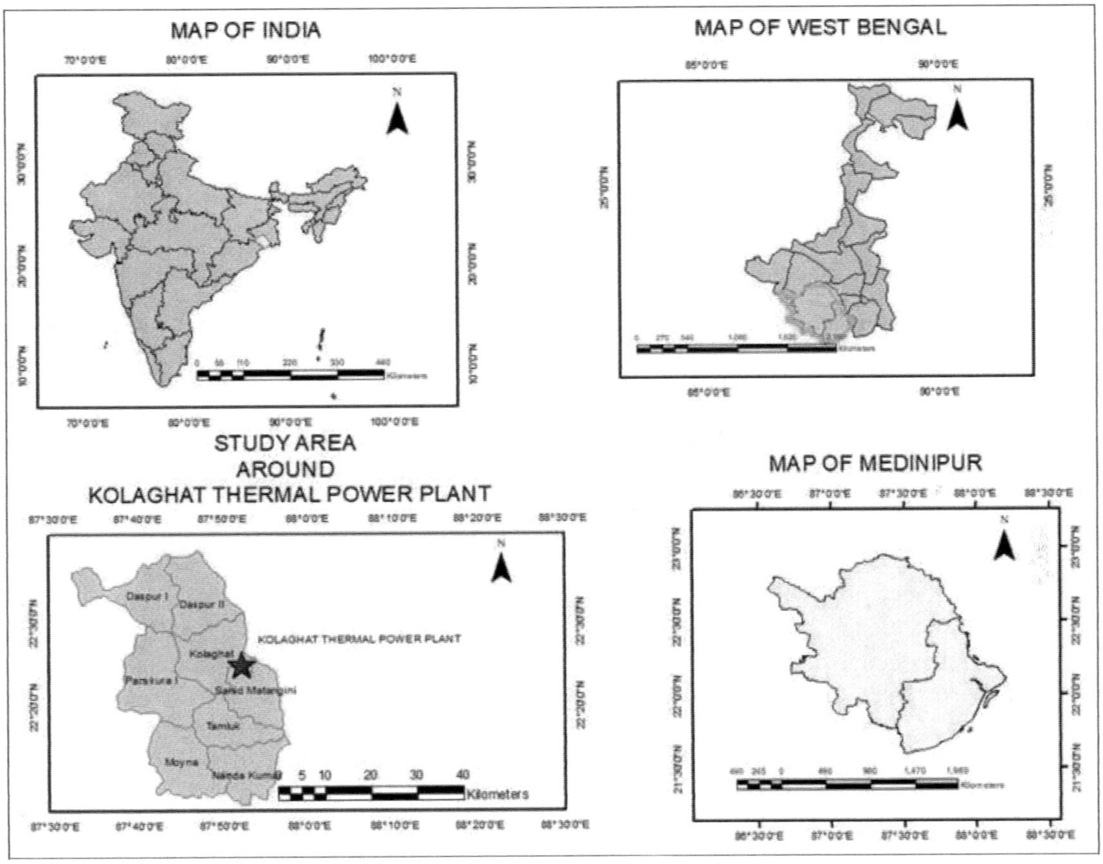


**Figure 1.** Location Map of Kolaghat Thermal Power Plant and surrounding blocks

## 3. Data Collection

All the data for the objective assessment has been obtained from the secondary sources. Following table provide details and sources of the respective data (table 1).

**Table 1.** Sources and details of the Data

| Data Type | Details | Source |
|---|---|---|
| Land Surface Temperature | Landsat 8 (OLI)acquired on 1st march 2019 | https://earthexplorer.usgs.gov/ |
| Ground water quality | Ground Water Year Book of West Bengal & Andaman & Nicobar Islands (2019-2020): April/May (2019-20). And from Water quality information system (WBPCB) | http://cgwb.gov.in/ http://emis.wbpcb.gov.in/ |
| Air Quality | Ambient Air Quality Monitoring Data For The Year 2019 | http://www.cpcbenvis.nic.in/air_quality_data.html |
| Non-communicable diseases, IPD and OPD | Health Management Information System 3 years data (2019 -2020, 2018 – 2019, 2017 – 2018) | https://nrhm-mis.nic.in/ |

## 4. Methodology

This paper aims to incorporate several environmental stressors and to build a single index for depicting the overall stress due to the thermal power plant's activities. These individual stressors have been selected subjectively based on their potential hampering environmental health and wellbeing. We have selected Air quality, Land surface temperature, drinking water quality, and non-communicable diseases status. The study also included a subjective assessment of residents' environmental and health perceptions, as well as occupational health in K.T.P.P. In the following section, we have discussed the selection, interpretation ad estimation of these stressors.

### *4.1. Environmental Stress and its dimensions*

Environmental stress is known as environmental event-forcing processes that influence an organism's existence and well-being and counteract organism activity. These stressors restrict the ecosystem's production, reproductive potential, and the sustainability of the ecosystem, and affect populations and communities of organisms and ecoscapes. [16]. Over time, the variety and complexities of these stressors have proliferated due to the growth of population and environmental exploitations [17,18]. Multiple stressors, such as organic pollution, thermal stress, and salinity alteration, are inducing significant and irreversible changes in coastal aquatic ecosystems as a consequence of anthropogenic activities e.g. urbanization, Industrialization, agriculture, fishing, coastal development, and anthropogenically induced global climate change [17]. Biological processes are disrupted, ecosystem functions are altered, and global and local biodiversity is decreased as a result of such stresses. In this study we have selected some prominent stressors to environment and people's health i.e. Ambient air quality, Ground water quality, Land Surface Temperature, and non-communicable diseases rate. All these dimensions have been aggregated to finally obtain the vulnerability and stress index.

### *4.2. Ambient air quality*

Air pollution has been identified as one of the important environmental stressor in the literature of several environmental and public health researches. Poor ambient air quality not only has detrimental effects upon physical and mental health of a human but also upon other biotic and abiotic components of the environment and further aggravating the stress. To represent air quality, we have used air

IOP Conf. Series: Earth and Environmental Science 1164 (2023) 012012 doi:10.1088/1755-1315/1164/1/012012

quality index. The air quality index is a single numerical number that indicates various parameters at the same time, making it simple to analyze and comprehend. The USEPA formed the AQI in 1976 as a way of communication between the public and government. Later, a number of other government bodies adopted and improved the approach. The National Air Quality Index established by India's Central Pollution Control Board is utilized in this work. The index is computed in two steps: calculating sub-indices of each parameter and aggregation of the sub-indices using CPCB's national ambient air quality standards. Weighted Additive, Root-Sum-Power, Root-Mean-Square, Min-Max Operator, and so forth are used for aggregation. Finally, a table is used to interpret the AQI results. Three key air performance indicators that are available at all of the sample locations have been chosen. These values are SO2, NO2, and PM10, all measured in g/m$^3$. These parameters are the most important parameters for air quality assessment found in the previous studies.

### 4.3. *Land surface temperature (LST)*

Heat stress has been explored in the past as an environmental threat that can affect people's productivity, health, and even existence. Heat stress not only has an impact on human health, but it is also a key contributor to crop failure, drought start and intensification, vegetative health degradation, and agricultural droughts[18,19]. Climate change is likely to increase the number of hot days and the incidence of heatwaves, resulting in heat-related health effects and death. Efforts to quantify human heat stress use various thermal stress indices, which are generally evaluated in continuous space using remote sensing technology[20].Land surface temperature model has been obtained from Landsat 8 OLI acquired on 1st march 2019 using following steps (figure 2). The Bands has been first processed for atmospheric correction before the calculation of LST.

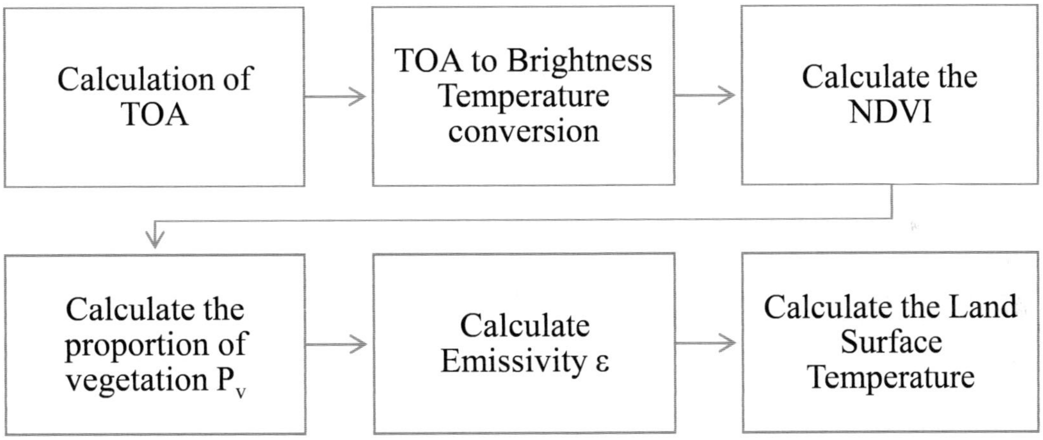


**Figure 2**. Methodology adopted for retrieving Land Surface Temperature [21]

### 4.4. *Ground water/ Drinking water quality*

Groundwater is one of the most valuable natural resources and it is widely utilized in India for drinking, household consumption, and agricultural uses. Access to clean drinking water is a basic human requirement. Contaminated surface water, industrial wastewater, household wastewater, and agricultural pollutants can pollute underground water. Untreated groundwater may harm human health directly. The water quality index (WQI) simplifies water parameters to an exclusive value to measure comprehensive water quality at a specific site and was developed to determine whether water samples are safe to drink.

In the current study, WQI has been computed in three phases utilizing the selected parameters i.e. PH, EC, Hardness, CA, Mg, Na, Alkanity, Cl, No3, So4, F, and TDS. We have used a weighted arithmetic water quality index technique for our investigation since it is simple and requires minimal data[22,23].

### 4.5. *Non-communicable diseases status*

Diseases or disabilities are the result of a combination of environmental exposure, genetic factors, and behavioural variables or one's lifestyle. Industrial inorganic pollution is directly related to environmental factors and mostly associated with pollutants rather with pathogens, hence the status of non-communicable diseases can be considered as the stress upon environment and health induced by Industrial pollution. This study emphasizes upon the impact of KTTP hence we have assessed the following non-communicable diseases.

- Respiratory (Asthma, COPD, Emphysema etc).
- Cardiovascular: Blood pressure and Heart related
- Gastrointestinal: Dyspepsia, IBS, Gastritis etc.
- Diabetes
- Neurological
- Allergic (rhinitis, asthma, skin or other)
- Skin related
- Cancer
- Psychological

To normalize the diseases cases with the population we have estimated diseases rate using total number of in and outpatient with the total non-communicable diseases cases. Finally, the non-communicable diseases rate has been estimated by averaging the normalised diseases cases for the years 2017–2018, 2018–2019, 2019–2020.

### 4.6. *Normalization and Aggregation of the sub-indices*

Because each variable in this research i.e. Ground Water Quality, Land Surface Temperature, Air Quality Index, Non- communicable disease has a distinct unit, they must all be brought into a consistent range (0-1) before they can be combined and this is referred to as normalisation [equation (1)].

$$\mathrm{Yn} = (\mathrm{Y} - \mathrm{YMin})/(\mathrm{YMax} - \mathrm{YMin}) \quad (1)$$

Yn = Normalized sub-indices, Y = Raw value of the sub-indices, YMax = Maximum value of Y, Minimum Value of Y

They have been put together after calculating the normalised values of each of the sub-indices using weighted mean [equation (2)]. Weights for each sub-indices has been subjectively determined by the author.

$$\frac{\sum_{i=1}^{n} wiXi}{\sum_{i=1}^{n} wi} \quad (2)$$

### 4.7. *Subjective Assessment: perception and occupational health*

Environmental and health perception of the inhabitants and the workers of the KTTP have been assessed using questionnaire survey. Perception study of the inhabitants residing within the extent of the study area has been divided into two parts i.e. perception and valuation of the inhabitants. Perception study has been structured with likert scale questionnaires and valuation study has been constructed using willing to Pay (WTP) or Contingency Valuation method (CVM). Opinion data of the workers has been collated with likert scale. Behavioural and attribute data has been collated using dichotomous and open-ended questions. All the questionnaires have been prepared after rigorous literature survey and after testing for validity and reliability i.e. face validity and content validity, Cronbach's Alpha coefficient test for reliability. We have performed non-parametric Kendal tau correlation to explore the relationship between "distance from the industry and air quality perceptions; water quality perceptions; people's income and their valuation, people's education and valuation. For calculating the environmental stress, we have selected four dimensions e.g. Ground water quality, Landsat Temperature, Air Quality, Non-Communicable Disease.

## 5. Result

The thermal condition indicates a gradual increase in temperature towards the west, especially towards Daspur I and II, Pasnkura I, and Kolaghat Blocks. The maximum and temperatures have been recorded in the area as 39°c in Paskura I block and a minimum of 27°c in the Nandakumar block respectively. Although local effects can be seen, it is limited to the vicinity of the plants. Hence, the majority of the heat stress cannot be attributed to the Thermal plant. The temperature recorded in the Thermal plant is 37.72°c which is gradually decreasing towards the surrounding area.

The Air Quality Index value in the study area ranges between 132 to 95 meaning the study area exhibits moderately polluted to satisfactory air. The air quality is getting better from the east to the west direction in the study area i.e. moderately polluted air quality can be observed near K.T.P.P. and it is getting better (figure 3). Hence, the influence of K.T.P.P on ambient air quality is recognizable. Blocks with better air quality were identified as the blocks having higher LST e.g. Daspur I, Paskura I, Moyna, and Nandakumar. Blocks of polluted air were identified as Kolaghat, Sahid Mathangini, and Daspur II. Groundwater quality in the area was found to be satisfactory i.e. <50. Spatial patterns of groundwater quality were found to be abrupt in the area, the northern part of the plant was observed to have a higher value of GQI indicating comparatively poor water quality. Sahid matangini block is observed to exhibit the best water quality and Kolaghat block has the highest GQI values (figure 4). Likewise, groundwater quality and non-communicable diseases rate also don't exhibit consistent spatial patterns nor any spatial association with the thermal plant. The highest NCD rate observed was 74% in Nanda Kumar, followed by Tamluk at 63%, Daspur-I having 13.2%, and Kolaghat with having NCD rate of 10. 41%.

Ergonomic risks identified among K.T.P.P's workers are electrical shocks, injuries from falls, lifting and handling machinery. In addition, skin diseases and allergies are common among the workers too. The majority of the workers have reported poor air and thermal conditions inside the plant and within the battery limit. Some reported soil pollution in the vicinity of the ash pond area.

The calculated vulnerability index score depicts a comprehensive picture of the region's stress level in terms of thermal stress, air quality, ground water quality, and the rate of non-communicable diseases. As per the index, Nanda Kumar, Tamluk, and Kolaghat blocks are under maximum stress and least stress identified in Sahid Matangini, Moyna, and Panskura I blocks (figure 5). However, air quality dominated the Index in terms of contribution compared to other sub-indices (figure 6).

**Table 2.** Correlation with Distance from K.T.P.P and Air, Water Quality Index

| **Distance from K.T.P.P and Air Quality** | **Correlation Coefficient** | **Sig. (2-tailed)** |
|---|---|---|
| Kendall's tau_b | -.436** | 0 |
| Spearman's rho | -.547** | 0 |
| **Distance from K.T.P.P and WQI** | **Correlation Coefficient** | **Sig. (2-tailed)** |
| Kendall's tau_b | -.216* | .033 |
| Spearman's rho | -.307* | .024 |

***. Correlation is significant at the 0.01 level (2-tailed). N = 54*

*Source: Primary field survey, 2021*

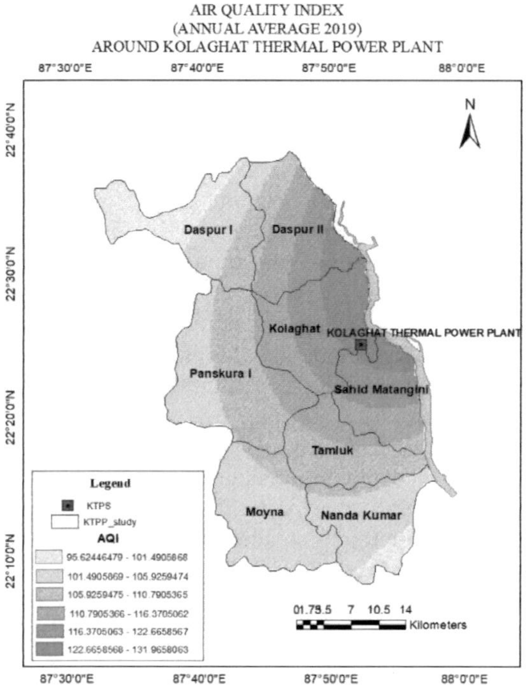


**Figure 3.** Spatial pattern of annual average air quality index (2019) around K.T.P.P.

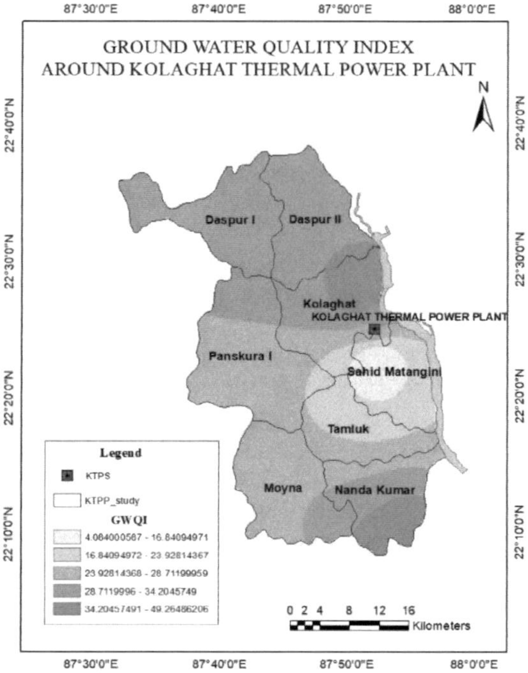


**Figure 4.** Spatial pattern of annual average Groundwater Quality index (2019) around K.T.P.P

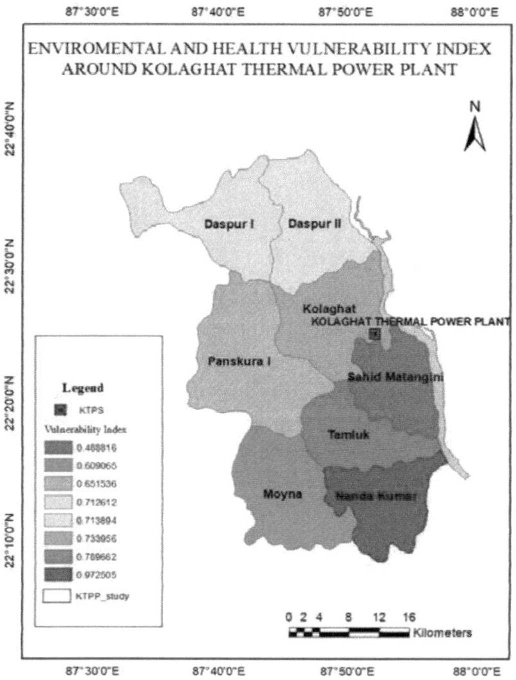


**Figure 5.** Environmental Health and vulnerability Index around K.T.P.P.

Results of subjective assessment reveal a statistically significant negative correlation between distance from K.T.P.P and Air Quality Index (table 2) indicating the impact of K.T.P.P upon ambient air quality. However, no significant correlation was found between distance from the plant and water quality and non-communicable diseases. In addition, inhabitants have reported seasonal (mostly in winter) fly ash deposition over many surfaces e.g., rooftop, yard, top of cars, etc. residing near the plants. A strong significant positive correlation has been observed between people's income level and their environment valuations, however, no significant correlation is found between people's education level and environmental valuations. The most common disease around thermal power plants reported by the inhabitants are skin problems, allergies, and eye irritation.

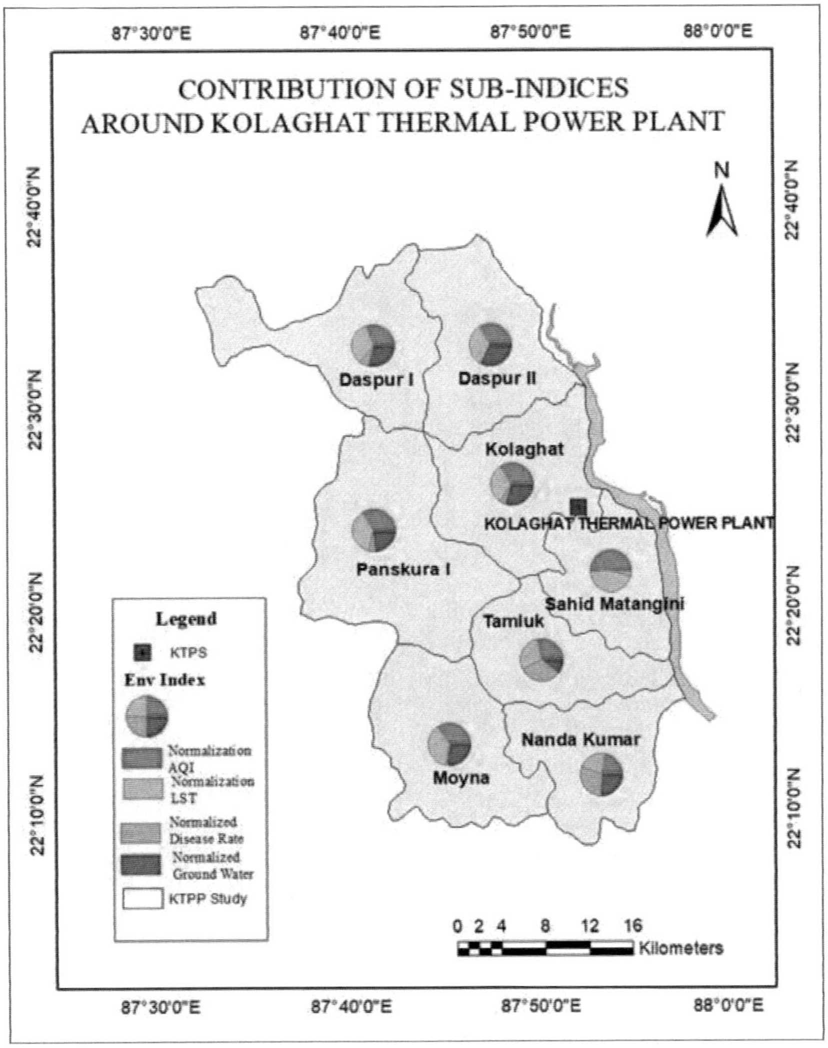


**Figure 6.** Contribution of Sub-indices in Environmental Health and vulnerability Indexaround K.T.P.P.

However, workers have reported regular operations and maintenance of pollution control systems installed in the plants. The plant was operational amid pandemic by following all safety rules and guidelines of Covid 19 i.e. social distancing, proper sanitizations, use of masks, and quarantining infected people. Approximately 20 % of the employees got affected by the Covid between March 2020to February 2021. Skin allergies and eye irritation affect the majority of people, according to perception surveys. Fatigue was reported by 18 percent of workers, while Respiratory Allergies and Irregular Heart Beats were reported by 11 percent of workers. Only a few workers are affected by any reportable ailment, such as a lung issue or high blood pressure.

**Table3.**Relation of Non-communicable Disease with Distance from the K.T.P.P

| | | | *Distance* | *NCD* |
|---|---|---|---|---|
| *Kendall's tau_b* | *Distance* | *Correlation Coefficient* | *1.000* | *-.184* |
| | | *Sig. (2-tailed)* | . | *.062* |
| | | *N* | *54* | *54* |
| | *NCD* | *Correlation Coefficient* | *-.184* | *1.000* |
| | | *Sig. (2-tailed)* | *.062* | . |
| | | *N* | *54* | *54* |
| *Spearman's rho* | *Distance* | *Correlation Coefficient* | *1.000* | *-.257* |
| | | *Sig. (2-tailed)* | . | *.061* |
| | | *N* | *54* | *54* |
| | *NCD* | *Correlation Coefficient* | *-.257* | *1.000* |
| | | *Sig. (2-tailed)* | *.061* | . |
| | | *N* | *54* | *54* |

*Source: Primary field survey, 2021*

## 6. Conclusion and Recommendations

The study suggests that the impact of the thermal plant is mostly limited to its vicinity and its impact cannot be observed on a larger scale. The most prominent influence in both subjective and objective assessments has been identified as fly ash emissions and air pollution. Although the association between non-communicable disease rates and industrial emission has not been found, poor thermal and ambient air quality conditions are indicated by the workers of the plant.

In addition, perception studies reported the prevalence of skin problems, allergies, and fatigue among workers. Based on these conclusions the following recommendations have been suggested. Since ash ponds have been identified as one of the primary causes of air, water, and soil pollution, they should be surrounded by green belts to reduce the negative impacts of the ashes in the air, water, and soil. The area lacks a stipulated amount of greenery as per the Ministry of Environment & Forests Environmental Impact Assessment Division i.e. 50 to 100m and consists of at least 3 tiers around plant boundary with a tree density of 2000 to 2500 trees per ha with a good survival rate of around 80%, shall be submitted. Over and above the green belt around the plant boundary, a green belt shall be developed around the ash pond [24]. We have suggested creating the necessary green belt by also incorporating phytoremediation plants. Suitable locations for green belts with phytoremediation plants have been suggested after assessing the locations of the ash ponds and the land use pattern in the surrounding areas.

To mitigate the harmful effects of the Ash ponds following recommendations have been suggested.

1. Creating 50 feet wide riparian buffer on both sides of the northern canal working as Hydraulic barrier for abating contamination from the ash ponds to the adjacent canal.
2. Finding out the suitable sites for phytoremediation within a 500meter battery limit by analyzing the land use pattern of the area.

Furthermore, we have suggested the region-specific phytoremediation techniques i.e. incorporating Phyto remedial plants for the proposed green belt and the riparian buffer (figure 7). For controlling of fly ashes that creates pollution in the environment by phytoremediation and plantations is required.

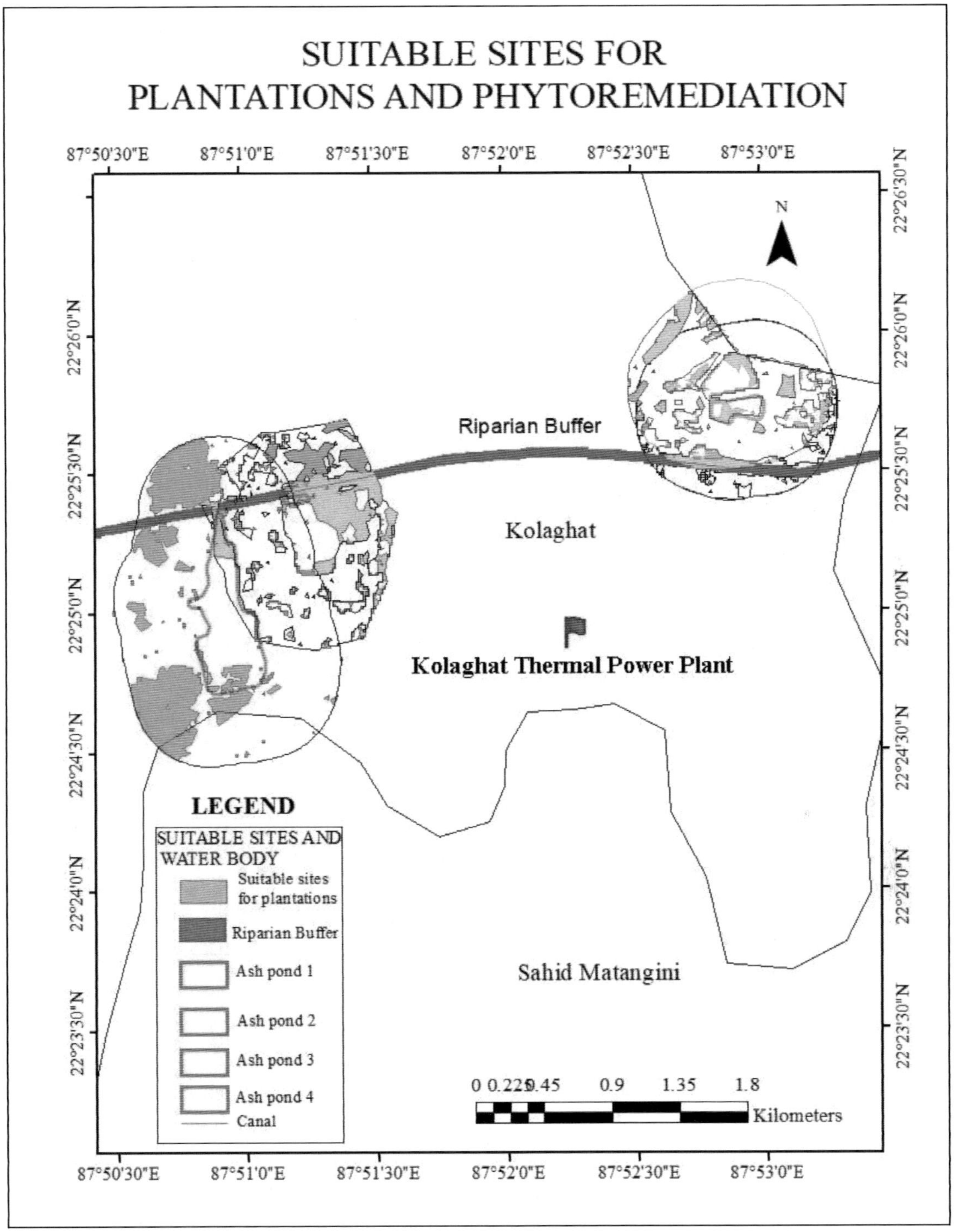


**Figure 7.** Suitable sites for plantations and phytoremediation around K.T.P.P

## References

[1] Devisscher T, Konijnendijk C, Nesbitt L, Lenhart J, Salbitano F, Cheng Z C, Lwasa S and van

den Bosch M 2019 SDG 11: Sustainable Cities and Communities – Impacts on Forests and Forest-Based Livelihoods *Sustainable Development Goals: Their Impacts on Forests and People*

[2] McFarlane R A, Barry J, Cissé G, Gislason M, Gruca M, Higgs K, Horwitz P, Huu Nguyen G, O'Sullivan J, Sahu S and Butler C D 2019 SDG 3: Good Health and Well-Being – Framing Targets to Maximise Co-Benefits for Forests and People *Sustainable Development Goals: Their Impacts on Forests and People*

[3] Sayer J, Sheil D, Galloway G, Riggs R A, Mewett G, MacDicken K G, Arts B, Boedhihartono A K, Langston J and Edwards D P 2019 SDG 15: Life on Land – The Central Role of Forests in Sustainable Development *Sustainable Development Goals: Their Impacts on Forests and People*

[4] Tripathi B 2018 Now, India is the third largest electricity producer ahead of Russia, Japan *indiaspend*

[5] Mukharji A 2021 Why India is on the brink of an unprecedented power crisis *BBC News*

[6] Bhattacharyya S C 1995 Thermal power generation and environment: A review of the Indian case *International Journal of Energy Research* **19** 185–98

[7] Munawer M E 2018 Human health and environmental impacts of coal combustion and post-combustion wastes *Journal of Sustainable Mining* **17** 87–96

[8] Boro J R 2020 *REPORT ON FLY ASH GENERATION AT COAL / LIGNITE BASED THERMAL POWER STATIONS AND ITS UTILIZATION IN THE COUNTRY FOR THE YEAR 2019 - 20* (NEW DELHI)

[9] Mondal I, Maity S, Das B, Bandyopadhyay J and Mondal A K 2016 Modeling of environmental impact assessment of Kolaghat thermal power plant area, West Bengal, using remote sensing and GIS techniques *Modeling Earth Systems and Environment* **2** 139

[10] Das J K 2014 Impact of Coal Ash Surrounding Thermal Power Stations: A Case Study in West Bengal. *Business Studies* **XXXV** & **XXX**

[11] Mandal A and Sengupta D 2005 Radionuclide and trace element contamination around Kolaghat Thermal Power Station, West Bengal - Environmental implications *Current Science (Bangalore)* **88**

[12] Dinda S 2014 Developmental project and its impact on adjacent river ecology: A case study of Kolaghat thermal power plant, West Bengal, India *International Journal of Geomatics* **5** 225–23

[13] Chakraborty S K 2021 Physicochemical Parameters and Their Seasonal Dynamics: Special Reference to Riverine Networks of South West Bengal, India *Riverine Ecology Volume 1* 477–521

[14] Jena S 2011 Environmental Impacts of Fly Ash Generated from a Coal Fired Power Plant in Indian Sub-continent *The 17th annual conference of the International Sustainable Development Research Society (ISDRS)* (New York, NY)

[15] Kanchan S, Kumar V, Yadav K, Gupta N and Arya S 2015 Effect of Fly Ash Disposal on Ground Water Quality near Parichha Thermal Power Plant, Jhansi: A Case Study *Current World Environment* **10** 572–80

[16] Freedman B 2018 *Environmental Stressors. Environmental Science.*

[17] Hewitt J E, Ellis J I and Thrush S F 2016 Multiple stressors, nonlinear effects and the implications of climate change impacts on marine coastal ecosystems *Global Change Biology* **22** 2665–75

[18] Bhuiyan C, Saha A K, Bandyopadhyay N and Kogan F N 2017 Analyzing the impact of thermal stress on vegetation health and agricultural drought – a case study from Gujarat, India *GIScience & Remote Sensing* **54** 678–99

[19] KENNEY W L, CRAIGHEAD D H and ALEXANDER L M 2014 Heat Waves, Aging, and Human Cardiovascular Health *Medicine & Science in Sports & Exercise* **46** 1891–9

[20] Song Y and Wu C 2018 Examining human heat stress with remote sensing technology *GIScience & Remote Sensing* **55** 19–37

[21] Sekertekin A and Bonafoni S 2020 Land Surface Temperature Retrieval from Landsat 5, 7,

IOP Conf. Series: Earth and Environmental Science 1164 (2023) 012012 doi:10.1088/1755-1315/1164/1/012012

and 8 over Rural Areas: Assessment of Different Retrieval Algorithms and Emissivity Models and Toolbox Implementation *Remote Sensing* **12** 294

[22] Brown R, McClelland N, Deninger R and Tozer R 1970 "A water quality index: do we dare?", Water & Sewage Works *A water quality index—crashing the psychological barrier. In Indicators of environmental quality* **117** 339–43

[23] Tiwari T N and Mishra M 1985 A preliminary assignment of water quality index to major Indian rivers *Indian Journal of Environmental Protection*

[24] Kumar V V., Swathy S, Radhika R, Khan M A F and Chakradhar B 2019 Green Belt Requirement for New and Expansion Projects for Obtaining Environmental Clearance in India *International Journal of Applied Environmental Sciences* **14** 475–87

# Monitoring Change in Urbanization and Green Space for Eastern Indian Cities in 30 Years-A Comparison between Kolkata and Bhubaneswar

**S Bose[1]*, A Mazumdar[1], S Basu[1]**

[1] Jadavpur University, Kolkata 700032, West Bengal, India

*Corresponding Author Email: bosesuddhasil@gmail.com

**Abstract.** Advancement in technology has made a new sphere of geospatial exploration. Application of geoinformatics in Landuse and Landcover study has made quantitative analysis most clear. Urbanization process is an inevitable factor in developing countries like India. Last 30 years eastern India has also been changed itself in terms of urban development. Kolkata and Bhubaneswar are two most important cities in eastern India where urban growth has taken place rapidly. Landuse and landcover pattern has also changed here with time. Urbanization process of both cities has been identified using satellite images taken from USGS. Applying satellite images, this paper has investigated about the urban green spaces that has been changed with time also depth of urbanization has been studied to find out an overall ecological balance in these cities. Different LULC classes has been chosen to define the ecology. Details about landscape matrices has been applied which are necessary to get the result. Urban Index (UI), Urban expansion index (UX) and Urban green index (UGI) has been used to calculate and compare results with both the cities. Geospatial analysis has been processed maintain time interval of 15 years to better understand the change that affected on land with time. Kolkata is one of the most urbanized cities as it has large history of urbanization whereas Bhubaneswar is also has developed with time, it has been selected as smart city for maintaining a holistic growth approach. Quantitative analysis and application of LecoS made the results more specific and appropriate to understand the actual relationship between growth of cities and greenspace towards sustainability.

**Keywords:** Urbanization, LULC, Urban Indices, Decadal Change, Urban Growth, Landscape Matrices, Greenspace.

## 1. Introduction

India is developing country [1]. Here with increasing population, urbanisation process has started taking place after the independence from British rule. Due to large size of India, Indian has different regional division. Eastern part of India, has major cities like Kolkata, Patna, Jamshedpur, Dhanbad, Ranchi, Bhubaneswar. many of them has observed significant urbanisation. The process of urbanisation can be monitored by remote sensing technology[2]. High resolution satellites have been used for near 50 years to take images of landforms and analysing the image, changes in Landcover and Landuse can identified [3]. There is much more scope to use different type of Urbanisation Indices to identify the change in urban area [4] with time and compare that with other cities [5]. Green space is also a serious concern for any city. Analysing satellite images, urban greenery can be identified and monitored. The changes in landscape and its relation within other landscapes can be described quantitatively using landscape ecology matrices [6] that will be effective for monitoring and the change in Urbanisation and green space more accurately. Kolkata and Bhubaneswar are one of the main important cities in Eastern India that has significance in India.

## 2. Study Area

19th century, there was predominance of British rules in eastern India [7], Kolkata was the capital of British India[8] and Bhubaneswar was also an important city of the state Orissa [9]. After the independence the importance of eastern part of India became inconsiderate. The capital of India shifted to Delhi after 1911.both the cities are the most important cities in eastern part of India.

### 2.1. *Kolkata*

Kolkata is situated on west bank of the river Hugli. It is the capital of West Bengal; it has a long history of Urbanisation of about 300 years. Kolkata city is administrated by Kolkata Municipal Corporation (KMC)[10]. It started its journey from 1876 as Calcutta Municipal Corporation. it consisting an area of about 207 $km^2$ with 144 wards, now a days. It is one of the populated cities in eastern India with more than 4 million people (census, 2011). Till 2012, the area of KMC was 186 $km^2$after 2012 it has increased to 207 $km^2$.former area has been considered here for the better understanding (figure 1).

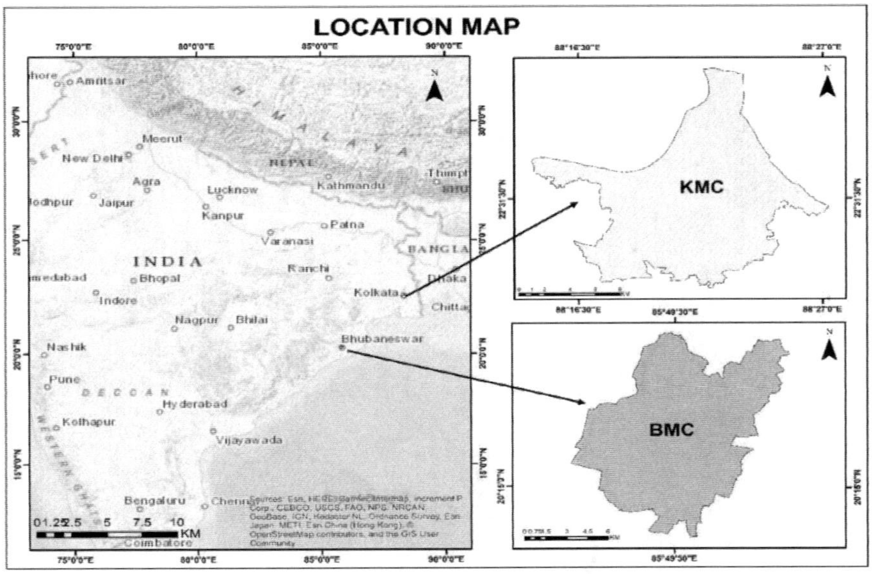


**Figure 1.** Location Map of Bhubaneswar and Kolkata Municipal Corporation.

### 2.2. *Bhubaneswar*

Bhubaneswar is situated in Orissa; it is the capital of Orissa. runs by Bhubaneswar Municipal Corporation (BMC) with area of 186$km^2$.It has 67 ward and 42 revenue village with population about 8.5 lakhs [11].Administration started its journey from 1948, and it has been considered to be a corporation from 1994 (figure 1).

## 3. Data Collection

The study of urbanisation remotely can be done best, using satellite images. Primary source of satellite image is ISRO for India, but due lack of availability, the satellite images has been collected from United states Geological survey's (USGS) official website USGS Earth explorer[12]. It is free of use

IOP Conf. Series: Earth and Environmental Science 1164 (2023) 012013 doi:10.1088/1755-1315/1164/1/012013

and accuracy is appreciable. Keeping in mind, the effect of the weather variability all the images are collected maintain a certain of the year. Keeping a gap of fifteen years, the data collection has started from 1990 and it ended in 2020 (figure 2). The year 2005 has been considered for a better understanding between these two years. (table 1)

**Table 1.** Data Acquisition from USGS Earth Explorer

| Location | Latitude/Longitude | Year | Path/Row | Spatial Resolution | Satellite | Bands |
|---|---|---|---|---|---|---|
| **KMC** | 22.5726,88.3639 | 1990 | 138/44 | 30m×30m | Landsat 05 Thematic Mapper | 1-5 |
| **BMC** | 20.2961,85.8245 | | 140/46 | | | |
| **KMC** | 22.5726,88.3639 | 2005 | 138/44 | | | |
| **BMC** | 20.2961,85.8245 | | 140/46 | | | |
| **KMC** | 22.5726,88.3639 | 2020 | 138/44 | | Landsat 08 Operational Land Imager | 1-7 |
| **BMC** | 20.2961,85.8245 | | 140/46 | | | |

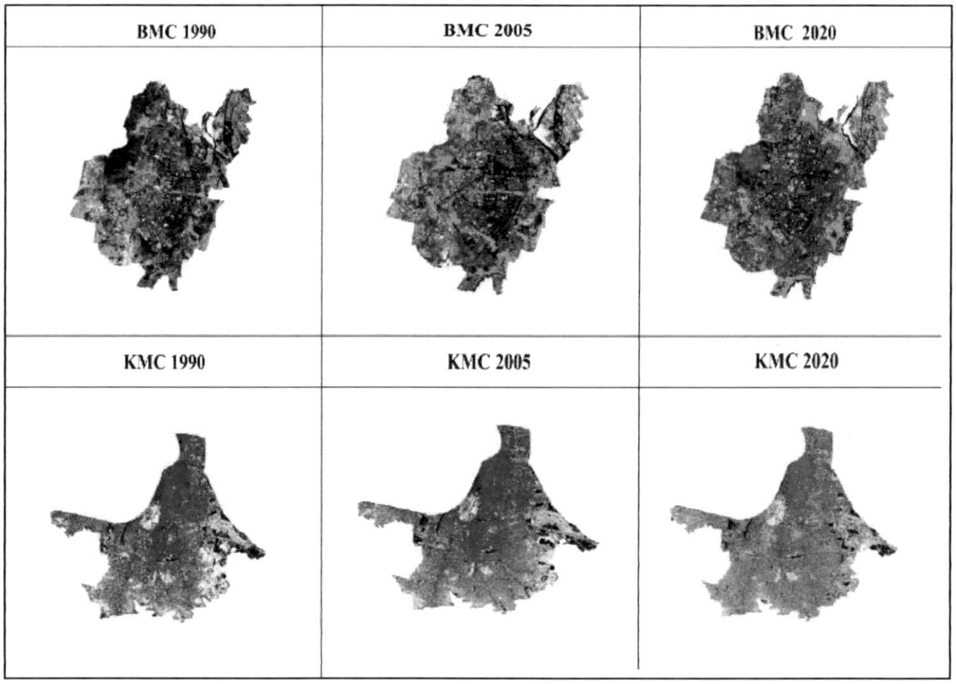


**Figure 2.**False Colour Composite Images Prepared for Classification

## 4. Methodology

This paper aims to monitor the changes between time with two different cities both administrated by the local urban bodies i.e. corporation and both cities having equal amount of area. Quantitative analysis of the landscapes has been performed for better analysis.30 years of timeframe has been selected to observe the changes in between.1990 has been selected as a base year from that year, observation has been started and 2020, has been selected as to end the timeframe and get more detailed about recent time. For better understanding of monitoring, 30 years has been split into another 15 years division.1990,2005 and 2020 has been considered as observing years in this paper (figure 3).

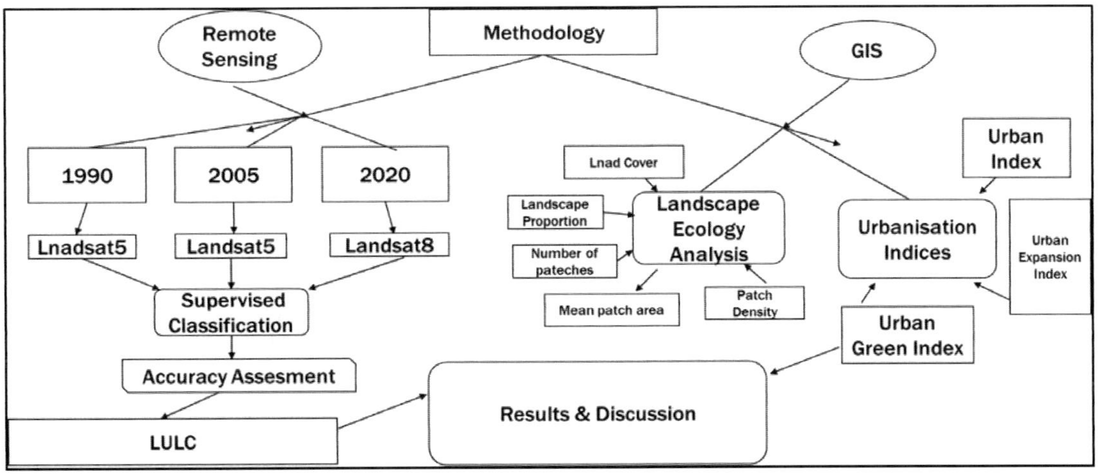


**Figure 3**. Methodology adopted for Monitoring Changes in Urbanisation and Greenspace of the cities

### 4.1. *Image pre-processing and processing*

This paper has utilised remote sensing and GIS process to monitoring the scenario for Urbanisation and green space. Satellite images have been collected which are already georeferenced. This image has been pre-processed for atmospheric correction. Image processing has been done on ERDAS Imagine 2014 software. (figure 2) Supervised classification process has been implemented to understand the Land use and Land cover change. All the image has been prepared with individually different training and testing sets. Parallelepiped has been selected for maximum likelihood. Thus, following the method, comparing with authentic base map the supervised image matched up to 80-90% in average for the abovementioned images. Accuracy assessment process has produced about 80to 90 percent accuracy for the images. Outcome of the images has been compared with ground truth data and it completed with positive outcome.

### 4.2. *Landuse and Landcover classification*

After the processing the images has been recoded with classes. Mostly conventional method has been followed, the Land use and land cover classification divided into water, greenery, wasteland for land cover and land use has been identified by high urban density and low urban density. Classes has been identified and selected based on the commonalities from two study areas (table 2).

**Table 2**.Land Cover and Landuse Classification Classes related to their Biome

| Landuse and Land Cover classes | Biome |
|---|---|
| Water | Lakes/Rivers/canals/wetlands |
| Vegetation | Greenery |
| Wasteland | Wasteland |
| High Builtup Density | Urban |
| Low Builtup Density | Urban |

### 4.3. *Landscape Ecology statistics parameter*

In python programming language, using two scientific libraries Scipy and Numpy, LecoS has been prepared in QGIS software as a plugin. It is an open source software and having free plugin. LecoS is to calculate basic and advanced landscape metrics and offers numerous functions to conduct landscape analysis. Up to now over 16 different landscape metrics are reinforced (table 3). LecoS comes with two different interfaces with core functions like the computation of landscape metrics having their own graphical interface, while more unconventional functionalities are only supported in the QGIS Processing toolbox [6].

**Table 3.** Different Landscape Matrices and Description

| Landscape composition matrices | Description |
|---|---|
| Number of patches (NP) | It is a patch-based calculation from the whole image. Number of patches has been counted from each of the class. Different classes showing different number of patches and with changing time changed patch numbers can identify changes in landuse and landcover |
| Mean Patch Area (MPA) | Arithmetic mean of different size of patches from each LULC classes has been calculated, with change of time change in patch area can be identified and it is useful for interpreting average change in classes |
| Patch Density (PD) | Patch density is calculated dividing the area of patches by the number of patches from each class. Change in density of the patch signifies change of area by class with time. |
| Landscape Proportion (LP) | Proportion of landscapes from the classified images according to LULC classification, shows the acquired land of each class from the total landscape. |

*4.4. Land Cover*

Landuse and Landcover classification prepared in tabular form to understand the quantitative analysis of different classes.

*4.5. Urbanisation Indices*

Understanding the change of urbanisation with different time several indices has been applied. Different values from the indices can be helpful to understand the comparison between two urbanised area abovementioned.

*4.5.1. Urban Index (UI).* It is the ratio between urban land and total land. Level of urbanisation be better identified with this index.in this paper Urban Land (UL) has been described as combined result of HBD (High Builtup Density) and LBD (Low Builtup Density).

$$UL = HBD + LBD \tag{1}$$

$$UI = \frac{UL}{TL} \times 100\% \tag{2}$$

Where UL defines urban land area combining high urban density and low urban density.TL represents Total land in defined area[4].

*4.5.2. rban Expansion Index ($UX_r$).* UX defines the change in urban landscape with change in time, here land area and time are two valuable factor which signifies changes in urbanisation with time. UX is a ratio between original urban area in specific time and change of urban area with changed time from the base time [4].

$$UX_r = \frac{UL_{t2} - UL_{t1}}{UL_{t1}} \times 100\% \quad (3)$$

Here, $UL_{t1}$ defines as, Urban Land of a specific time and $UL_{t2}$ defines the changed time, $UX_r$ represents the urban Expansion Ratio, as it is a ratio it has no specific unit to describe.[13]

*4.5.3. Urban Green Index (UGI).* This Index has been devised to identify the proportion of Urban Green Space (UGS) in comparison with simultaneous urban development [13]. In this paper, UGS is composed of calculating vegetation and water area.

$$UGS = (Water + Vegetation)/2 \quad (4)$$

The equation for Urban Green Index is,

$$UGI = \frac{UGS_{t2} - UGS_{t1}}{(HDB_{t2} + LDB_{t2}) - (HDB_{t1} + LDB_{t1})} \quad (5)$$

Here, UGS=amount of green space, HDB=High Built-up Density, LDB=Low Built-up Density.

## 5. Result and Discussion

### 5.1. Classification Results

Landuse and land cover images has been produced with changing time for both the cities. The quantitative tabular values have been analysed here.

*5.1.1. Land Cover.* The classified LULC maps of the cities is shown in figure 4.BMC area having proportion of water contained area in the city remained same for over 30years. Where as in KMC area, water contained area has been decreased to half almost. In BMC area, about 33% of vegetation from the total area was vegetated where as in 2005 it has decreased to 25% of the total land, now in 2020 it has been increased to near about 40% due to taking vegetation as a serious issue by the administration. Whereas in case of KMC acquiring area of vegetated land has been decreased consistently. In 1990, it was about 40% of the total land where as in 2020 it has been decreased to 15%.in the term of wasteland, values showing a significant situation of land encroachment due to urbanisation process. Bhubaneswar had an area of 91.83km$^2$ in 1990, but with time about 40 km$^2$ area of wasteland had been transformed into another class of land. Now in 2020 just about 2km$^2$ area has been left to treated as wasteland., as Kolkata has a long history of urbanisation so in 1990 it had a minimum area of wasteland of 2.5km$^2$, with time it has been decreased to lowest. A few meters area has been banished as wasteland (table 4).

**Table 4.** Landuse and Landcover Classification values in Sq.Km.

| Classes | 1990 | | 2005 | | 2020 | |
|---|---|---|---|---|---|---|
| | BMC | KMC | BMC | KMC | BMC | KMC |
| Water | 4.27 | 7.53 | 4.30 | 6.54 | 4.57 | 3.75 |
| Vegetation | 59.94 | 77.07 | 45.58 | 58.37 | 74.72 | 26.21 |
| Wasteland | 91.83 | 2.54 | 51.63 | 0.0018 | 1.88 | 0.0009 |
| HBD | 1.31 | 71.5 | 63.52 | 55.28 | 71.90 | 58.93 |
| LBD | 29.81 | 25.65 | 22.13 | 65.1 | 34.09 | 96.40 |

*5.1.2. Landuse.* In this paper, land use has been divided into two classes for a better understanding the density of urban area. Bhubaneswar, as under urbanisation process gone high in urban built up density, specifically in between 1990 to 2005, it has increased drastically in area.HBD increased moderately between 2005 to2020.Incase of Kolkata, it is significant to observe that 1990, HBD was about 72km2 but with time it is decreasing slowly. In 30 years of time span in BMC area, Low Built up Density has remained the same, but increasing with time as slow pace. In case of Kolkata municipal area, low built up density is increasing at a moderate to high pace (table 4).

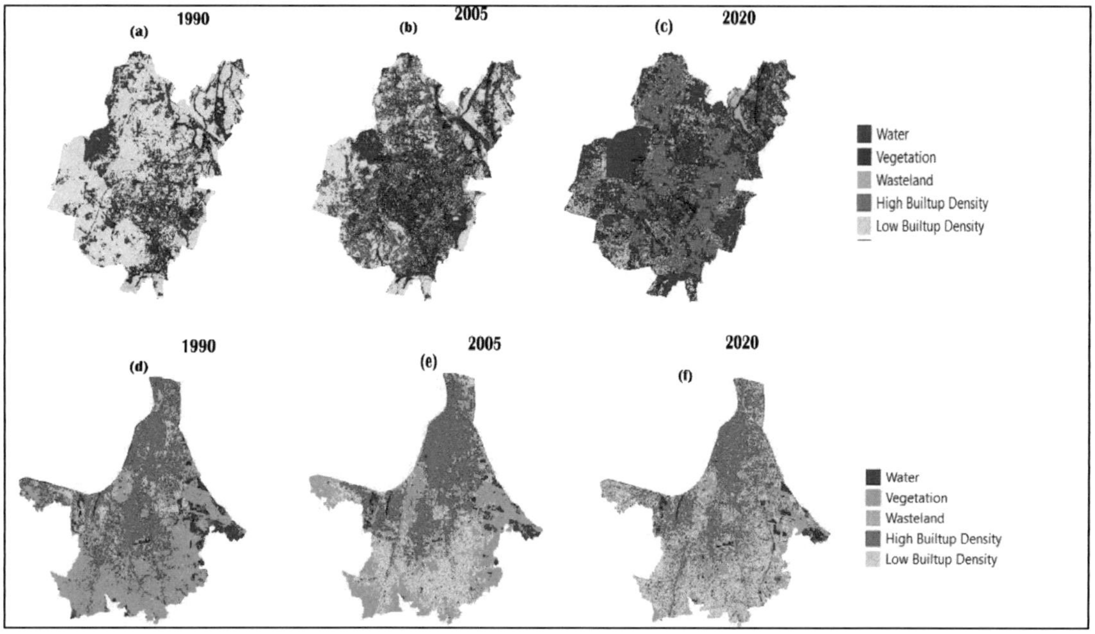


**Figure 4.** Classified Images according to Landuse and land cover classes:(a, b, c) showing images of Bhubaneswar and (d, e, f) showing images of Kolkata.

In both cities' urbanisation process has left its marks prominently but urban built up density are different for the different city. Here period and other economic and social factors have an importance on the effect of urbanisation.

*5.2. Landscape Matrices*

A landscape matrix has been done on LecoS. It is pixel-based calculation. Number of patches (NP) showing the counts of patches in each class. Incensement in number of patches signifies the fragmentation of land in their respective class and decreased number of patches shows the concentration of each LULC class (figure 5). In Bhubaneswar the number of patches increased for water area, that means the available water land has been divided into small parts because of encroachment for other classes, at first number of patches has been amplified then it got reduced so it a picture of changing pace in development of a city, from 1990 to 2005 Bhubaneswar has a growth with high pace, in 2005 to 2020 it has slowed down. in case of Kolkata number of patches for water remained alike in 30 years. Count of vegetation got increased and low built up density got decreased. huge number of changes in count of HBD can be seen in 2020, that means density of the of high built up density got fragmented.

Mean patch area (MPA) massively decreased in context of wasteland from 1990 to 2005, it shows that with time average area of wasteland got compromised with other feature classes in Bhubaneswar. Kolkata had an average area of wasteland and vegetation both got decreased with time, which signifies

IOP Conf. Series: Earth and Environmental Science 1164 (2023) 012013 doi:10.1088/1755-1315/1164/1/012013

all that lands has been transformed into other land. in case of Kolkata that lands have been changed into Urban land. Lower urban builtup density radically increased in Kolkata.

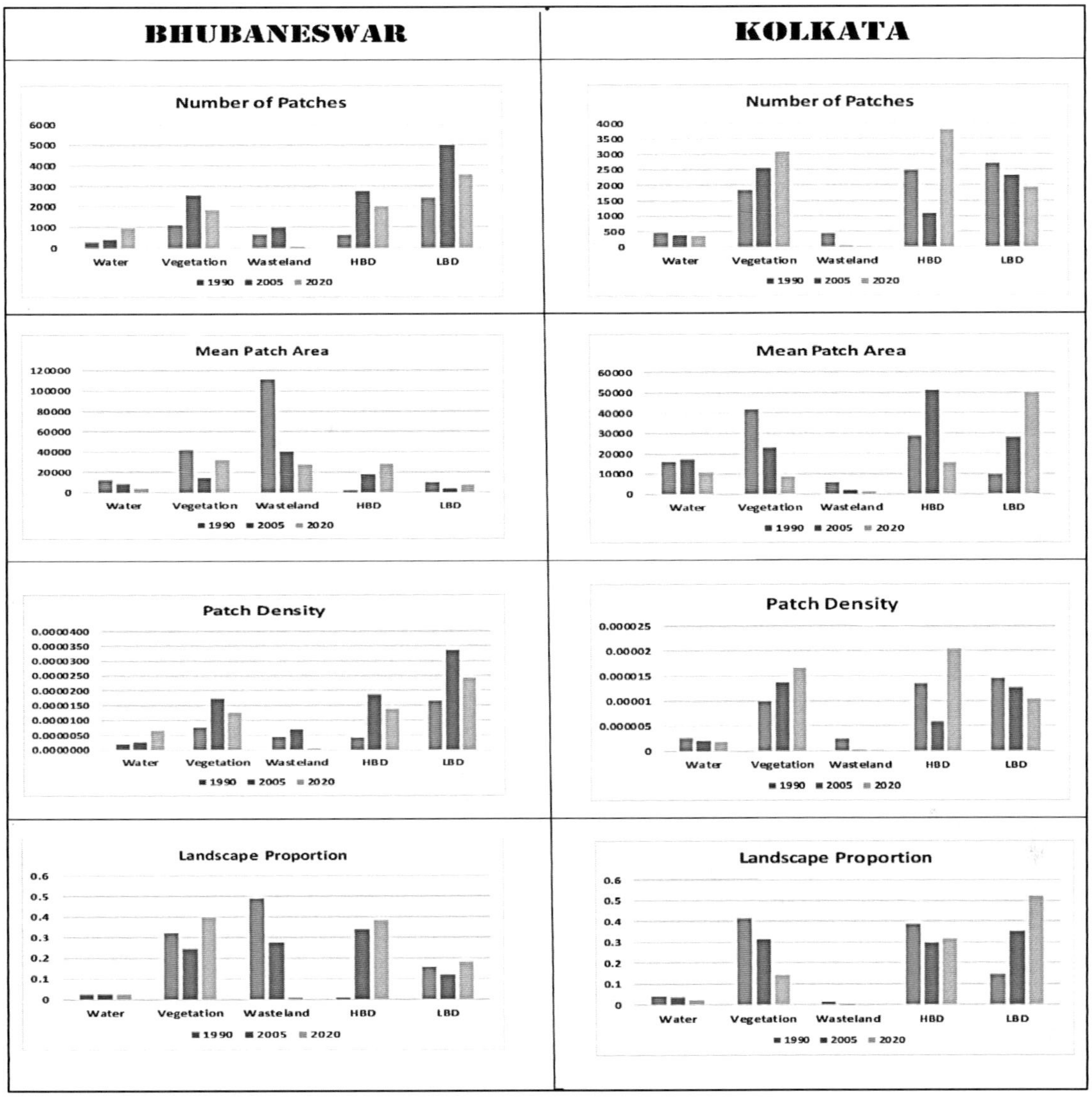


**Figure 5.** Diagram showing values of different Landscape Matrices

In terms of Patch Density (PD) for Bhubaneswar, all increased between 1990 to 2005 and later got checked in next 15 years except for wasteland. In the period of 2005 to 2020 wasteland got decreased density wise.

Kolkata has faced mixed situation with different types of changes in different Landuse classes. High urban density increased in 30 years, because as it much as it was under high urban density that places got dense more than before. But LBD got decreased with time because with time its count got decreased. It signifies that Kolkata has witnessed denser HBD area but less dense in LBD with time.

The landscape ecology gets cleared with the use of Landscape Proportion (LP) matrices, In Bhubaneswar proportion of wasteland got decreased whereas, HBD and LBD increased in last 30 years. In situation of Kolkata, mainly proportion of greenery decreased hugely and LBD changed its proportion drastically (figure 5).

### *5.3. Urbanisation Indices*

Urbanisation indices has been prepared for better understanding of land which are especially engaged to urban services. Urbanisation Index, Urban Expansion Ratio and Urban Green Index will define more accurate results on scenario of urban landscape. These indices are based on HBD, LBD and UGS those are composed for executing the result (figure 6).

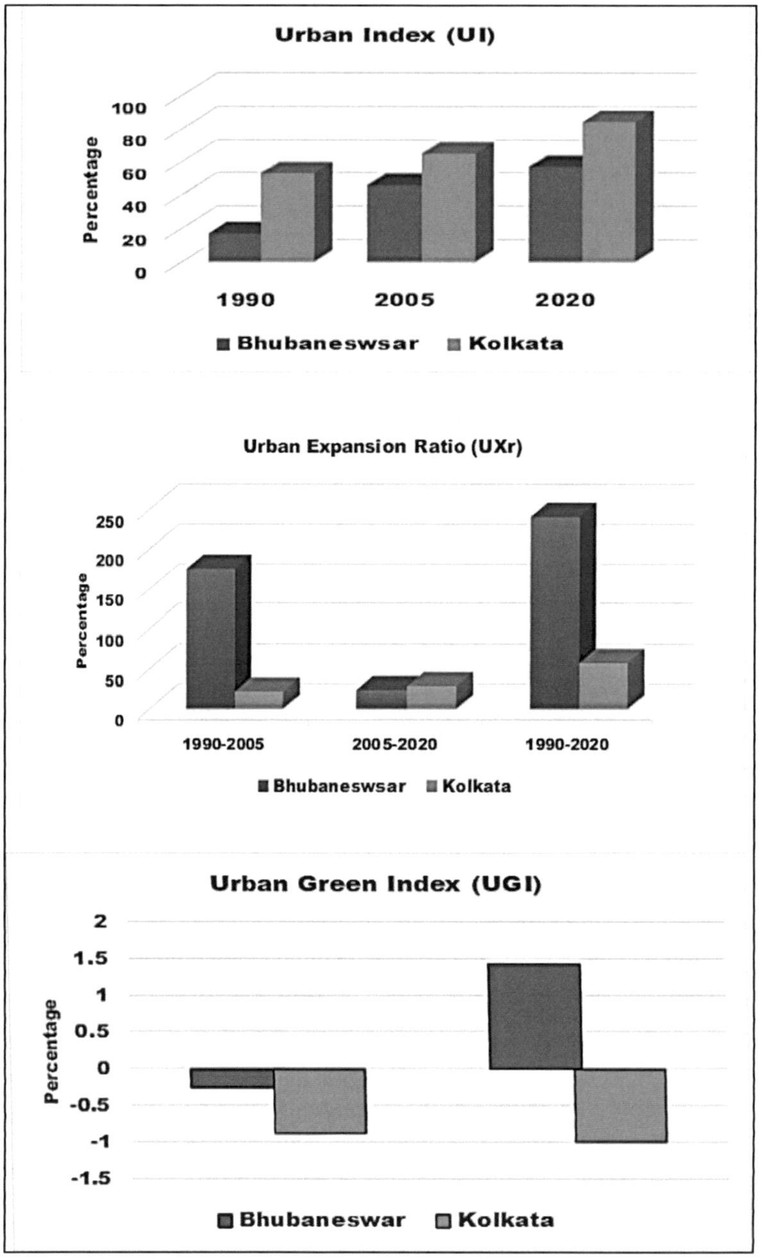


**Figure 6.** Diagram showing Values of different Urbanisation Indices.

*5.3.1. Urban Index (UI).* In 1990, there was 17% of urban land in Bhubaneswar.in 2005 it changed to 46% of urban land and with another 15 years it holds 57% of urban land in total area. Kolkata had 33% of total urban land in 1990, in next fifteen years it increased into 65% of total urban land and in 2020, 84% land are converted as urban land.

*5.3.2. Urban Expansion Ratio ($UX_r$).* Expansion ratio showing, Bhubaneswar has increased its urban area 175% in between 1990 to 2005 and in next fifteen years it has increased to 23%. Bhubaneswar has witnessed more than 240% growth in urban area in last 30 years.it is a very positive sign of development in terms of urbanisation of a city. Kolkata had a long history of establishment, so in 1990 to 2005 it has increased to 22% and in next fifteen years it has acquired 29% of urban land and within the timeframe of thirty years it has a growth of 58% from total land.

*5.3.3. Urban Green Index (UGI).* Percentage urban greenery can be calculated with this index. Bhubaneswar has lost 0.26% of urban green space in between 1990 to 2005 and positively increased 1.44% green space in between 2005 to 2020.Kolkata has lost 0.88% in first fifteen years of timeframe and another 1% in next fifteen years.

## 6. Conclusion

Kolkata and Bhubaneswar are two important cities in eastern India. These cities have been compared using remote sensing and GIS in this paper. Both cities gone through the process of urbanisation in time span of thirty years. Kolkata has a history of urbanisation more than 300 years so here scope of increment is less, where Bhubaneswar is experienced drastic urbanisation in last 30 years, growth rate of Bhubaneswar is about 250% in between 1990 to 2020, whereas Kolkata has witnessed about 60% of urbanisation. Process of urbanisation encroached greenery. It can be seen clearly. there is need of maintain environmental sustainability. Municipal corporation for both cities should be focused on urban greenery. Landscape matrices showing that, land with water substances should be resisted from fragmenting into parts. Water substantial land should be much more concentrated. A positive perspective showing that high urban built up is decreasing with time and low built up density is increasing, that showing the local governing body has maintained to give free spaces to the localities and more planned construction of urban area is being performed. In this paper, by collecting satellite data remote sensing and GIS methods has been performed to monitor the cities remotely and observe the changes in land pattern with passage of time. All the work, on this paper are based on secondary data and job has been done remotely. Here lies the scope of doing such work on every city with frequent time and find out detailed changes and take actions into it. A new era of computerisation has opened an infinite path that leads to more scientific and economic analysis that will be fruitful to maintain sustainability. This paper tried to make a mark on that way.

## References:

[1] Khalid A M, Sharma S and Dubey A K 2021 Concerns of developing countries and the sustainable development goals: case for India *Int. J. Sustain. Dev. World Ecol.* **28** 303–15
[2] Bhatta B 2009 Analysis of urban growth pattern using remote sensing and GIS: A case study of Kolkata, India *Int. J. Remote Sens.* **30** 4733–46
[3] De Almeida C M, Vieira Monteiro A M, Câmara G, Soares-Filho B S, Coutinho Cerqueira G, Lopes Pennachin C and Batty M 2005 GIS and remote sensing as tools for the simulation of urban land-use change *Int. J. Remote Sens.* **26** 759–74
[4] Yunfeng H, Yifang B, Qian Z and Jiyuan L 2009 The trajectory of urbanization process in the Yangtze River Delta during 1990 to 2005 *2009 Jt. Urban Remote Sens. Event*
[5] Molinaro R, Najjar M K, Hammad A W A, Haddad A and Vazquez E 2020 Urban Development Index (UDI): A comparison between the city of Rio de Janeiro and four other global cities *Sustain.* **12**
[6] Jung M 2015 LecoS-A python plugin for automated landscape ecology analysis

[7] Johnson B 2021 The East India Company and its role in ruling India - Historic UK
[8] Kolkata - Capital of British India | Britannica
[9] BSCL :: Bhubaneswar Smart City Limited
[10] KMC 2017 Official Website of Kolkata Municipal Corporation
[11] Census and 2011 2011 Bhubaneswar Municipal Corporation *Census Popul. 2020 data*
[12] EarthExplorer
[13] Haas J, Furberg D and Ban Y 2015 Satellite monitoring of urbanization and environmental impacts—A comparison of Stockholm and Shanghai *Int. J. Appl. Earth Obs. Geoinf.* **38** 138–49

# Spatio-temporal Analysis of Environmental Criticality: Planned Versus Unplanned Urbanization

**R Mallik[1], K Dikkila Bhutia[1], S Roy[1], M Nandi[1], P Dash[2], K Mukherjee[1*]**

[1] Department of Geography, School of Science, Adamas University, Kolkata-700126, India
[2] Department of Geography, Khejuri College, Baratala, West Bengal-721431, India

*Corresponding Author Email: kasturi.mukherjee@adamasuniversity.ac.in

**Abstract.** The economic boom in the Indian cities is causing rapid urban growth, mostly in an unplanned fashion. The growth of urban built-up is primarily taking place by replacing vegetation and other low radiative surfaces, increasing the magnitude and spatial extent of heat concentrations. The resulting phenomenal increase in temperature, known as Urban Heat Island (UHI), raises environmental criticality. Satellite remote sensing provides a breakthrough for monitoring the spatiotemporal variations of UHI by estimating Land Surface Temperature (LST) and surface biophysical parameters. The objective of this work is to compare the changing pattern of LST that resulted from urban growth and associated biophysical characteristics in a planned city (Kalyani city) and an unplanned city (Barasat city) in West Bengal, India from 2005 to 2019. Using Landsat data, the study retrieved summer LST along with the prepared vegetation index (NDVI) and built-up index (NDBI) for the years 2005 and 2019. The Environmental Criticality Index (ECI) was calculated for the periods from LST, NDVI, and NDBI datasets. The long-term (1988- 2019) LST has been derived using the cloud computation technique to analyze the trend. Over the years, though the average LST of Kalyani is relatively high from Barasat, a rapid increase in LST is noticed for Barasat city. Between 2005 – 2019, the rapid unplanned growth in Barasat city has not only increased the LST but also raised the concern for environmental criticality as compared to Kalyani City. The correlation of LST with NDVI and NDBI suggests that urban heating is significantly controlled by the surface characteristics that need to modify through proper planning for urban sustainability. This study may assist planners, administrators, and researchers in decision-making.

**Keywords:** Land Surface Temperature, Urban Planning, Planned and Unplanned cities, Environmental Criticality Index, NDVI

## 1. Introduction

Urban areas can be described as the cradle phase of human civilization and culture as it provides employment opportunities, educational facilities, transport communication facility, and increases the standard of living [1]. Worldwide, especially in developing nations, major advancements in human civilization in terms of facilities, infrastructure, and employment opportunities have been noticed in urban areas. The concentration of economic activities and opportunities in the urban hotspots has created a pull factor for migration. The rapid transformation of places into cities through urbanization and development, resulting in a large amount of population influx is one of the major concerns for various environmental problems [2]. According to the United Nations (UN), the rate of urbanization i.e., the percentage of people living in cities is expected to reach 70% by 2050, although it was 50% in 2008 [3]. However, as a major part of the human population is inhabiting urban areas, it often raises various environmental problems and has a huge impact on climate, both, at local and global scales [4].

The transformation of the human economy from an agriculturally based to an industrial one led to alternate large acres of both green areas and farmland into buildings, roads, factories, etc. The alteration of land use practices has significantly affected the water budget and heat budgets of the cities. As natural processes are interlinked, with the change in surface cover characteristics, surface-atmosphere interactions have altered which led to a change in the urban microclimate. The significant impact of such conversion of green areas and bare soils into impervious concrete and asphalt structures is the gradual warming of cities, known as "Urban Heat Islands" (UHI) [5,6,7]. The structures constructed with dark surfaces (asphalt) have low albedo and are capable to increase thermal energy by absorbing more solar radiation [4]. In addition, the heat created from factories, air conditioners, automobiles, etc. all combinedly contribute to the formation of UHI [9,10,11,12]. The adverse impact of urban micro-climate change can be minimized through an in-depth study of the UHI phenomenon, leading to a proper guide for adopting a suitable urban planning strategy [13,14].

The study of the UHI phenomenon involves an analysis of Land Surface Temperature (LST) focusing on the role of anthropogenic activities in the radiation budget. The spatial pattern of LST is largely controlled by urban sprawls as studied by many researchers [15,16] The type and magnitude of urban sprawling manifest the rate of urban growth, as well as, the nature of urban planning adopted in a city. Hence, the pattern of LST varies for planned and unplanned cities [17]. Environmental problems are more acute in unplanned cities as compared to planned cities. The planned cities consist of proper allocation for green space, open ground, widen streets, gaps between buildings, etc. Moreover, appropriate building code is strictly followed. Therefore, the internal generation of heat by restricting longwave radiation is limited. Unfortunately, the scenario is reversed for unplanned cities. Illegal encroachment, unscientific alteration of surface cover, crisis for space, overpopulation, and crowding, corruption, etc. limits the proper growth of cities. As the number of unplanned cities is higher in developing countries, the problems of environmental pollution and the UHI phenomenon are critical.

The remote sensing data obtained from space-borne satellites have a major application in urban climate studies [18]. Freely available remotely sensed data from multiple sensors and GIS technology facilitate researchers a wide opportunity to study the UHI phenomenon at various scales in contrast to the classical method that depends on intensive field data collections [19,20]. Remote sensing satellites enable the estimation of LST which has a variety of uses in environmental and ecological studies and spatial planning. Many researchers have exploited the potential of satellite data to investigate urban thermal environments with different specific purposes like examining the spatiotemporal patterns of LST, the relation of LST with surface characteristics, building structure, and urban sprawl, forecasting of UHI under different urbanization scenarios, etc. [21-27]. The automatic data-capturing capabilities of sensors with high spatiotemporal coverage have overridden the limitations of manual data collection for monitoring urban microclimate. Several advantages of satellite remote sensing have attracted researchers for multi-fold applications in environmental issues.

Remote sensing sensors like Landsat Thematic Mapper (TM), Enhanced Thematic Mapper Plus (ETM+), Operational Land Imager (OLI), etc. are capable to capture LST information using thermal infrared (TIR) bands [28]. In addition, multiple TIR bands of Moderate-resolution Imaging Spectroradiometer (MODIS) sensors provide a good opportunity to capture the UHI phenomenon at various spatial scales [29,30,31,32,33,34,35,36]. The radiant energy emitted directly from the land surface is captured by the sensor in Thermal Infrared (TIR) spectral region that is often used to derive LST [37,38,39,40]. The difference in LST between the urban built-up and non-built-up surface shows the intensity of the UHI effect [41].

Researchers have been interested in understanding the relationship between land surface temperature (LST) and surface characteristics. In order to do so, various studies have been conducted that use different approaches to analyze this relationship. One approach that has been used in several studies is to examine the relationship between LST and vegetation cover, using vegetation indices such as the normalized difference vegetation index (NDVI). These studies have aimed to understand how LST is affected by the presence and density of vegetation on the surface [42,29,43,14,44]. Another study investigated the impact of different land cover features on land surface temperature (LST) in the Indian cities of Panaji and Tumkur. To do so, the study utilized several indexed-based surface parameters, including the soil-adjusted vegetation index (SAVI), enhanced built-up and bareness index (EBBI), and modified normalized difference water index (MNDWI). The purpose of the study was to understand how LST is influenced by different land cover features in the Indian context. [45]. A study conducted in Madhya Pradesh, India has analyzed the relationship between land surface temperature (LST) and two remote sensing indices: normalized difference vegetation index (NDVI) and normalized difference built-up index (NDBI). The study used Landsat-8 data to assess the relationship between LST and these two indices in the region. [46]. Researchers have also investigated the pattern of thermal environment and UHI phenomena of the Sriniketan-Shantiniketan Planning area, West Bengal [46]. A study has been conducted to examine the impact of changes in land use and land cover (LULC) on land surface temperature (LST) in the Barrackpore subdivision. The study specifically focused on how changes in LULC have affected LST in this region [47]. Although urban planning has a vast array of applications, there are few studies that examine the effects of urban development on LST in planned and unplanned cities.

The current study compares the spatiotemporal pattern of LST in planned and unplanned cities with a case study of Kalyani and Barasat cities, respectively, and examines how it relates to surface characteristics and environmental criticality. The study has calculated LST, NDVI, and NDBI, from open-source satellite data. Between 2005 and 2019, the spatial distribution and association of LST and surface parameters (NDVI and NDBI) were investigated. The environmental Criticality Index (ECI) was calculated based on LST and NDBI and correlation was performed. The study may be beneficial to policymakers, administrators, and urban planners to take suitable strategies for the further development of the cities.

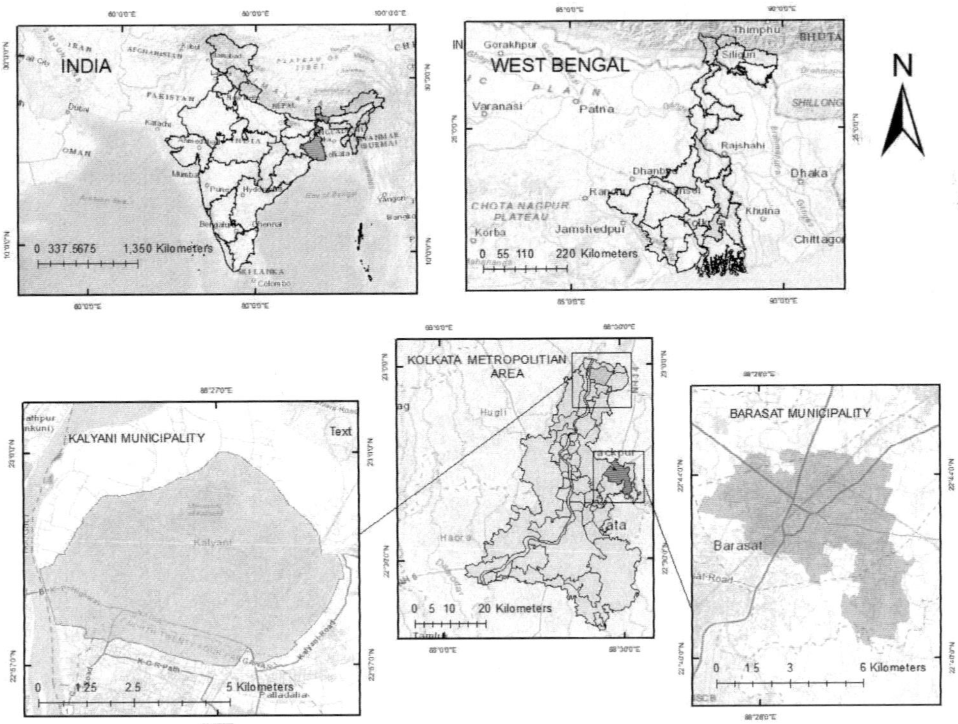


**Figure 1.** The geographical location of Kalyani and Barasat municipalities

## 2. Study Area

The Barasat and Kalyani cities which are located within the periphery of the Kolkata Metropolitan Area (figure 1), are selected in the present study to make a comparison of the effect of planned and unplanned urban growth on LST. Barasat city is the district headquarter of North 24 Parganas district, West Bengal. The Barasat Municipality is bordered by the Bidhannagar subdivision in the south, the Nadia district in the north, Gaighata, Baduria, Bashirhat, and Haroa blocks in the east, and the Barrackpore subdivision in the west. The municipality is located between 22°40′58″ to 22°44′04″ N latitudes and 88°27′00″ to 88°31′30″ E longitudes, covering an area of 31.6 sq. km. The city lies in the Gangetic delta at an average elevation of 11 meters, on almost flat terrain. The region enjoys a tropical climate with an annual rainfall of 1,579 mm, experiencing dry winter and a hot-humid summer. The maximum temperature reaches in May at about 40°C and a minimum temperature of 10°C in January.

The administrative area of the city is divided into 32 wards at present, increasing from 15 wards in 1986. The growth of the city has significantly increased after it was declared the administrative headquarters of the North 24 Parganas district in 1986. As per the Census of India (2011), 2,83,443 people inhibit in Barasat city. However, the population has increased gradually, especially after the Liberation War of Bangladesh in 1971 [48]. The city experienced a huge influx of emigrant population as it is well connected by roads and trains with the areas adjacent to the Bangladesh border like Basirhat, Hasnabad, etc. Noticeably, the population increased by 125% in 2011 from 1991[49]. This unplanned urbanization has led to the loss of natural landscapes and habitats, as natural areas are developed for residential, commercial, and industrial purposes.

Kalyani city is situated 38 Km far north of Barasat in the Nadia district, West Bengal. It is created through planning in 1950 by the Former Chief Minister Dr. Bidhan Chandra Roy. Earlier, it was an American WW2 Military Base, used for Hump Operations. The Kalyani municipality, established in 1995, covers an area of 29.24 sq. km, geographically located between 22.9747° N and 88.4337° E. It lies on the Eastern bank of the river Hooghly, over the Bengal Basin with an average elevation of 11m, consisting of fertile alluvial soil. Meteorologically, Kalyani also experiences a similar climate to Barasat with an annual rainfall of 1250mm. Temperatures in Kalyani reach highs in April-May, averaging to 35-40°C during summer and 10°C-18°C in the winter. As Kalyani is a planned city, most of the important administrative buildings are situated in the city center area. The rapid growth of residential apartments is seen with plots starting from 2-6m away from roads. It mostly consists of pucca buildings with kachha in the slum belts [50]. As per the Census of India (2011), Kalyani has a population of 100,620. Noticeably, though the geographical area of Barasat is greater than Kalyani by 2 sq. km, the size of the population in Barasat is more than double as compared to Kalyani. There has been a 22% rise in population from 2001-2011 in Kalyani. Kalyani is directly connected by trains to Sealdah, Ranaghat, and Krishnagar and by roads to Kolkata and Bongaon.

## 3. Data and Methods

This study utilized freely available satellite images from three different sensors: Landsat 5 (TM), Landsat 7 (ETM+), and Landsat 8 (OLI/TIRS). These images were from path 138 and row 44. The purpose of using these images was to gather data for the study. The satellite images have been collected and processed in two ways. Firstly, the Landsat 5 images of 2005 and Landsat 8 images of 2019 were directly downloaded from the United States Geological Survey (USGS) website and processed in image processing software to derive surface biophysical characteristics and spatial distribution LST for summer, which further used to calculate environmental criticality. Secondly, the data archive of Landsat 5, 7, and 8 (1988 to 2019) has been accessed and processed using Google Earth Engine to derive the long-term trends in the average summer LST of the cities. The broad methodology followed in this study is illustrated in figure 2 and described in the following sub-sections.

### 3.1 Estimation of surface biophysical parameters

*3.1.1. Extraction of Built-up Area.* The urban growth from the year 2005 to 2019 has been identified by estimating changes in the built-up area of the cities.

IOP Conf. Series: Earth and Environmental Science 1164 (2023) 012014 doi:10.1088/1755-1315/1164/1/012014

*3.1.2.* The built-up areas were derived from the satellite images of 2005 and 2019 using a supervised classification algorithm. As only the built-up area is concerned for this purpose, the training samples for urban built-ups were chosen carefully for the training and classification of the images. The output land use map was converted into binary classes assigning value '1' to the built-up class and '0' to the other classes. The spatiotemporal change and corresponding growth in the built-up area between 2005 and 2019 were analyzed by Barasat and Kalyani cites separately.

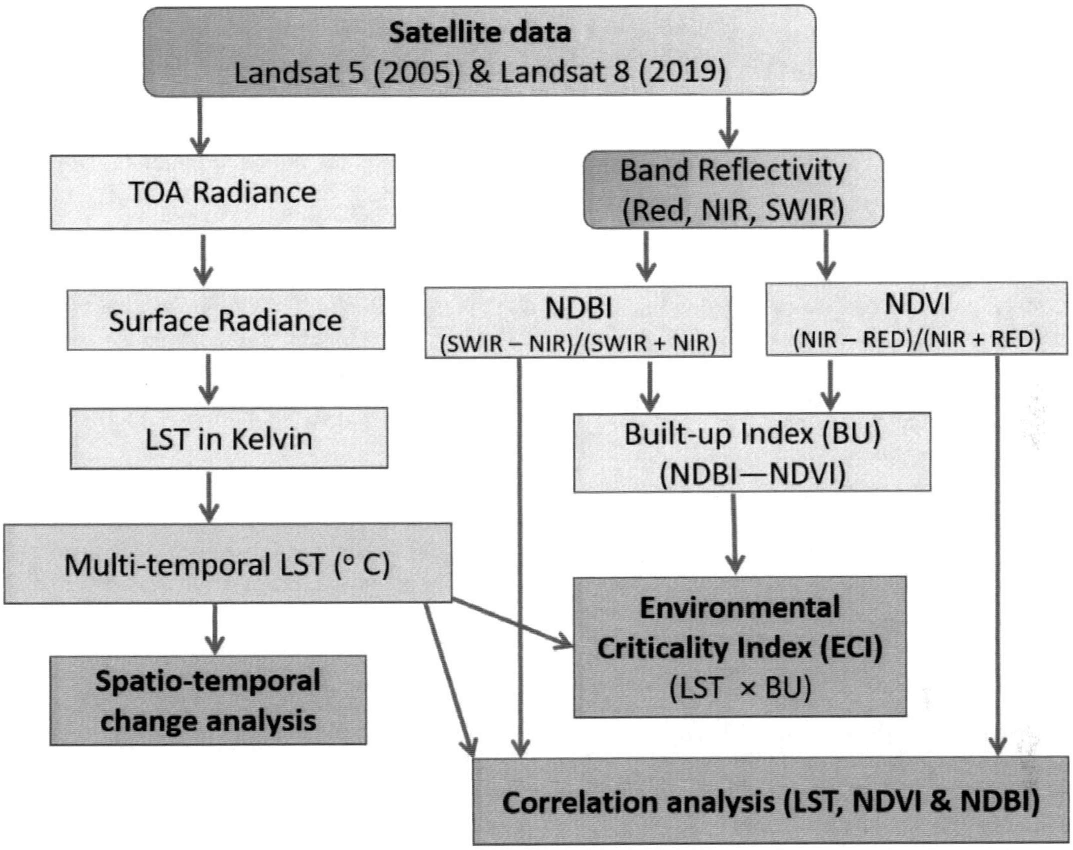


**Figure 2.** Methodological flowchart

*3.1.3. Surface vegetation cover (NDVI).* In urban areas, the transformation of natural surface cover, especially vegetation cover, into built-up cover may be responsible for the UHI phenomenon. Therefore, in this study, two surface characteristics parameters i.e., NDVI and NDBI have been prepared from satellite data. The vegetation indices derived from readily available satellite imagery are often used as important surface parameters in many environmental studies. As the reflectance of vegetation maximum in the Near Infrared (NIR) spectral region as compared to the visible spectrum, is expressed as follows [51]:

$$NDVI = (NIR - Red) / (NIR + Red) \quad (1)$$

The NDVI was calculated from Landsat 5 data using band 4 (NIR) and band 3 (Red), while band 5 (NIR) and band 4 (Red) were used for Landsat-8 images. Within the value range of NDVI between -1 and +1, the value close to +1 exhibits dense vegetation cover, while values in minus show water bodies, and values close to 0 show bare surface.

*3.1.4. Built-Up Index (NDBI).* NDBI is popularly used as a proxy for urban settlement. Over the urban areas, the reflectance is usually high in Short-Wave Infrared (SWIR) band and very low in the NIR band. Therefore, SWIR and NIR bands are used to calculate NDBI using the following equation:

$$NDBI = (SWIR - NIR) / (SWIR + NIR) \tag{2}$$

For NDBI calculation, band 5 (SWIR) and band 4 (NIR) of Landsat 5 data, and band 6 (SWIR) and band 5 (NIR) of Landsat-8 data were used in this study. The value of NDBI varies between -1 and +1, where a higher NDBI value represents the built-up area and a low value represents vegetation and water. However, the built-up area was identified from NDBI using a threshold derived after many iterations.

### *3.2. Estimation of Land Surface Temperature (LST)*
The estimation of LST from multispectral data involves several steps. First, the conversion of DN values to at-sensor spectral radiance [52,53] and then the conversion of the at-sensor spectral radiance to at-sensor brightness temperature by using the inverse of the Planck Function. It is supposed that the earth's surface acts as a black body that has an emissivity value of '1' [54]. Finally, the brightness or radiant temperature is changed into the kinetic temperature or surface temperature. The basic algorithm is not described in this paper, but further information for retrieving LST from Landsat TM, ETM+, and OLI/TIRS can be obtained from the author's earlier work [55,56].

### *3.3. LST Retrieval through Cloud Computation*
Several studies have derived LST from Landsat data and proposed algorithms to retrieve LST from Landsat series [39,56,57,59]. Such methods may not be suitable to handle a large number of data sets as it requires high computational efficiency and time-taking. [58] proposed an approach to retrieve LST in Google Earth Engine (GEE) for Landsat data series using Statistical Mono-Window (SMW). GEE allows users to process big sizes of remote sensing data online under the limited computational resources of the community based on online data archives and programming scripts like Java. In this study, a freely available Java script for GEE prepared by [58] has been used to retrieve LST data of the summer season from Landsat 5, 7, and 8 for the period 1988 – 2019. Based on the suitability of the availability under less cloud cover conditions, each scene from either Landsat 5, 7, or 8 has been processed in GEE for the study domain. Then, the calculated LST for each scene was averaged for the study domains i.e., Barasat and Kalyani city boundaries. Finally, the average annual LST for the summer period was calculated by computing the average of each subset average for the months of April and May throughout the years from 1988 to 2019. This data was used to analyze the long-term trends in the LST of Barasat and Kalyani, separately. The detailed methodology of LST retrieval using the SMW algorithm in GEE is discussed in [58].

### *3.4. Correlation among LST, NDVI, and NDBI*
The influence of surface biophysical parameters on LST has been investigated by analyzing the correlations of LST with NDVI and NDBI for selected samples across the study site at various buffer distances from the city core. Firstly, buffer areas were created at 500m, 1000m, 1500m, 2000m, 2500m, and 3000m radius around the city center. The overlapping area between two consecutive buffer zones was removed. Then, a minimum of 20 random sample points was generated for each segment in GIS the platform. Next, the values of NDVI, NDBI, and, LST for each sample point were extracted for the year 2019. Finally, the relationships were investigated through scatterplot and Pearson's correlation (also called Pearson's R). 'R' is a measure of the closeness of association of the points in a scatterplot to a linear regression line based on those points, calculated as follows:

$$R = \frac{n(\Sigma xy) - (\Sigma x)(\Sigma y)}{\sqrt{[n\Sigma x^2 - (\Sigma x)^2] - [n\Sigma y^2 - (\Sigma y)^2]}} \tag{3}$$

### *3.5. Environmental Criticality Index (ECI)*

ECI is a prerequisite that helps in the identification of regions derogating from ambient environmental conditions [14] calculated ECI by dividing LST with NDVI, both parameters are stretched at a 1-255 value range. The present study followed the approach of [60] to calculate ECI by multiplying the Built-Up (BU) index with LST. The BU index was calculated by subtracting NDVI from NDBI. The formulas are as follows:

$$BU = NDBI - NDVI \quad (4)$$

$$ECI_{(LST\text{-}BU)} = LST \times BU \quad (5)$$

## 4. Results and Discussion

### *4.1. Change in the built-up area*

The spatial distribution of the built-up area has been derived to capture the spatial growth of urban settlements between 2005-2019 in Barasat and Kalyani (figure 3). As compared to 2005, Barasat city experienced rapid urban growth at the core in 2019 growth; while in Kalyani, the growth of settlement was observed surrounding the existing settlement of 2005, especially in the southwest of the city. The percentage of built-up area to the total area in Barasat city was compatibly less than in Kalyani in 2005, though the statistics get reversed in 2019 (figure 4).

Between the periods (2005 to 2019), the built-up area in Kalyani increased by 14.36% of the total area from 47.41% (2005) to 54.97% (2019). Noticeably, Barasat City has seen a total built-up area growth of 20.58% from 42.16% (2005) to 65.56% (2019). The analysis showed that Barasat has gained more built-up area than Kalyani throughout the years 2005-2019. The rapid increase of built-up area in Barasat is mostly attributed due to unplanned urbanization and illegal migrations as Barasat is situated close to the Bangladesh border.

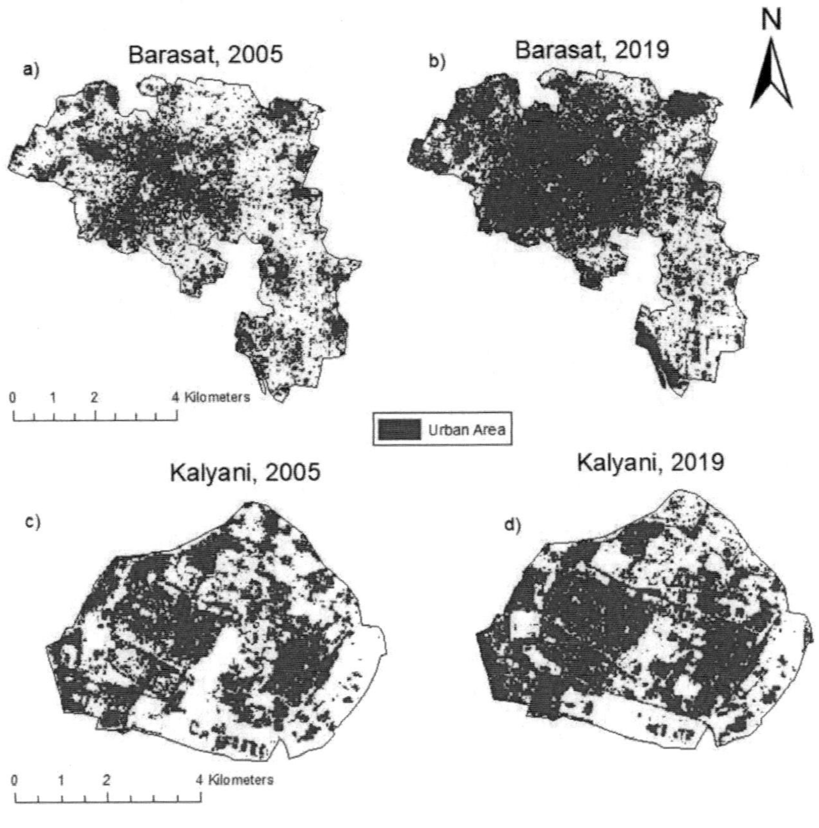


**Figure 3.** Map displaying the distribution of built-up (Kalyani, Barasat)

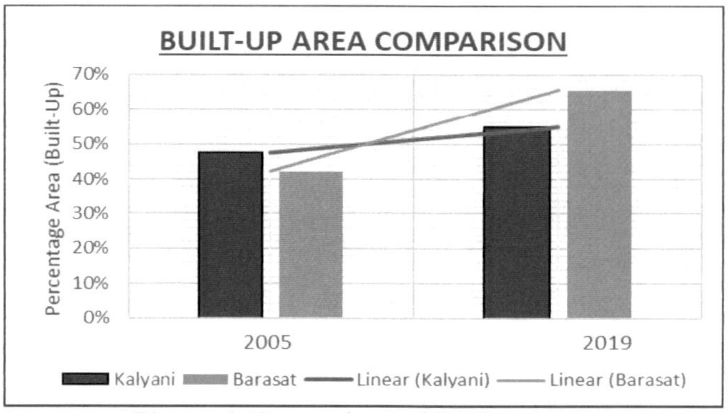


**Figure 4.** Comparison of Built-up area.

### *4.2. Analysis of surface biophysical parameters*

The surface biophysical parameters of the cities have been analyzed by exploring the pattern of NDVI and NDBI. The spatial distribution of NDVI and NDBI have been illustrated in figure 5 and figure 6, respectively, and summary statistics are in table 1. In Barasat city, NDVI ranges from -0.09 to 0.52 with a maximum value in 2005. The low NDVI value is observed at the city core, and the extent of low NDVI has increased in 2019 as compared to 2005 (figure 5a, 5b). The average NDVI value of Kalyani city is more than 0.3 which is slightly higher than Barasat. A major decrease in NDVI is noticed in the southwestern part of Kalyani (figure 5c, 5d). The expansion of areas with low NDVI values in 2019 is attributed due to urban expansion in the cities. The growth of cities that mostly happens haphazardly destroys urban green belts. Moreover, the cities' lowlands, wetlands, and bare grounds are often converted into built-up without considering their significance to keeping the balance of the natural environment.

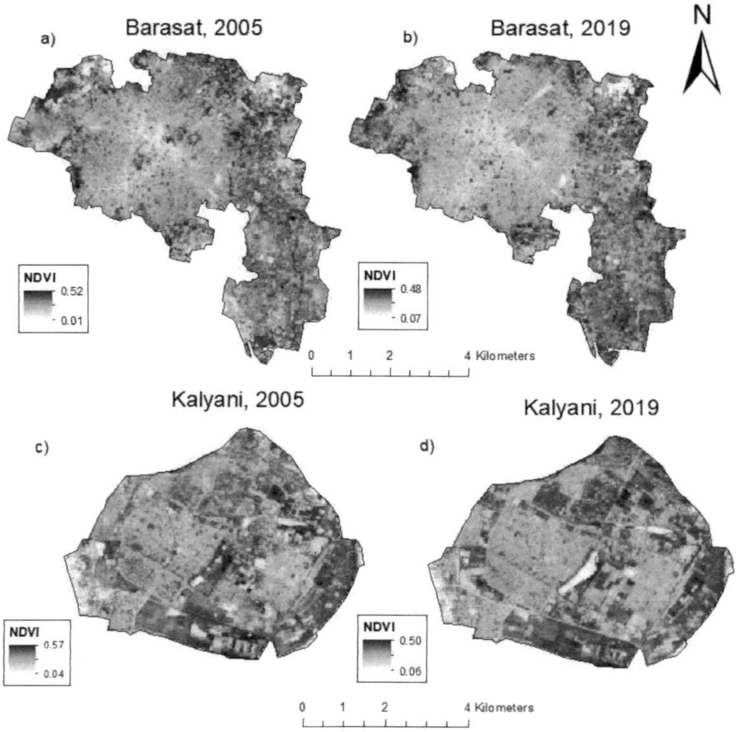


**Figure 5.** NDVI distribution over 2005,2019 (Barasat, Kalyani)

**Table 1.** Summary statistics of Barasat and Kalyani for the years 2005 and 2019

| | | Barasat | | | | Kalyani | | | |
|---|---|---|---|---|---|---|---|---|---|
| | Year | MAX | MIN | AVG | STD | MAX | MIN | AVG | STD |
| NDVI | 2005 | 0.52 | -0.09 | 0.27 | 0.08 | 0.57 | -0.18 | 0.32 | 0.09 |
| | 2019 | 0.48 | 0.2 | 0.27 | 0.07 | 0.5 | -0.07 | 0.3 | 0.08 |
| NDBI | 2005 | 0.33 | -0.23 | 0.04 | 0.06 | 0.32 | -0.22 | 0.04 | 0.06 |
| | 2019 | 0.27 | -0.3 | -0.08 | 0.05 | 0.23 | -0.32 | -0.14 | 0.05 |

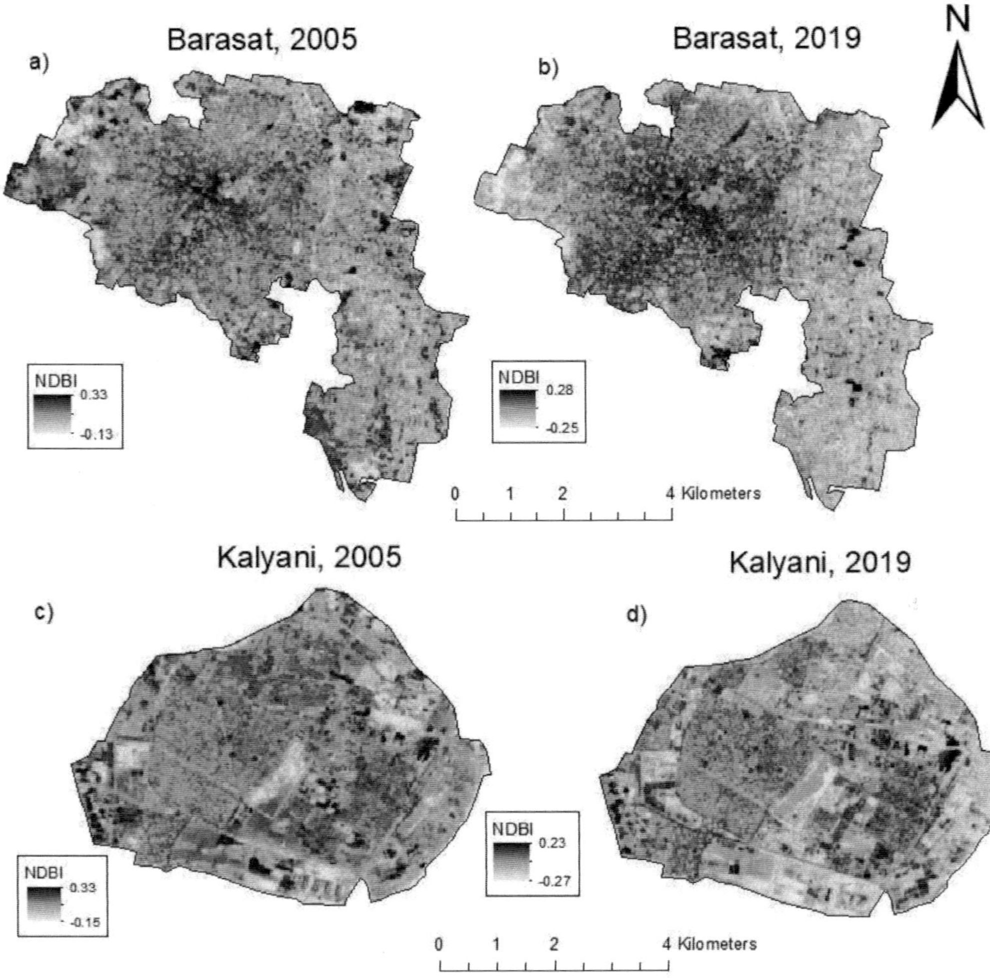


**Figure 6.** NDBI distribution of Barasat and Kalyani (2005, 2019)

The spatiotemporal change in NDBI as illustrated in figure 6 shows a close match with the pattern of NDVI distribution, but in an inverse way (figure 5). Areas with high built-up have extremely low NDVI values. In Barasat, the drastic increase of NDBI and decrease of NDVI in the city core from 2005 to 2019 is due to an increase in impervious surfaces i.e., residential buildings, concrete pavements, commercial zones, etc., reflecting the growth of the city in an unplanned fashion.

IOP Conf. Series: Earth and Environmental Science 1164 (2023) 012014 doi:10.1088/1755-1315/1164/1/012014

The distribution of settlement in the center of Kalyani city is not compact as compared to Barasat as evidenced by figures 6c, and 6d. Being a planned city, a large number of agricultural fields, vegetation cover, open space, playgrounds, lakes, etc. are located in Kalyani city. Therefore, it is hard to find a compact city center, rather settlements are more or less evenly distributed and separated from each other by green space in several pockets or colonies across the city.

*4.3. Spatio-temporal change in LST*

In this study, the LST has been derived and analyzed for the summer. The long-term pattern of summer LST has been estimated for the period 1988-2019 through the google earth engine. The average yearly summer LST calculated for the Barasat and Kalyani is shown in figure 7. For both cities, the overall trend of LST from 1988 to 2019 is showing an increase in temperature, though the trend is remarkably progressive from 1997 to 2019. Over the years, average LST fluctuates for both cities, though the magnitude of LST in the peaks is high for Kalyani as compared to Barasat (figure 7). Interestingly, the difference in average LST between the cities is greater before 2005 and closely matches afterward. The trend line shows a steady rise in temperatures for Kalyani, whereas Barasat showed a comparatively steep rise in LST. The rapid urbanization in Barasat city after 2005 may be attributed to a sharp increase in LST in comparison to the Kalyani.

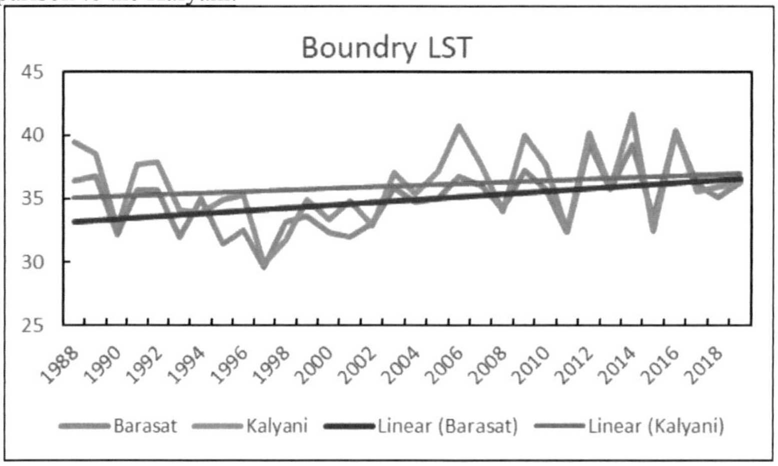


**Figure 7.** Long-term changes in average summer LST from 1988 to 2019 in Barasat and Kalyani

In order to make a comparative analysis in the spatial patterns of the UHI that may be contributed by the expansion of urban growth, the study calculated LST from satellite data of the years 2005 and 2019. The raster layers of calculated LST have been reclassified by the mean-standard deviation method that has been applied by many researchers [61,62]. In this method, the LST regime has been classified into five LST zones based on mean and standard deviation as given in table 2. The spatial distribution of the various LST zones in Barasat and Kalyani for 2005 and 2019 is shown in figure 8. The corresponding area of each LST zone for these two years is summarized in table 3.

**Table 2.** The classification scheme used in LST (Yangzi et al., 2019; Chu et al., 2020)

| Zone | Range | Description |
|---|---|---|
| 1 | T<Mean - 1 Standard deviation (SD) | Low-Temperature Zone |
| 2 | Mean - 1 SD <T< Mean-0.5 SD | Secondary Moderate Temperature Zone |
| 3 | Mean - 0.5 SD <T< Mean +0.5 SD | Moderate Temperature Zone |
| 4 | Mean + 0.5 SD <T < Mean + 1 SD | Secondary High-Temperature Zone |
| 5 | T>Mean + 1 Standard deviation | High-Temperature Zone |

**Table 3.** Values derived for each of the 5 zones according to classification scheme

| Zone | Description | Barsat | | Kalyani | |
|---|---|---|---|---|---|
| | | 2005 | 2019 | 2005 | 2019 |
| 1 | Low-Temperature Zone | <22.346 | <28.514 | <23.825 | <29.321 |
| 2 | Secondary Moderate Temperature Zone | 22.346 to 22.659 | 28.514 to 28.985 | 23.039 to 23.432 | 29.321 to 29.666 |
| 3 | Moderate Temperature Zone | 22.659 to 23.285 | 28.985 to 29.928 | 23.432 to 248 | 29.666 to 30.358 |
| 4 | Secondary High Temperature Zone | 23.285 to 23.598 | 29.928 to 30.4 | 24.218 to 24.611 | 30.358 to 30.703 |
| 5 | High Temperature Zone | >23.598 | >30.4 | >24.611 | >30.703 |

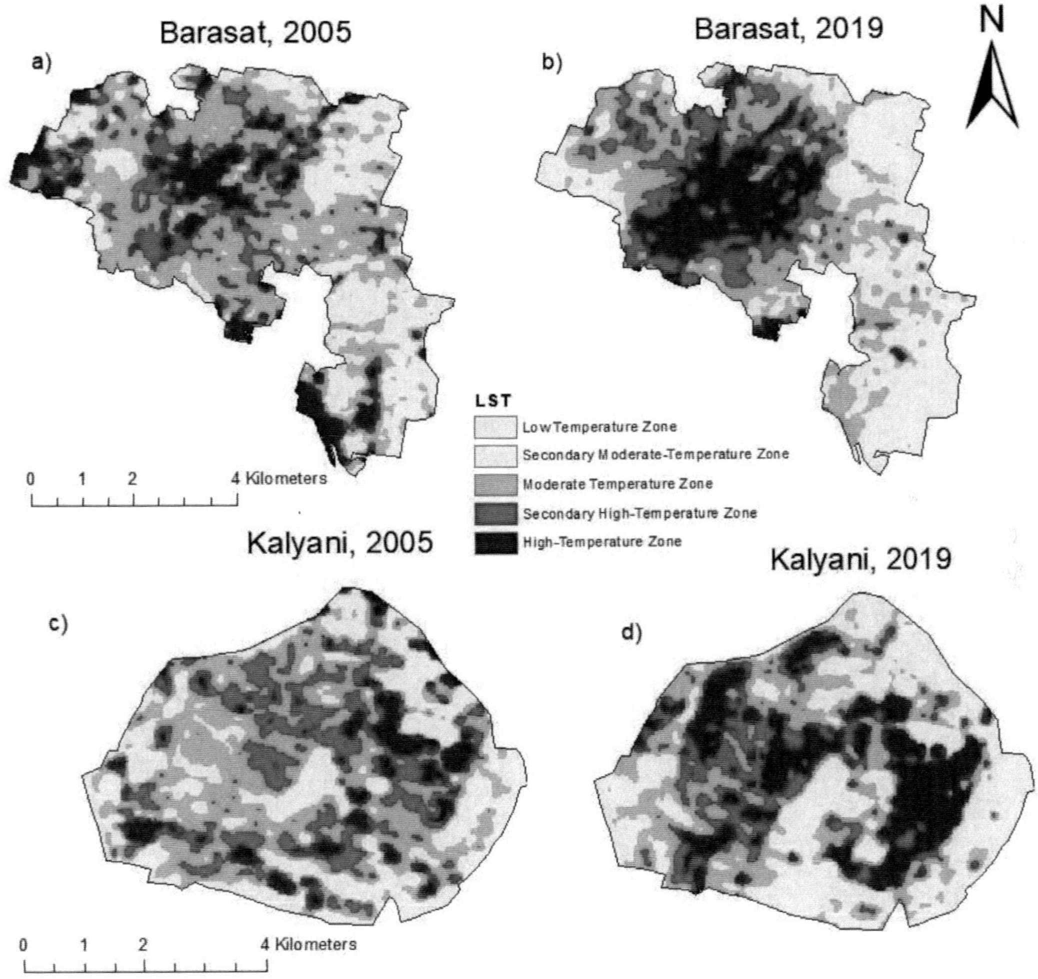


**Figure 8.** Spatial pattern of LST zones of Barasat in 2005 (a), 2019 (b) and of Kalyani in 2005 (c), 2019 (d)

In Barasat, the city core reflects a high-temperature zone that increased in spatial extent from 2005 to 2019 (figure 8a, 8b). From the earlier analysis, it is found that the core exhibit lower NDVI values and high NDBI values (figure 5 and figure 6). Additionally, the growth of the high LST zone in the core closely resembles the growth of urban built-up from 2005 to 2019 (figure 3). Overall, the extent of the high-temperature zone has increased by 8% from 14.13% in 2005 to 22.83% in 2019 table 3. Noticeably, 41.43% of the area in 2005 lies in the moderate LST zone which is close to the mean LST. However, the scenario has changed in 2019 indicating that the regional variation of LST has increased as compared to 2005 as extensive urban growth has occurred. In the eastern part of the city, the low LST reflects its association with high NDVI and low NDBI values. The existence of vegetation, open space, water bodies, and cropland in this part may be attributed to low LST. Thus, this analysis reflects the role of surface biophysical characteristics to control surface temperature.

Kalyani city also showed a significant increase in the extent of the high LST zone in 2019 (22.83% area) as compared to 2005 (02.65% area) (figure 8c, 8d). With the increase in compactness and growth of settlements in two major pockets of the city have attributed to such change in LST. From table 3 it is evident that the extent of low and high LST zones has increased in 2019 as compared to 2005. Such changes have caused a reduction in the spatial coverage of secondary moderate, moderate, and secondary high-temperature zones. Likely to Barasat, Kalyani also reflects an increase in spatial heterogeneity in LST, but comparatively in less magnitude. However, this analysis shows that with the increase in population pressure and consequent built-up growth, the planning strategy of Kalyani city become in effective controlling urban LST.

### *4.4. Relationship of LST with NDVI and NDBI*

The association between surface biophysical parameters and the surface temperature has been identified through correlation analysis of LST with NDVI and NDBI for the year 2019. The scatterplot between LST and NDVI for Barasat and Kalyani (figure 9) shows the inverse relations i.e., an increase in LST with a decrease in NDVI value. The correlation coefficient value for Barasat is 0.77 and for Kalyani is 0.71 which indicates a good relationship between the variables. This finding is very indicative to support previous analysis that surface biophysical characteristics significantly control the changes in LST. The unplanned growth of the urban area by removing green spaces is largely responsible for the temperature rise in Barasat city. Likely to Barasat, the reduction of vegetation cover in Kalyani from 2005 to 2019 may result in the rise of surface temperatures (see LST range in table 2).

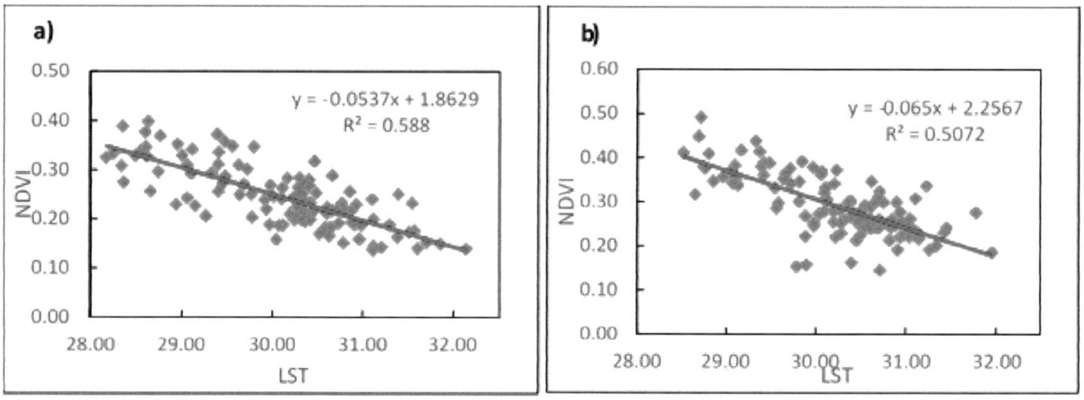


**Figure 9.** Correlation between LST and NDVI in (a) Barasat and (b) Kalyani for the year 2019

The correlation between the LST and NDBI of Barasat and Kalyani in the year 2019 is represented by the scatter plot (figure 10). The figure clearly shows a positive correlation with a good correlation coefficient of 0.76 for Barasat and 0.74 for Kalyani. The correlation analysis confirms that the surface the temperature will increase with the increase of compactness and growth of built-up area unless proper planning is adopted mandating space for vegetation and a wide gap between settlements. The analysis signifies that the increase of built-up compactness, especially in unplanned cities like Barasat, the urban growth may lead to amplifying the trapping of heat and aggravating the UHI phenomenon.

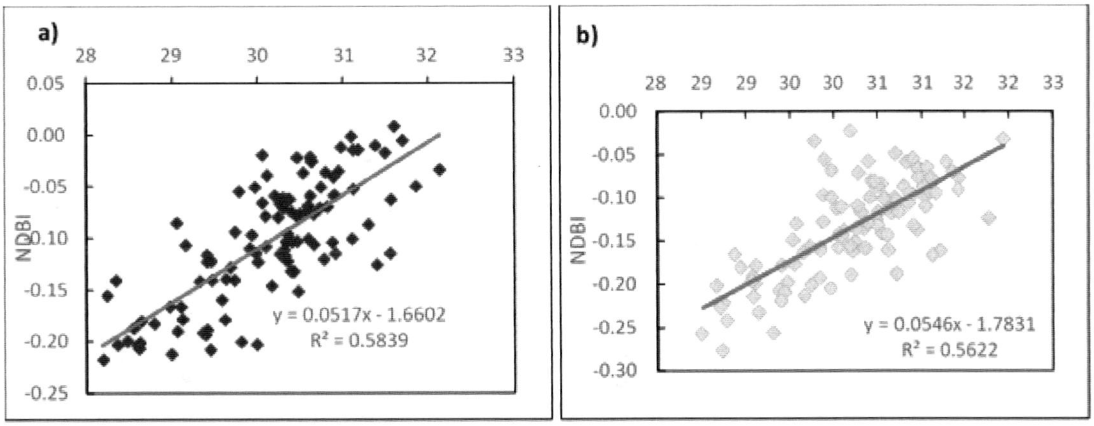


**Figure 10.** Correlation between LST and NDBI of 2019 in Barasat (a) and Kalyani (b)

### *4.5. Environmental Criticality*

In this study, the environmental criticality of the cities has been calculated as ECI using the built-up index and LST of the years 2005 and 2019. The spatial distribution and changes in ECI between 2005 and 2019 of Barasat (figure 11a, 11b) and Kalyani (figure 11c, 11d) cities indicate that environmental criticality has increased over the periods. Likely to LST, the major increase in ECI in Barasat city is noticed in the city core from 2005 to 2019. The spatial extent of the high and very high ECI zones in 2019 has increased as compared to 2005. Figure 11a and figure 11b shows similarity in results with the pattern of NDBI. It is found that the pixels with high ECI values also match with a high NDBI value and low NDVI value. However, in the southern part of Barasat, the increase in the extent of the low criticality zone in 2019 from 2005 is due to an increase in the deviation of LST from urban to non-urban areas (like vegetation, cropland, etc.) in 2019. Such similarity was also noticed in the southern and northern parts of Kalyani city in 2019 (figure 11d).

The spatial pattern of ECI in Kalyani for the years 2005 (figure 11c) and 2019 (figure 11d) shows a significant increase in the spatial extent of the very high ECI zone between the periods. This may be attributed due to the increase of urban sprawl in the city in 2019 as compared to 2015. However, the change in the extent of high and very high ECI zones between the periods is not significant in Kalyani unlike in Barasat. Such differences between the cities are might be due to the difference in the adaptation of urban planning. Though Kalyani was developed as a planning city, the efficiency of planning is seeming as less impactful in controlling urban microclimate (specific temperature) and environmental criticality under the current urbanization scenario.

## 5. Conclusion

This paper tried to make a comparative study between the planned and unplanned cities on spatiotemporal change in LST in relation to surface biophysical characteristics (NDVI and NDBI), and environmental criticality derived from the satellite images. The estimated summer LST for the period of 1988 – 2019 has shown an increasing trend for, both, Barasat and Kalyani. Overall, the magnitude of LST derived from satellite images is higher in Kalyani than in Barasat. Noticeably, the percentage of built-up area in Kalyani was more as compared to Barasat before 2005. Afterward, rapid urban growth in Barasat lead to an increase in a built-up area and resulted in the rise of LST. The study found that from 2005 to 2019, rapid and unplanned urban development occurred in Barasat at the city core replacing vegetation and open spaces. The spread of urban sprawling is also evidenced by the change in NDVI and NDBI analyzed for these periods. On the other hand, Kalyani also showed a growth in the built-up area and the corresponding change in NDVI and NDBI. However, the extent of urban growth is comparatively less in Kalyani in comparison to Barasat.

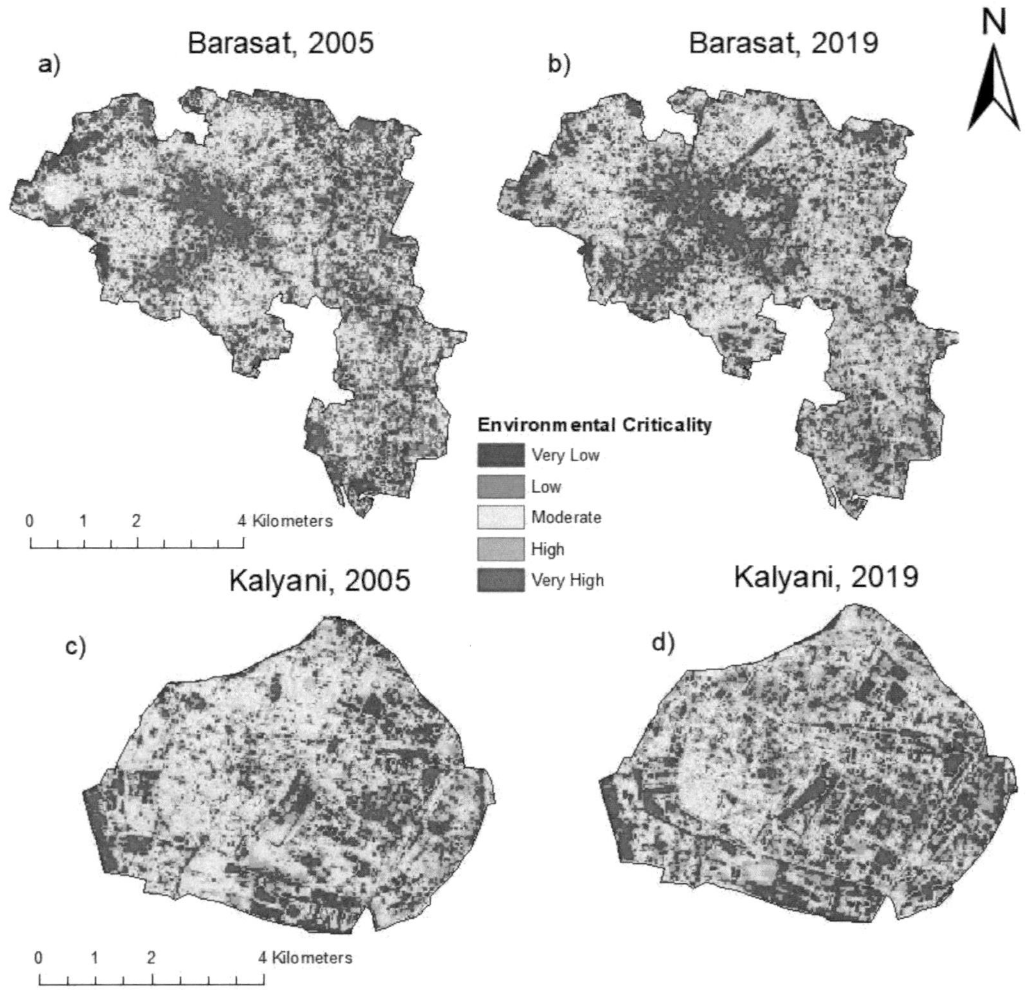


**Figure 11.** Spatial pattern and changes in environmental criticality of Barasat in 2005 (a), 2019 (b), and of Kalyani in 2005 (c), 2019 (d)

The correlation analysis of LST with surface characteristics showed that LST has a good negative relation with vegetation cover (NDVI) and good positive relation with built-up (NDBI) for both cities. The analysis of environmental criticality revealed a dramatic change in ECI at the core of Barasat city. As the city core is well connected with road and rail networks and concentrated with urban facilities, therefore occupied by concrete buildings rapidly. The findings suggest that the issues of LST and environmental criticality in Barasat city are raising at an alarming rate. The study found the raising concern of the UHI phenomenon needs special attention from the planner for the future development of Barasat city. Similarly, though Kalyani is developed as a planned city, the increasing population pressure and corresponding urban sprawling also aggravated the UHI phenomenon gradually. The increase in the spatial extent of high ECI rings the alarms for future situations. However, the pattern of changes in LST and ECI between the cities from 2005 to 2019 are reflecting how unplanned urban growth intensifies the UHI phenomenon and environmental criticality as compared to planned development. Therefore, this article has successfully established the different patterns of thermal environments for planned and unplanned urban areas.

The planned and unplanned cities, both, in India, are witnessing rapid urban growth that must have a good economic significance but need control for future development by not overlooking the environmental aspects. Proper planning is a prerequisite for preventing and mitigating environmental extremes. Regions facing less than or high-temperature extremes can be counteracted with the presence of green spaces, and city morphology with the allowance of more vegetation, while these conditions are given impetus by the unchecked built-up growth. The implementation of appropriate bylaws through proper urban planning can strictly regulate illegal construction, tree removal, etc., and will be beneficial for improving the urban environment, as well as, satisfying the local needs for space and infrastructure. This study is not only beneficial for the urban planners, administrators and urban developers of Barasat and Kalyani cities, but also helps researchers and others to investigate in another part of the globe.

**References**

[1] Sujathamma, C. (2019) 'Environmental issues of urban areas and wellness' *IJSDR* **4**(5) 579-582

[2] Tregonning, G., Barr, S., Dawson, R. and Ranjan, R., 2019. A multi-objective spatial optimization framework for sustainable urban development. In *2nd International Conference on Natural Hazards and Infrastructure ICONHIC2019*. Newcastle University.

[3] Secretariat, U.N., 2007. World population prospects: The 2006 revision and world urbanization prospects: The 2007 revision. Population Division of the Department of Economic and Social Affairs of the United Nations Secretariat. Available from http://esa. un. org/unup.

[4] Mills, G., Cleugh, H., Emmanuel, R., Endlicher, W., Erell, E., McGranahan, G., Ng, E., Nickson, A., Rosenthal, J. and Steemer, K., 2010. Climate information for improved planning and management of mega cities (needs perspective) *Procedia Environmental Sciences* ***1*** 228-246.

[5] Oke, T.R., 1973. City size and the urban heat island *Atmospheric Environment (1967)* **7**(8) 769-779.

[6] Dousset, B. and Gourmelon, F., 2003. Satellite multi-sensor data analysis of urban surface temperatures and landcover *ISPRS journal of photogrammetry and remote sensing* **58**(1-2), 43-54.

[7] Kim, H., 1990. *A method of classification for multisource data in remote sensing based on interval-valued probabilities* (Doctoral dissertation, Purdue University).

[8] Comarazamy, D.E., Gonzalez, J. and Luvall, J.C., 2010, December. Investigation of Long-Term Impacts of Urbanization and Global Warming in a Coastal Tropical Region. In *AGU Fall Meeting Abstracts* 2010 B21E-0369.

[9] Guhathakurta, S. and Gober, P., 2007. The impact of the Phoenix urban heat island on residential water use *Journal of the American Planning Association* **73**(3) 317-329.

[10] Khan, S.M. and Simpson, R.W., 2001. Effect of a heat island on the meteorology of a complex urban airshed *Boundary-Layer Meteorology* **100**(3) 487-506.

[11] Sailor, D.J. and Lu, L., 2004. A top–down methodology for developing diurnal and seasonal anthropogenic heating profiles for urban areas *Atmospheric environment* **38**(17) 2737-2748.

[12] Weng, Q., 2010. Remote sensing and GIS integration. McGraw-Hill Professional Publishing.

[13] Tian, M., Wang, P. and Sun, W., 2010. A Review of Retrieving of Land Surface Parameters Using the Land Surface Temperature-Vegetation Index Feature Space *Advances in Earth Science* **25**(7) 698-705.

[14] Senanayake, I.P., Welivitiya, W.D.D.P. and Nadeeka, P.M., 2013. Remote sensing-based analysis of urban heat islands with vegetation cover in Colombo city, Sri Lanka using Landsat-7 ETM+ data. *Urban Climate* **5** 19-35.

[15] Son, N.T. and Thanh, B.X., 2018. Decadal assessment of urban sprawl and its effects on local temperature using Landsat data in Cantho city, Vietnam *Sustainable cities and society* **36** 81-91.

[16] Carneiro, E., Lopes, W. and Espindola, G., 2021. Linking Urban Sprawl and Surface Urban Heat Island in the Teresina–Timon Conurbation Area in Brazil *Land* **10**(5) 516.
[17] Kadhim, N., Mourshed, M. and Bray, M., 2016. Advances in remote sensing applications for urban sustainability *Euro-Mediterranean Journal for Environmental Integration* ***1***(1) 1-22.
[18] Chen, X.L., Zhao, H.M., Li, P.X. and Yin, Z.Y., 2006. Remote sensing image-based analysis of the relationship between urban heat island and land use/cover changes *Remote sensing of environment* **104**(2) 133-146.
[19] Loveland, T.R. and Irons, J.R., 2016. Landsat 8: The plans, the reality, and the legacy *Remote Sensing of Environment* **185** 1-6.
[20] Roy, D.P., Wulder, M.A., Loveland, T.R., Woodcock, C.E., Allen, R.G., Anderson, M.C., Helder, D., Irons, J.R., Johnson, D.M., Kennedy, R. and Scambos, T.A., 2014. Landsat-8: Science and product vision for terrestrial global change research *Remote sensing of Environment* **145** 154-172.
[21] Zhang, X., Zhong, T., Feng, X., Wang. K. 2009. Estimation of the relationship between vegetation patches and urban land surface temperature with remote sensing *International Journal of Remote Sensing*. **30** (8) 32-45.
[22] Srivastava, P.K., Majumdar, T.J. and Bhattacharya, A.K., 2009. Surface temperature estimation in Singhbhum Shear Zone of India using Landsat-7 ETM+ thermal infrared data *Advances in space research* **43**(10) 1563-1574.
[23] Ma, Y., Kuang, Y. and Huang, N., 2010. Coupling urbanization analyses for studying urban thermal environment and its interplay with biophysical parameters based on TM/ETM+ imagery *International Journal of Applied Earth Observation and Geoinformation* **12**(2) 110-118.
[24] Qin, X., Wu, X., Chiew, Y. and Li, Y., 2012. A green roof test bed for stormwater management and reduction of urban heat island effect in Singapore *British Journal of Environment and Climate Change* **2**(4) 410-420.
[25] Mozafari, G. and Narangifard, M., 2015. The Vegetative Cover Deterioration Due to the Droughts on the Mulla Sadra Watershed and Application of Remote Sensing Techniques in Its Monitoring.
[26] Fu, P. and Weng, Q., 2016. A time series analysis of urbanization induced land use and land cover change and its impact on land surface temperature with Landsat imagery *Remote sensing of Environment* **175** 205-214.
[27] Ibrahim, F. and Rasul, G., 2017. Urban land use land cover changes and their effect on land surface temperature: Case study using Dohuk City in the Kurdistan Region of Iraq *Climate* **5**(1) 13.
[28] Yu, C. and Hien, W.N., 2006. Thermal benefits of city parks *Energy and buildings* ***38***(2) 105-120.
[29] Lu, D. and Weng, Q., 2006. Use of impervious surface in urban land-use classification *Remote sensing of environment* **102**(1-2) 146-160.
[30] Pu, R., Gong, P., Michishita, R. and Sasagawa, T., 2006. Assessment of multi-resolution and multi-sensor data for urban surface temperature retrieval *Remote Sensing of Environment* **104**(2) 211-225.
[31] ZHENG, Z.F., LIU, W.D. and WANG, Y.C., 2006. The distributive character of urban heat island effect in the Beijing region *Journal of Nanjing Institute of Meteorology* **5** 694-699.
[32] Momeni, M. and Saradjian, M.R., 2007. Evaluating NDVI-based emissivities of MODIS bands 31 and 32 using emissivities derived by Day/Night LST algorithm *Remote Sensing of Environment* **106**(2) 190-198.
[33] Benali, A., Carvalho, A.C., Nunes, J.P., Carvalhais, N. and Santos, A., 2012. Estimating air surface temperature in Portugal using MODIS LST data *Remote Sensing of Environment* **124** 108-121.
[34] Mallick, J., Rahman, A. and Singh, C.K., 2013. Modeling urban heat islands in heterogeneous land surface and its correlation with impervious surface area by using night-time ASTER satellite data in highly urbanizing city, Delhi-India *Advances in Space Research* **52**(4) 639-655.

[35] Lazzarini, M., Marpu, P.R. and Ghedira, H., 2013. Temperature-land cover interactions: The inversion of urban heat island phenomenon in desert city areas Remote Sensing of Environment **130** 136-152.
[36] Improved surface temperature estimates with MASTER/AVIRIS sensor fusion Grigsby S.P., Hulley G.C., Roberts D.A., Scheele C., Ustin S.L., Alsina M.M. (2015) *Remote Sensing of Environment* **167** 53-63.
[37] Yuan, F. and Bauer, M.E., 2007. Comparison of impervious surface area and normalized difference vegetation index as indicators of surface urban heat island effects in Landsat imagery *Remote Sensing of environment* **106**(3), 375-386
[38] Effat, H.A. and Hassan, O.A.K., 2014. Change detection of urban heat islands and some related parameters using multi-temporal Landsat images; a case study for Cairo city, Egypt *Urban Climate* **10** 171-188.
[39] Ahmadi, H., Arabi, R. and Fatahi, L., 2015. Thermal Behavior of Green Roofs in Different Climates. *Current World Environment* **10**(Special Issue) 908.
[40] Mazidi, A. and Hoseini, F.S., 2015. Effects of changing land use and land cover on the heat island in urban area of Yazd using remote sensing data *Geography and Development Iranian Journal 13*(38) 1-12.
[41] Deilami, K., Kamruzzaman, M. and Liu, Y., 2018. Urban heat island effect: A systematic review of spatio-temporal factors, data, methods, and mitigation measure *International journal of applied earth observation and geoinformation* **67** 30-42.
[42] Weng, Q., Lu, D. and Schubring, J., 2004. Estimation of land surface temperature–vegetation abundance relationship for urban heat island studies *Remote sensing of Environment* **89**(4) 467-483.
[43] Amiri, R., Weng, Q., Alimohammadi, A. and Alavipanah, S.K., 2009. Spatial–temporal dynamics of land surface temperature in relation to fractional vegetation cover and land use/cover in the Tabriz urban area, Iran *Remote sensing of environment* **113**(12) 2606-2617.
[44] Kumar, D. and Shekhar, S., 2015. Statistical analysis of land surface temperature–vegetation indexes relationship through thermal remote sensing *Ecotoxicology and environmental safety* **121** 39-44.
[45] Ramaiah, M., Avtar, R. and Rahman, M., 2020. Land Cover Influences on LST in Two Proposed Smart Cities of India: Comparative Analysis Using Spectral Indices *Land* **9**(9) 292.
[46] Malik, M,S., Shukla, J, P., & Mishra, S., 2019. Relationship of LST, NDBI and NDVI using Landsat-8 data in Kandaihimmat Watershed, Hoshangabad, India.
[47] Das, M. and Das, A., 2020. Assessing the relationship between local climatic zones (LCZs) and land surface temperature (LST)–A case study of Sriniketan-Santiniketan Planning Area (SSPA), West Bengal, India *Urban Climate* **32** 100591.
[48] Mukherjee, K. and Das, P., 2018. Modelling the Relationship between Urban Growth Modes and the Thermal Environment-A Case Study of the Barasat Municipality, West Bengal. Journal of Geography, Environment and Earth Science International, 1-19.
[49] Mukherjee, K., 2019, February. Delineating fringe by global diversity value: A case study on Barasat municipality, India. In 2019 International Conference on Big Data and Computational Intelligence (ICBDCI) 1-12. IEEE.
[50] Tarafdar, A. and Patnaik, U., Research Article Effect of Vegetative Cover on Urban Heat Island Effect.
[51] Van de Griend, A.A. and OWE, M., 1993. On the relationship between thermal emissivity and the normalized difference vegetation index for natural surfaces *International Journal of remote sensing* **14**(6) 1119-1131.
[52] Markham, B.L. and Barker, J.L., 1985. Spectral characterization of the Landsat Thematic Mapper sensors *International Journal of Remote Sensing* **6**(5) 697-716.

[53] Sobrino, J.A., Li, Z.L., Stoll, M.P. and Becker, F., 1996. Multi-channel and multi-angle algorithms for estimating sea and land surface temperature with ATSR data *International Journal of Remote Sensing* **17**(11) 2089-2114.
[54] Comarazamy, D.E., González, J.E., Luvall, J.C., Rickman, D.L. and Mulero, P.J., 2010. A land–atmospheric interaction study in the coastal tropical city of San Juan, Puerto Rico *Earth Interactions* **14**(16) 1-24.
[55] Mukherjee K, Ray R. An Analytical Study on the Relationship between LULC Dynamics and Land Surface Thermal Environment-A Case Study on Barasat Municipal Area, North 24 Parganas, West Bengal, India. International Journal of Remote Sensing & Geoscience. 2017; **6**(2).
[56] Parastatidis, D., Mitraka, Z., Chrysoulakis, N. and Abrams, M., 2017. Online global land surface temperature estimation from Landsat *Remote sensing* **9**(12) 1208.
[57] Malakar, N.K., Hulley, G.C., Hook, S.J., Laraby, K., Cook, M. and Schott, J.R., 2018. An operational land surface temperature product for Landsat thermal data: Methodology and validation. *IEEE Transactions on Geoscience and Remote Sensing* **56**(10) 5717-5735.
[58] Ermida, S.L., Soares, P., Mantas, V., Göttsche, F.M. and Trigo, I.F., 2020. Google earth engine open-source code for land surface temperature estimation from the Landsat series. *Remote Sensing* **12**(9) 1471.
[59] Ahmadi, M., Dadashi, A. (2017). 'The Identification of Urban Thermal Islands based on an Environmental Approach, Case Study: Isfahan Province *Geography and Environmental Planning* **28**(3) 1-20.
[60] Sasmito, B. and Suprayogi, A., 2018, June. Spatial Analysis of Environmental Critically due to Increased Temperature in The Built Up Area With Remote Sensing. In *IOP Conference Series: Earth and Environmental Science* **165**(1) 012011.
[61] Yang, Z., Chen, Y., Wu, Z., Qian, Q., Zheng, Z. and Huang, Q., 2019. Spatial heterogeneity of the thermal environment based on the urban expansion of natural cities using open data in Guangzhou, China *Ecological Indicators* **104** 524-534.
[62] Chu, L., Oloo, F., Bergstedt, H. and Blaschke, T., 2020. Assessing the link between human modification and changes in land surface temperature in Hainan, China using image archives from google earth engine *Remote Sensing* **12**(5) 888.

# Potential Effect of the Changing Urban Climate on Non-Communicable Diseases: A Case Study on Barasat City, India

**R Chhetri**[1], **K Mukherjee**[1*], **P Dash**[2]

[1]Department of Geography, School of Basic and Applied Sciences, Adamas University, Kolkata-700126, India
[2] Department of Geography, Khejuri College, Baratala, West Bengal-721431, India

*Corresponding Author Email: kasturi.mukherjee@adamasuniversity.ac.in

**Abstract.** Rapid land use change in urban areas creates a unique climate, which risks the health of its inhabitants. The unique urban climate is attributable to a phenomenon recognized as the Urban Heat Island (UHI). Under such circumstances, the urban regions become hotter than their rural surroundings, creating an island of higher temperatures. Human exposure to excessively warm heat islands, particularly in humid climates like India, is becoming an increasingly serious public health issue. This paper briefly analyses the formation of surface urban heat islands (SUHI), identified by estimating the Land Surface Temperature (LST) and also its potential impacts on the occurrence of heat-related Non-Communicable Diseases (NCD) within a city. Barasat city of India has been chosen to demonstrate this work. LST has been estimated from the Landsat data using a remote sensing platform. It has been observed that the high LST in Barasat city perfectly followed an emitting shape in descending order from the city center to the suburban areas. A primary household survey was conducted in 2018 in the month of March-April, where 200 samples were collected. The information has been collected from the samples regarding their occupation, average hours of exposure to high heat zones, and the NCDs they have suffered from. It has been found that heat cramps, heat stress, heat strokes, and respiratory diseases are common for people exposed to the high heat zone. Although NCDs increase the socioeconomic burden in developing cities, NCD surveillance is limited in India. This study will assist in formulating a better urban neighborhood considering the city's population exposed to UHI.

**Keywords**: Urban climate, Urbanization, Land Surface Temperature, Non-Communicable Diseases

## 1. Introduction

The land is among the most important resources of the earth on which all kinds of human activities are performed. Land surfaces are utilized by human beings for different purposes like producing food, providing shelter, recreation, extracting, and processing materials which give rise to the concept of "land use". "Land cover" refers to the forest, water bodies, rock, or soil that covers the land [1]. The worldwide recent trend of LULC changes has been dominated by the loss of agricultural land, forest areas, and waterways into the urban landscape [2] which is the pre-condition of urbanization. But this conversion of the previous, vegetated non-urban land into urban built-up has been recognized as a foremost factor behind the loss of biodiversity, disruption of ecological processes, climatic change, disturbances in biogeochemical cycles, etc. [3].

The hasty change in the form and function of structures in urban areas creates its own unique climate, which risks the health of its inhabitants. In the last few decades, researchers have also paid attention to the engulfment of different LULCs by expanding built-up, causing irreversible thermal environmental changes [4]. The land surface temperature (LST) is rising both quantitatively and spatially as a consequence of increased urbanization at the expense of the low radiative surface. Due to an incident identified as the Urban Heat Island (UHI), this process changes the urban climate. The phenomena of UHI have first explored in the city of London and illustrated in the 1810s by a British scientist, named Luke Howard.

Atmospheric heat islands, such as the CLHI and BLHI, signify hotter urban air. The term "SUHI" describes the urban surfaces' generally greater temperature when contrasted to the nearby rural areas. While surface UHIs are severer during the day, atmospheric UHIs increase at night [5,6]. SUHI is identified by estimating Surface Temperature (LST) [7]. The radiating skin temperature of the land is referred to as LST. It is a crucial factor in the physics of the earth's surface. In numerous scientific investigations in the areas of hydrology, ecology, and global dynamics, it plays a significant role [8]. At both the local and global levels, thermal infrared (TIR) remote sensing offers a distinctive way of attaining LST data. The land surface emits a large percentage of the energy spotted by the sensor in this spectral range [9]. Any surface's thermal inertia determines the kind of temperature the surface experiences. The variability of thermal response increases with urban impervious surfaces' expansion due to the sensible heat flux's huge extent. Thus, surface urban heat island (SUHI) is one of the significant implications of LULC change with urbanization, which increases imperviousness and induces vegetation displacement [10].

According to a US effect evaluation of the number of deaths and energy consumed for hot and cold periods, the effect of the UHI on heat-related causalities was estimated to be an increase of 1.1 deaths per million people. Heat island exposure is increasing the risk of chronic non-communicable diseases (NCDs) like heat stress, stroke, cardiovascular disease, some cancers, respiratory distress, etc. in various Indian cities also [11]. Most diseases are brought on by prolonged or frequent exposure to high temperatures. Blood rushes to the skin's surface as the body attempts to cool itself under extreme heat, and less blood penetrates the brain, muscles, and other organs, interfering with physical and mental function and, in rare cases, having fatal repercussions. NCDs were historically considered a problem in developed countries. But the burden of such diseases has amplified rapidly in most regions of the world [11]. The investigation has focused so far in India on communicable diseases such as malaria, dengue, and tuberculosis. Very little evidence on the urban climate change influences on health is available. However as anticipated, there will probably be more cases of excessive heat exposure and the resultant heat strokes in the future [12]. Between 2012 and 2030, it is estimated that India will spend US$6.2 trillion on treating various NCDs, including cancer, chronic respiratory illnesses, cardiovascular diseases, and mental health disorders [13]. Despite the growing impact of NCDs on public health in India, the NCD mitigation and eradication strategy is impeded by a lack of adequate monitoring and reliable data [14]. This reminds greater academic attention to effective NCD surveillance and understating its correlation with changing urban climate. This article presents a framework for understanding the changing urban climate and evaluates the linkages with NCDs in Barasat city of India.

## 2. Study area

One of West Bengal's oldest towns, Barasat, has been selected for this research as the study area (figure 1) The town rapidly became a residential neighborhood during the 1947 partition of India. This happened as a result of the massive inflow of Hindu immigrants from former Pakistan. The city continued to flourish after Bangladesh was formed in 1971. Its significance has increased massively since it was designated as the district headquarters of North 24 Parganas in 1986, and it has begun to be more significantly impacted by accelerated urbanization. Between 1991 and 2001 it is concerning that the population of Barasat increased by 125.5%, a rate far exceeding that of India and West Bengal's urban population growth (figure 2). There are a number of challenges brought on by this surge of urbanization and rapid population growth. The massive, unplanned conversion of land into urban space created issues and led to significant surface warming [15].

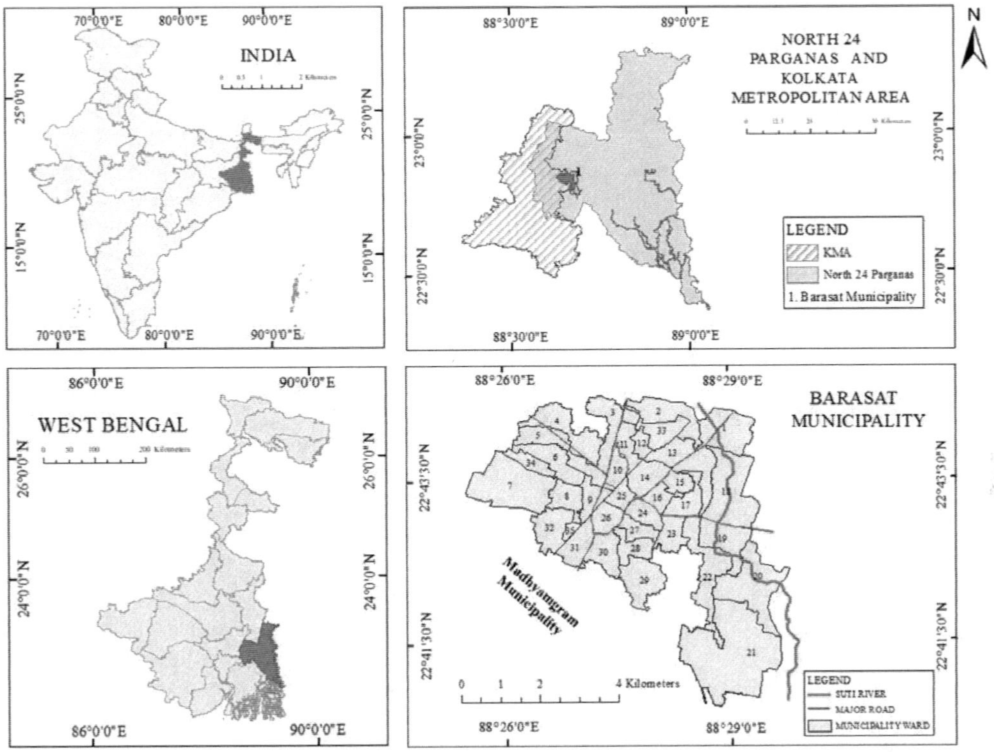


**Figure 1.** A map of the study area

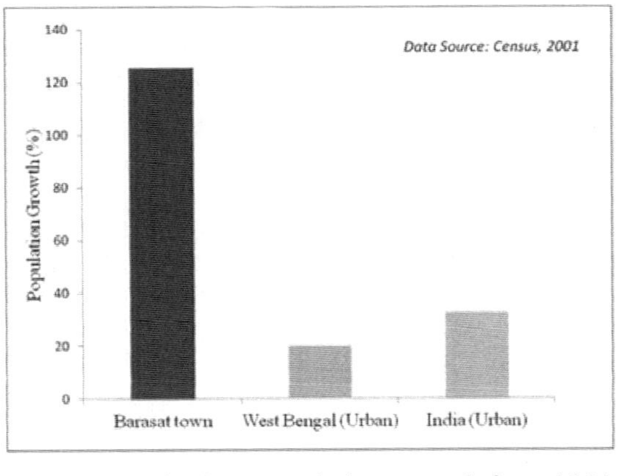


**Figure 2.** The rate of urban population growth from 1901 to 2001

### 3. Data Used

The entire Landsat archive is now available free of cost for scientific research, which is a treasure of information used for assessing the dynamics of LST [16]. Table 1 contains information about the sensors employed in this study.

**Table 1.** Details of secondary data used

| Data | Period |
|---|---|
| Landsat 8 OLI &TIRS | 2014-04-22, 2017-04-14 |
| Landsat 7 ETM | 2001-04-26 |
| Landsat 5 TM | 1989-04-25, 1995-04-02<br>2005-04-13, 2006-05-02<br>2007-04-03, 2010-04-11 |

## 4. Methodology

### *4.1. LST estimation*

For estimating LST from Landsat 7 and Landsat 8, the Mono-window and Split-window algorithms have been used respectively. This process has been elaborately discussed in the author's earlier work [17].

### 4.2. *LST validation*

Ground-based thermal emissivity has been measured in October 2017 using an infrared thermometer. Association between the ground-based thermal emissivity and satellite-driven LST is evaluated by Student's t-test at a 0.05 significance level. The disparity between the sensors' resolution (30 m) and the spatial dimensions of the field measurements (1 m) is the main issue that researchers have when validating remote sensing data with ground truth observations. The conformity of ground data measurements and satellite data is useful when the site of interest is homogeneous, but it is challenging to find the right test site for validation. In literature, bare soil, vegetation, and water body have been used as the most appropriate sites [18] for validation. In this study, thermally homogenous sites have been identified, like vegetated areas, built-up areas, and water bodies that have an uninterrupted large areal extension (>1 $km^2$). As the field verification for the satellite images of 1989, 2001, and 2014 is not possible in this research, the reliability of satellite data on LST estimation is validated for the year 2017 only.

### *4.3. Ground measurements*

On a day without any clouds, surface temperature measurements were made on the aforementioned LULC classes using an infrared thermometer of the MT-5 model [19]. With a 1.5% accuracy, the MT-5 Digital Infrared Thermometer intercepts surface temperature in a single channel (8–14 m) in the emissivity range of 0.01-0.99 [20]. On October 27, 2017, numerous observations were made at the study site to record the spatiotemporal heterogeneity of the surface temperature within the specified LULC classes.

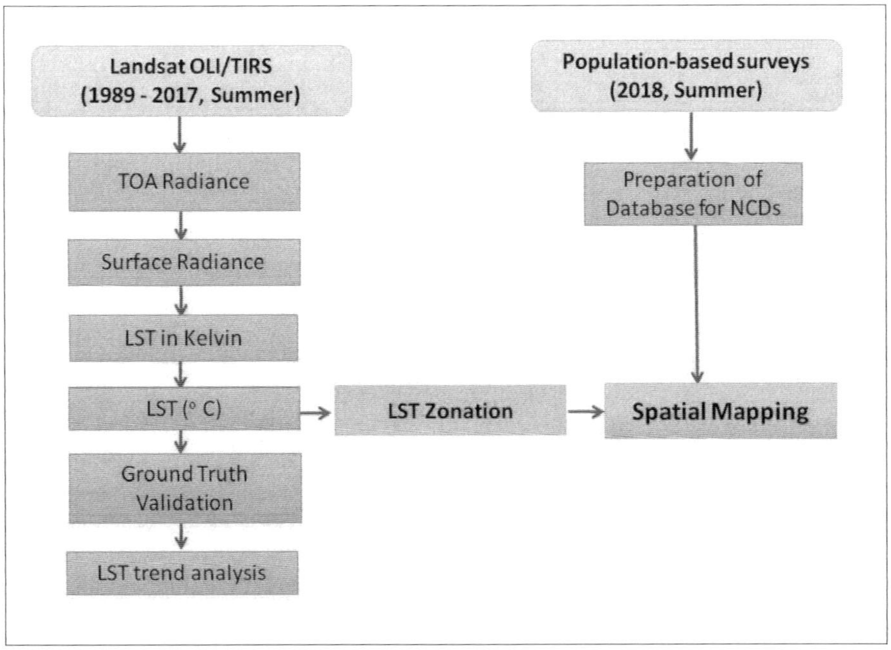


**Figure 3.** The methodology followed in this study

### *4.4. Trend analysis of surface temperature*

To comprehend the summer temperature trend from 1989 to 2017, the Mann-Kendall test was adapted to analyse the trend of LST data. This test which is usually known as Kendall's τ statistic denotes test statistics of distribution given by Kendall is a popular non-parametric test for evaluating the identification of non-linear trends [21]. This method employs rank-based analysis, which is effective at decomposing the impact of extremes and robust when using skewed variables [22]. Under the null hypothesis H0, the deseasonalized data (x1, x2,....xn) in MKT are assumed to be a sample of n independently distributed random variables. The alternative hypothesis H1 states that the distributions of xk and xj are not identical for all k, j ≤ n with k ≠ j. Equations (2) and (3) allow for the calculation of the test statistic S, whereas equation (1) computes variances as follows [23]:

$$S = \sum_{k=1}^{n-1} \sum_{j=k+1}^{n} \mathrm{sgn}(x_j - x_k) \tag{1}$$

$$\mathrm{sgn}(x_j - x_k) = \begin{cases} +1 \text{ if } (x_j - x_k) > 0 \\ 0 \text{ if } (x_j - x_k) = 0 \\ -1 \text{ if } (x_j - x_k) < 0 \end{cases} \tag{2}$$

$$\mathrm{Var}(S) = [n(n-1)(2n+5) - \textstyle\sum_t t(t-1)(2t+5)]/18 \tag{3}$$

Where t is the size of a specific tie, and the sum of all t equals the total number of ties. Equation (4) is used to calculate the standard normal variate (z) when the sample size (n) is bigger than 10:

$$z = \begin{pmatrix} \frac{S-1}{\sqrt{\mathrm{Var}(S)}} & \text{if } S>0 \\ 0 & \text{if } S=0 \\ \frac{S+1}{\sqrt{\mathrm{Var}(S)}} & \text{if } S<0 \end{pmatrix} \tag{4}$$

Sen's estimator: To determine the trend in a time series calculated as true slope, a basic non-parametric approach is used (change per unit time) [24]. Equation (5) estimates the slope for the N pairs of data as follows:

$$Q_i = \frac{x_j - x_k}{j-k} \quad for\ i = 1, \ldots, N \tag{5}$$

where, $x_j$ and $x_k$ are, respectively, data values at times j and k (j>k). The median of these N values is used to calculate Qi, Sen's estimator of slope. When N is odd, $Q_{med}$ = Q(N+1)/2, and when N is even, $Q_{med}$ = [QN/2 + Q(N+2)/2]/2, respectively. However, a two-sided test at the 100(1-)% confidence interval is used to determine the significance of $Q_{med}$.

*4.5. Spatial mapping of UHI*

The raster layer of derived LST of summer 2017 was further reclassified into five categories based on the mean-standard deviation as the reclassified maps are more suitable for the identification of elevated temperature zones [25-26]. From each LST raster map, the mean (μ) and standard deviation (σ) of the study domain has been calculated. Then considering half of the standard deviation as range of the class groups, the raster dataset has been reclassified into five LST zones as described in table 2. The 'High-temperature zone' corresponds to the heavily urbanised area having high thermal conductivity can be defined as the heat island.

**Table 2**. Classification criteria of LST following mean-standard deviation method

| LST Zone | LST Classification Method | Zone Description |
|---|---|---|
| 1 | $T < \mu - \sigma$ | Low-temperature zone |
| 2 | $\mu - \sigma < T < \mu - 0.5\,\sigma$ | Secondary moderate-temperature zone |
| 3 | $\mu - 0.5\,\sigma < T < \mu + 0.5\,\sigma$ | Moderate-temperature zone |
| 4 | $\mu + 0.5\,\sigma < T < \mu + \sigma$ | Secondary high-temperature zone |
| 5 | $T > \mu + \sigma$ | High-temperature zone |

*4.6. Preparation of the NCD database*

The Non-communicable disease data were collected by using a population-based survey method. Questionnaires on health and demography have been prepared and the survey was conducted with GPS in the month of March - April, 2018. Primary data have been collected from all the five LST zones of the city by adapting stratified random sampling techniques. A total of 200 natives were surveyed taking 5-6 respondents have been selected from each ward. Respondents have been selected from each ward based on random sampling with minimum of 1 hour of sun exposure. Thorough the survey the authors have tried to understand the duration of exposure to different LST zones and the occurrence of heat related diseases. The entire methodology has been illustrated in figure 3.

## 5. Result and Discussion

*5.1. LST validation*

The predicted LST produced from Landsat data is verified using the ground surface temperature measured in the field with an IR thermometer. It is noteworthy to mention that as the field data before 2015 were neither available nor collected, therefore the result is validated only for the Landsat 8 data of October 2017. The summary statistics of the validation results are shown in Table 3. The student's t-test indicates that the satellite data-derived LST and the field measurements do not differ significantly. This indicates that Landsat data can be successfully used to derive LST for the study domain.

**Table 3.** Validation of estimated LST by Student's t-test

| Statistical parameters | Measured LST (°C) 27th October, 2017 | Satellite data-derived LST (°C) 23rd October, 2017 | Student's t |
|---|---|---|---|
| Mean | 30.24 | 28.25 | 1.56* |
| SD | 5.97 | 0.69 | |
| Variance | 35.68 | 0.48 | |
| N | 22 | 22 | |

* Significant at α = 0.05

**Table 4.** Summary statistics of LST (°C) derived for summer season

| Year | Maximum | Minimum | Range | Mean | Standard Deviation |
|---|---|---|---|---|---|
| 1989 | 21.69 | 17.92 | 3.77 | 19.27 | 0.36 |
| 1995 | 22.54 | 17.33 | 5.21 | 18.89 | 0.48 |
| 2001 | 24.78 | 19.96 | 4.82 | 21.92 | 0.63 |
| 2005 | 25.89 | 20.83 | 5.06 | 22.97 | 0.62 |
| 2006 | 24.78 | 20.54 | 4.24 | 22.31 | 0.7 |
| 2007 | 27.26 | 21.11 | 6.15 | 23.6 | 0.84 |
| 2010 | 28.07 | 22.25 | 5.82 | 24.67 | 0.83 |
| 2014 | 35.17 | 26.24 | 8.90 | 31.1 | 0.97 |
| 2017 | 36.44 | 27.54 | 8.93 | 29.78 | 1.06 |

### *5.2. Analysis of the trend of LST*

Figure 4 presents the results of the spatial variability of LST in degrees Celsius (°C) throughout the summers from 1989 to 2017. There is a significant spatial variation of LST values within the study area. The moderate to high-temperature spots are clustered in the middle part of the city which is covered with built-up. This high LST zone is forming the Urban Heat Island (UHI). Figure 4 clearly indicates that UHI in the Barasat municipality has evolved an emitting form that descends from the city centre to outlying areas.

The summary statistics of LST for the mentioned periods (1989–2017) are provided in table 4. The municipality's mean temperature has increased significantly over the previous few decades, which is an indication that the urban land surfaces are gradually warming up. The maximum and minimum LST during the summer is about 27.4°C and 21.5°C, consecutively. The highest and lowest LST were 21.89°C and 17.92°C, respectively, in 1989, whereas the corresponding figures for 2017 are 36.44°C and 27.54°C. The lowest temperature recorded was 27.54 °C in 2017, which is significantly warmer than the highest temperature recorded, 21.69 °C, in 1989. 2017 is the warmest over past 28 years. From the table 4, it has been clear that there is a rising trend of the highest, lowest and mean LST from 1989 to 2017.

It is noteworthy that the range of LST (highest~lowest) is also gradually increasing from 1989 to 2017, which signifies intensification of SUHI phenomenon. With $R^2$ of 0.84 and a p-value of 0.001 (figure 5), the summer LST (1989–2017) demonstrated a distinct upward trend that was significant at the 95% confidence level. The rising LST range patterns are evidence of the changing urban climate. The standard deviation of LST has gradually increased, which suggests that LST variability has increased.

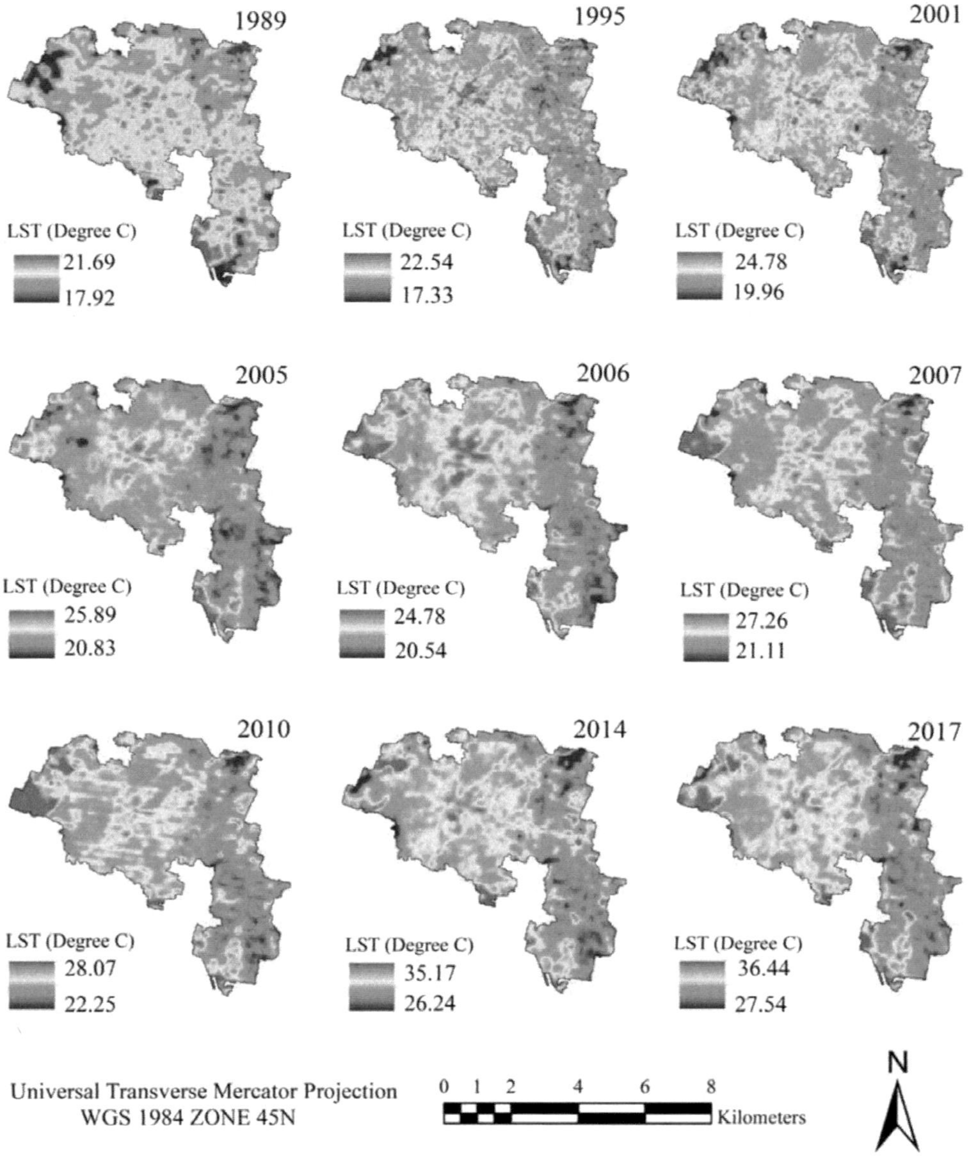


**Figure 4.** Spatial pattern of LST during summer season

*5.3. Spatial mapping to identify heat risk*

The many types of land surfaces in a city cause temperature to vary, which causes equivalent variances in the population's potential exposure to heat. The compact and contiguous urban region in the city centre contains the warmest zone (high temperature zone, secondary high-temperature zone), known as the UHI, where the mean standardized LST is always higher than the zonal mean (figure 6). The low-temperature zone, secondary moderate-temperature zone, and moderate-temperature zone correspond to water bodies, vegetated areas, and crop lands, respectively. The high temperature zone and secondary high temperature zone of surface temperature (UHI) roughly correspond to the densely populated areas of Barasat city. Because of the existence of roads, buildings, and other impermeable surfaces, the densely populated area typically has a higher solar radiation absorption rate as well as a higher thermal capacity and conductivity when compared to the nearby non-urban areas. The bare soil in the peripheral areas of the city is also linked to higher LST due to low thermal inertia of it. It quickly gets heated throughout the day time and quickly cools at night.

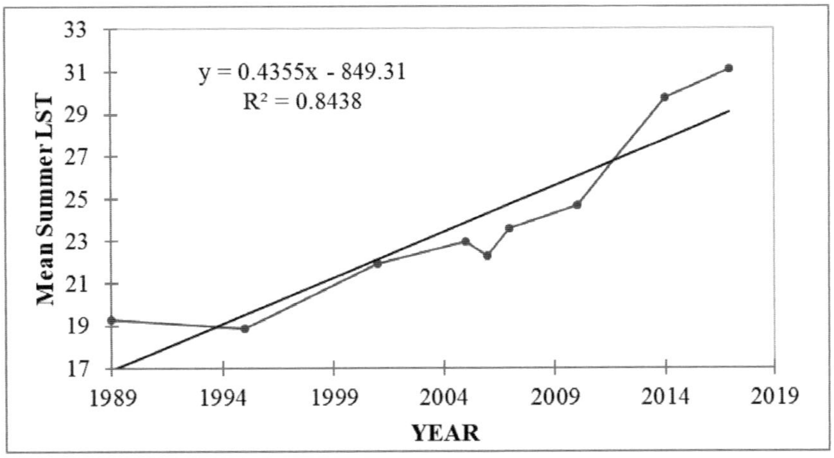


**Figure 5.** Trend of yearly mean summer LST

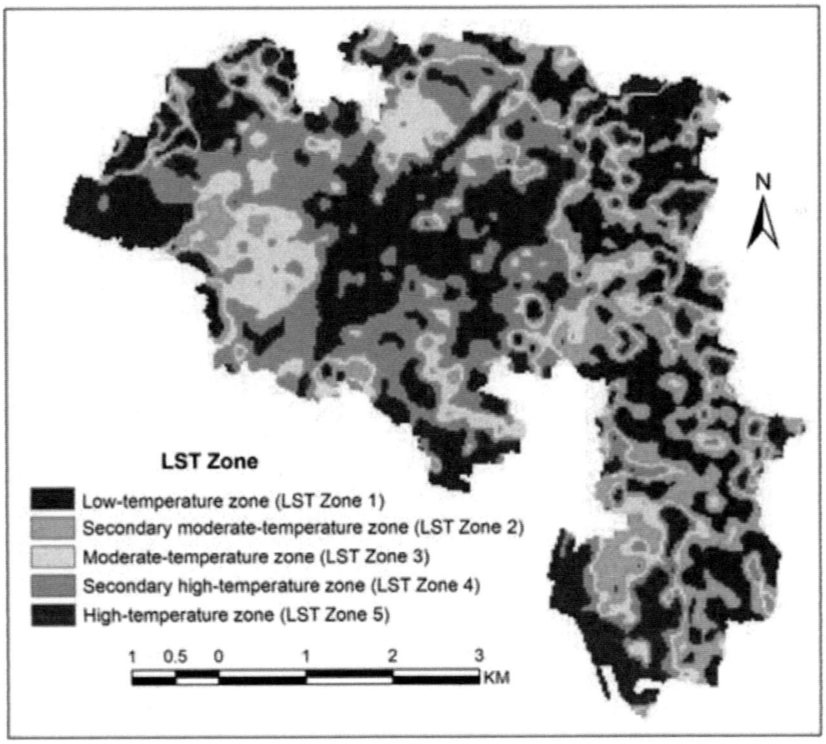


**Figure 6.** LST Zonation

During the field survey, it has been found that in the case of fewer than 3 hours of exposure in this UHI maximum recorded was heat exhaustion followed by heat cramps and loose motion. In the case of 3-6 hours of exposure in this zone, the maximum recorded was heat exhaustion followed by heat cramps and loose motion. Heat exhaustion was the most common symptom after 7-10 hours of exposure, followed by skin allergies, depression, and loose motion. Heat exhaustion was the most common symptom, followed by skin allergies and loose motion after 11-14 hours of exposure. Heat exhaustion has been the most frequent problem after 11-14 hours of exposure, followed by heat cramps, skin allergies, depression, and loose motion. This study attempted to identify the most frequent non-communicable diseases (NCDs) linked to various LST Zones. A questionnaire survey gathered field data on NCDs from each LST zone. The percentages of respondents commonly suffer from different NCDs and found working or staying in particular LST zones with a minimum exposure time of 6 hours are shown in figure 7.

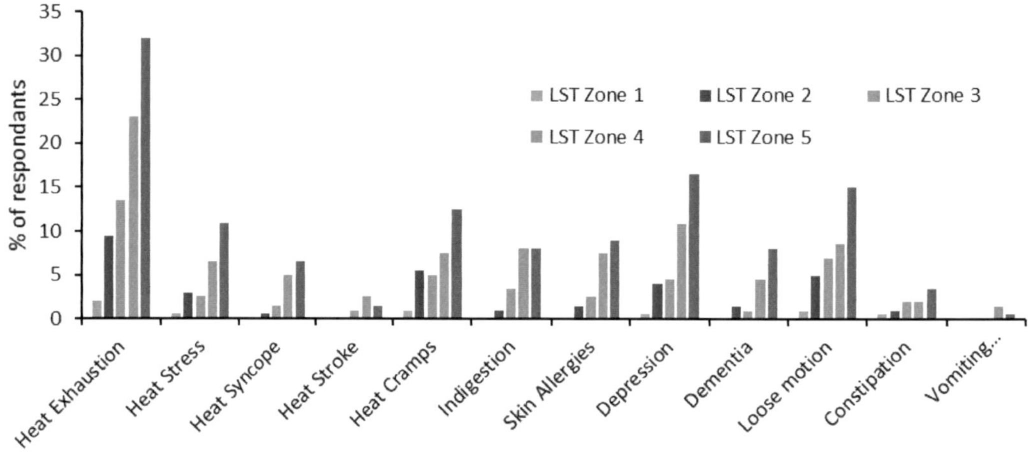


**Figure 7.** Graph showing percentages of respondents having a particular health problem working/staying in different LST zones

The figure illustrates that the significant responses were recorded under each NCD category from the persons closely associated with LST Zone 5, followed by LST Zone 4. The percentages of positive responses for each NCD are limited for those exposed in Zone 1. Figure 7 also suggests that heat exhaustion is the most occurred NCD of people across all LST Zones. The other common NCDs are depression, heat cramps, and loose motion, especially for respondents from LST Zone 5. The most significant percentages of people suffer from heat exhaustion compared to other NCDs, maximally 32% in LST Zone 5 and 23% in Zone 4. People working outdoors in open shops and driving vehicles are found very sensitive to NCDs compared to others. The overall analysis suggests a significant impact of LST on human health, especially those exposed to high-temperature zone for long periods.

## 6. Conclusion

This research highlights the potential impact of elevated LST on the occurrences of non-communicable diseases. A little database is available in our country for understanding the non-communicable diseases among the masses. At the time of the survey, it has also been found that both public and private hospitals of Barasat do not have any data records for NCDs. Because of such infrastructural and behavioral constraints, this study has only incorporated the population-based survey which consists of the demographic and health survey. One advantage is that the population-based survey has helped to overcome the distortion of any health information. As Barasat is lying close to the city of Kolkata and it is developing rapidly, many urban developments are expected in near future. In the process of urbanisation, the city has already led to the rise in temperature which has an impact on increasing the prevalence of non-communicable diseases. People who have been exposed long to different temperature zones are surveyed. Almost all had suffered from any one of the heat-related diseases. It is true that without population influx and urbanisation the world could have not been developed. However, it might be argued that enhancing urban neighbourhoods is the key to maintaining the high rates of urban expansion. This study will assist in formulating a framework for making a better urban neighbourhood considering the vulnerable population of the city.

**Reference**

[1] Turner M G 1990 Spatial and temporal analysis of landscape patterns *Landsc. Ecol.* **4**(1) 21–30
[2] Long H, Tang G, Li X and Heilig G K 2007 Socio-economic driving forces of land-use change in Kunshan, the Yangtze River Delta Economic Area of China *J. Environ. Manage.* **83** 351–64
[3] López E, Bocco G, Mendoza M, Duhau E 2001 Predicting land-cover and land-use change in the urban fringe. *Landsc. Urban. Plan.* **55**(4) 271–285
[4] Berling-Wolff S, Wu J 2004 Modeling urban landscape dynamics: A case study in Phoenix, USA *Urban Ecosyst.* **7**(3) 215-240
[5] Yuan F, Bauer M 2007 Comparison of impervious surface area and normalized difference vegetation index as indicators of surface urban heat island effects in Landsat imagery *Remote Sens. Environ.* **106**(3) 375-386
[6] Roth M, Oke T, Emery W 1989 Satellite-derived urban heat islands from three coastal cities and the utilization of such data in urban climatology *Int. J. Remote Sens.* **10**(11) 1699-1720
[7] Yuan F, Wu C, Bauer M 2008 Comparison of Spectral Analysis Techniques for Impervious Surface Estimation Using Landsat Imagery *Photogramm Eng. Remote Sensing* **74**(8) 1045-1055
[8] Li Z, Tang B, Wu H, Ren H, Yan G, Wan Z et al. 2013 Satellite-derived land surface temperature: Current status and perspectives *Remote Sens. Environ.***131** 14-37
[9] Sobrino J, Jimenez-Munoz J, Soria G, Romaguera M, Guanter L, Moreno J et al. 2008 Land Surface Emissivity Retrieval from Different VNIR and TIR Sensors. *IEEE Trans. Geosci. Remote Sens.* **46**(2) 316-327
[10] Brabec E, Schulte S, Richards P 2002 Impervious Surfaces and Water Quality: A Review of Current Literature and Its Implications for Watershed Planning *J. Plan. Lit.* **16**(4) 499-514
[11] Kjellstrom T, Butler A, Lucas R, Bonita R 2009 Public health impact of global heating due to climate change: potential effects on chronic non-communicable diseases *Int. J. Public Health* **55**(2) 97-103
[12] IPCC 2007 [Internet] Available from: https://www.ipcc.ch/2007/ [Accessed on 16 January 2021]
[13] WHO 2016 [Internet]. World Health Organization. Available from: https://www.who.int/news-room/fact-sheets/detail/noncommunicable-diseases [Accessed on 16 January 2022]
[14] Kroll M, Phalkey R, Kraas F 2015 Challenges to the surveillance of non-communicable diseases – a review of selected approaches *BMC Public Health* **15**(1) 1-12
[15] Mukherjee K 2020 Delineating fringe by rough-set theoretic approach: a case study on Barasat city, India *J. Phys. Conf. Ser.* **1579**(1) 012026
[16] Chander G, Markham B 2003 Revised landsat-5 TM radiometric calibration procedures and postcalibration dynamic ranges *IEEE Trans Geosci Remote Sens* **41**(11) 2674-2677
[17] Mukherjee K, Das P 2018 Modelling the Relationship between Urban Growth Modes and the Thermal Environment - A Case Study of the Barasat Municipality, West Bengal *Journal of Geography, Environment and Earth Science International* **17**(2) 1-19
[18] Coll C, Caselles V, Galve J, Valor E, Niclos R, Sanchez J et al. 2005 Ground measurements for the validation of land surface temperatures derived from AATSR and MODIS data *Remote Sens. Environ.* **97**(3) 288-300
[19] Burnham R, McKinley R, Vincent D 2006 Three Types of Skin-Surface Thermometers *Am J Phys Med Rehabil* **85**(7) 553-558
[20] Salisbury J, D'Aria D 1992 Infrared (8–14 µm) remote sensing of soil particle size *Remote Sens. Environ.* **42**(2) 157-165
[21] Kendall M 1975. *Multivariate analysis*. Charles Griffin & Company Ltd.

[22] Partal T, Kahya E 2006 Trend analysis in Turkish precipitation data *Hydrol. Process.* **20**(9) 2011-2026
[23] Hirsch R, Slack J 1984 A Nonparametric Trend Test for Seasonal Data With Serial Dependence *Water Resour. Res.* **20**(6) 727-732
[24] Sen P 1968 Estimates of the Regression Coefficient Based on Kendall's Tau *J Am Stat Assoc.* **63**(324) 1379-1389
[25] Yang Z, Chen Y, Wu Z, Qian Q, Zheng Z, Huang Q 2019 Spatial heterogeneity of the thermal environment based on the urban expansion of natural cities using open data in Guangzhou, China *Ecol. Indic.* **104** 524-534
[26] Chu L, Oloo F, Bergstedt H, Blaschke T 2020 Assessing the link between human modification and changes in land surface temperature in Hainan, China using image archives from google earth engine *Remote Sens.* **12**(5) 888

IOP Conf. Series: Earth and Environmental Science 1164 (2023) 012016 doi:10.1088/1755-1315/1164/1/012016

# Assessing the Socio-Economic Setup of Ravangla, Sikkim, and the Vulnerability of its Community using Geoinformatics

**S De[1], D Routh[1], S Hazra[1], P Kandar[1], S Sarkar[1], S Mitra[1], T Bhadra[1,*], R Sarkar[1], M Baral[1], A Rai[1]**

[1] Department of Geography, Adamas University, West Bengal, India

[*]Corresponding Author Email: tuhinbhadra.au@gmail.com

**Abstract.** The geoinformatics platform integrates remotely sensed data and Geographic Information Systems (GIS). It has a vast range of applications. The technique is widely used not only to study the physical aspects of a region but also to analyse the social and economic setup. The objective of this study is to understand the socio-economic setup of Ravangla, which is a small tourist town located in South Sikkim, India. The primary data have been collected from the field survey and analysed using GIS tools. The social indices such as social structure, housing condition, and the condition of basic facilities such as health, education, transport, market, and others have been analysed in the present study. GIS has been used to study the spatial distribution of these social and economic parameters. The technique has also been used to assess the socio-economic vulnerability of the community living in Ravangla. The Composite Vulnerability at the community level has been calculated based on socioeconomic parameters and a vulnerability map has been prepared. Based on the Composite Vulnerability Index, the households in the study area have been classified as Highly, Moderately, and Less Vulnerable. The GIS techniques have been used to amalgamate the socioeconomic components of vulnerability to recognize the most vulnerable households in Ravangla. The study will help planners, policymakers, and administrators to take necessary steps for mitigating socio-economic problems in the study area.

**Keywords:** Composite Vulnerability Index, Economic Vulnerability, Geoinformatics, Ravangla, Sikkim, Social Vulnerability.

## 1. Introduction

Geoinformatics is a branch of science that deals with acquiring, processing and handling spatial data. It combines remote sensing (RS) and geographical information systems (GIS) for spatial analysis which has huge applications in different fields. The technique is widely used not only to study the physical aspects of a region such as land-use land-cover mapping, weather forecasting, disaster management, and so on but also to analyse the social and economic setup like spatial distribution of facilities available in a region. It also helps in the planning and policy-making processes. The geographic properties of socioeconomic information of a region are completely different from its physical information [1], but they are interdependent. The socioeconomic setup has been used with other parameters like demographic, biophysical, etc. in several studies to solve the community-level problem with the application of geoinformatics. A geospatial technology-based integrated ecological and socioeconomic framework was used by Tali et al. to rank the medicinal plants in Jammu and Kashmir based on their priority [2]. Similarly, in another study, the GIS technique was used to map the malaria-endemic areas in the eastern part of Uttar Pradesh, India by integrating geographical information, socioeconomic setup and epidemiology of the region [3]. GIS is a useful tool for regional development studies as it integrates social and environmental data in the same platform.

The United Nations (UN) defines vulnerability as "the conditions determined by physical, social, economic and environmental factors or processes that increase the susceptibility of an individual, a community, assets or systems to the impact of hazards" [4]. A study by Maleki et al. reported vulnerability assessments of rural households in Iraq [5]. The geoinformatics platform was used to calculate the vulnerability index (VI) based on the Intergovernmental Panel on Climate Change (IPCC) guideline under the study [5]. Another study by Samanta et al. estimated the composite vulnerability of the Indian Sundarbans, where the socioeconomic and physical parameters were integrated using a geoinformatics platform [6]. The study by Senapati and Gupta assessed the vulnerability of the fishing community in the village of Mumbai, India where the climate change scenarios were considered [7]. The frequency of extreme weather occurrences will increase as a result of climate change [8]. A good number of studies have been done by researchers all over the world on the vulnerability of communities due to climate change. Studies have also been carried out on the vulnerability of the population in urban areas due to heat exposure [9-12]. People living in temperate climates are vulnerable to heat waves causing health issues. A study by Malakar and Mishra shows the vulnerabilities of eleven cities in India based on the vulnerability index, space index, infrastructure index, social index, and hazard index [13]. The study reveals that cities like Jaisalmer and Jodhpur are the most vulnerable, whereas metropolitan cities (except Chennai) are less vulnerable. Studies have also been done to assess the vulnerability of a community where indigenous knowledge and GIS are integrated for natural resource management.

Existing works on Ravangla, Sikkim, focus on the socio-economic setup of this region. Chakraborty and Chakma's work on the economy and social development in Sikkim explores that the occupation of the rural people in Sikkim is agriculture and it is based on organic farming techniques [14]. Sharma and Rohatagi's study on the rural development of Sikkim analyses the socio-economic data by using qualitative and quantitative techniques [15]. Another study by Sharma and Chettri, on cultural aspects of Sikkim, describes the house type, agricultural practice, culture of this area, electricity, water supply, and other socio-economic setups [16]. The literacy status of Sikkim has also been studied by researchers [17]. No study has been done to assess the vulnerability of the community of Ravangla. From this perspective, the present study emphasizes the socio-economic vulnerability at the community level in this region. The objective of this study is to understand the socio-economic setup of Ravangla and to assess the socio-economic vulnerability of the community living in this town. The GIS platform has been used in the study to prepare spatial maps of the vulnerable community. The study will help planners, policymakers, and administrators to take necessary steps for mitigating socio-economic problems in Ravangla.

## 2. Study Area

The Indian state of Sikkim is located in the eastern part of the country (figure 1). The state is bordered by Nepal, China, and Bhutan to the east, north, and west. West Bengal, an Indian state, is located in the south of Sikkim. Sikkim is divided into four districts, i.e. East, West, North, and South Sikkim (figure 1). South Sikkim is located in the southernmost part of the state. The altitude of South Sikkim ranges between 400 to 2000 metres and the climate of the region is mostly temperate. Namchi, Ravangla, Melli and Jorethang are the major urban centres in the region. Most of the industrialization in Sikkim takes place in South Sikkim as the state is geologically stable, the occurrence of landslides is comparatively less, has a good transport network and availability of flat land. South Sikkim is also famous for tea cultivation. Ravangla is a small tourist town in South Sikkim (figure1) and is located between Pelling and Gangtok. The altitude of the Ravangla is approximately 2440 metres. Upper parts of the Ravangla experience snowfall during winter months.

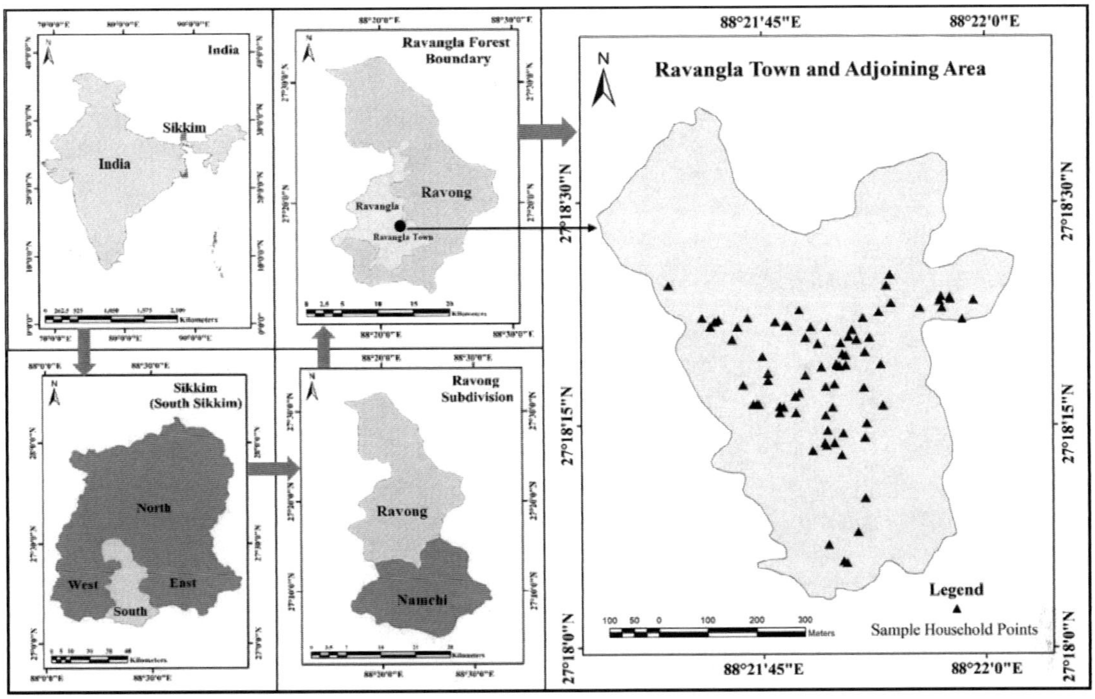


**Figure 1.** Study Area.

## 3. Methodology

The primary data related to the social and economic setup of Ravangla, have been collected from 100 households (figure 1) during the field survey. The social indices such as social structure, housing condition, and the condition of basic facilities such as health, education, transport, etc have been analysed in the present study. Composite Vulnerability at the community level has been calculated based on socioeconomic parameters (figure 2).

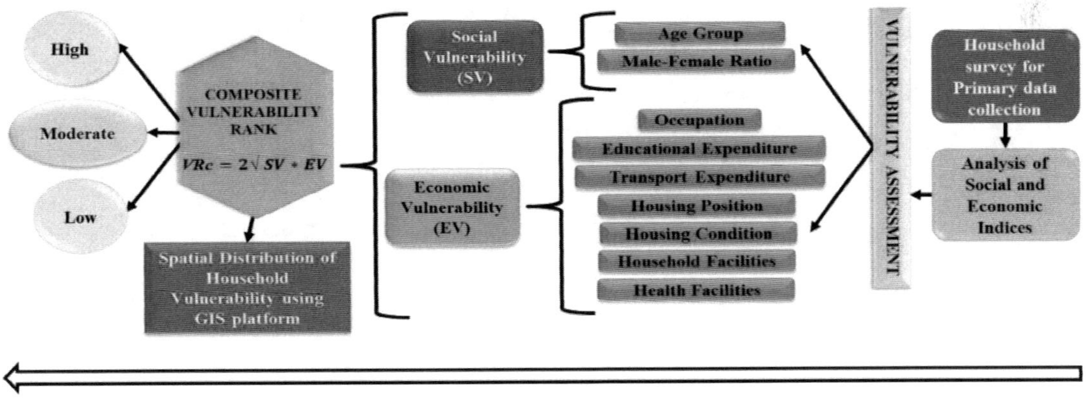


**Figure 2.** Methodology of Vulnerability Assessment.

In the present study, economic parameters, like household facilities occupation, transport expenditure, education expenditure, health facilities, etc, and social parameters like age group, male-female ratio, etc. of Ravangla have been taken to calculate the composite vulnerability index (figure 2). Vulnerability Rank (table 1) has been assigned to each parameter based on stakeholder consultation.

IOP Conf. Series: Earth and Environmental Science 1164 (2023) 012016 doi:10.1088/1755-1315/1164/1/012016

Table 1 shows the parameter-wise vulnerability ranking categories, where, rank 1 represents low vulnerability, rank 2 and rank 3 represent moderate and high vulnerability respectively. After the detailed analysis, vulnerability maps have been prepared. Based on the Composite Vulnerability Index the households in the study area have been categorized as Highly, Moderate, and Less Vulnerable. The GIS platform has been used to amalgamate the socioeconomic components of vulnerability for recognizing the most vulnerable households in Ravangla.

**Table 1.** Vulnerability Ranking of some Social and Economic Parameters.

| *Vulnerability Components* | *Parameters* | *Descriptions* | *Vulnerability Ranks* |
|---|---|---|---|
| **Social Vulnerability (SV)** | Age Group | Families having <30% old or child member | Rank 1 |
| | | Families do not fit in Rank 1 and 3 | Rank 2 |
| | | Families having >30% old or child member | Rank 3 |
| | Male-Female Ratio | Families having <50% woman | Rank 1 |
| | | Families do not fit in Rank 1 and 3 | Rank 2 |
| | | Families having >50% woman | Rank 3 |
| **Economic Vulnerability (EV)** | Occupation | Govt. employment/Service sector/Teaching/Business | Rank 1 |
| | | Driver/Shopkeeper/Repairman/Dairy farming/Mechanic | Rank 2 |
| | | Farmer/ Daily labour/Construction worker | Rank 3 |
| | Family Expenditure (for Education and Transport) | Education: >8000 Rs, Transport: >4000 Rs | Rank 1 |
| | | Education: 4000-8000 Rs, Transport: 2000-4000 Rs | Rank 2 |
| | | Education: <4000 Rs or free, Transport: <2000 Rs | Rank 3 |
| | Health Facilities | Easy access to ambulance/doctors/treatment centre | Rank 1 |
| | | Not easy access to ambulance/doctors /treatment centre | Rank 2 |
| | | No access to ambulance/doctors/treatment centre | Rank 3 |
| | Housing Position | Roadside | Rank 1 |
| | | Valley | Rank 2 |
| | | Slope | Rank 3 |
| | Housing Condition | Owner, Pucca house, >8 rooms, >4 floors | Rank 1 |
| | | Govt. Quarter, Semi-pucca house, 4-8 rooms, 2-4 floors | Rank 2 |
| | | Tenant, Kutcha house, <4 rooms, <2 floors | Rank 3 |
| | Household Facilities | Electricity, Water supply, T.V., Computer, Personal Transport | Rank 1 |
| | | Electricity, Water supply, T.V. | Rank 2 |
| | | Electricity, Water supply | Rank 3 |

* Rank 1: Low vulnerability; Rank 2: Moderate vulnerability; Rank 3: High vulnerability

IOP Conf. Series: Earth and Environmental Science 1164 (2023) 012016 doi:10.1088/1755-1315/1164/1/012016

## 4. Results and Discussion

The community-level vulnerability in Ravangla has been assessed based on some social and economic parameters. The age structure and male-female ratio are analysed to assess social vulnerability. Around 49 per cent population of Ravangla are male and 51 per cent are female. The maximum population of the region belongs to the age group between 16 to 30 years. The study shows that the maximum number of households is highly vulnerable according to the age group and male-female ratio (figure 3).

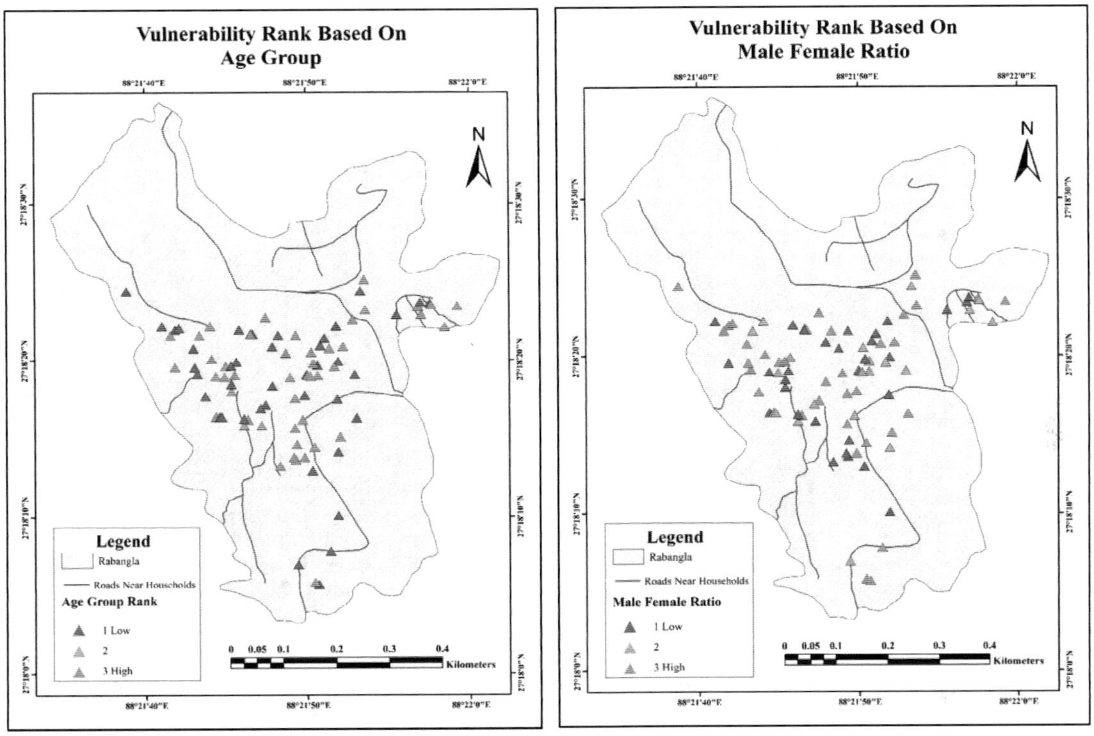


**Figure 3.** Vulnerability Rank based on the age group and the male-female ratio.

The economic vulnerability has been assessed based on occupation structure, educational expenditure, transport expenditure, health care facilities, housing position, housing condition, and household facilities (table 1). There is a noticeable difference in the occupation status among the male and female populations. The share of the male population among the total population engaged in any occupation is around 76 per cent and the share of the female population is very less i.e. around 24 per cent. Most of the working population in this town are engaged in secondary and tertiary activities. Approximately 25 per cent of people have their own business followed by 12 per cent and 10 per cent are shopkeepers and Government employees respectively. People, who are engaged in tertiary activities are comparatively less vulnerable in terms of economic condition, whereas the people engaged in secondary and primary activities are moderately and highly vulnerable respectively. According to the occupation structure, the number of moderately vulnerable households is higher in this region (figure 4). The education status among males and females is almost similar, the share of male and female population having an education is around 51 per cent, and 49 per cent respectively. The education expenditure is quite high in Ravangla, around 34 per cent of surveyed population pay 5000 to 10000 rupees per month. This group is economically less vulnerable compared to others (figure 4). Around 31 per cent of surveyed population pay 1000 to 5000 rupees per month for their child's education. This group is moderately and highly vulnerable in terms of economic status (figure 4). Based on educational expenditure 50 per cent of surveyed households are highly vulnerable.

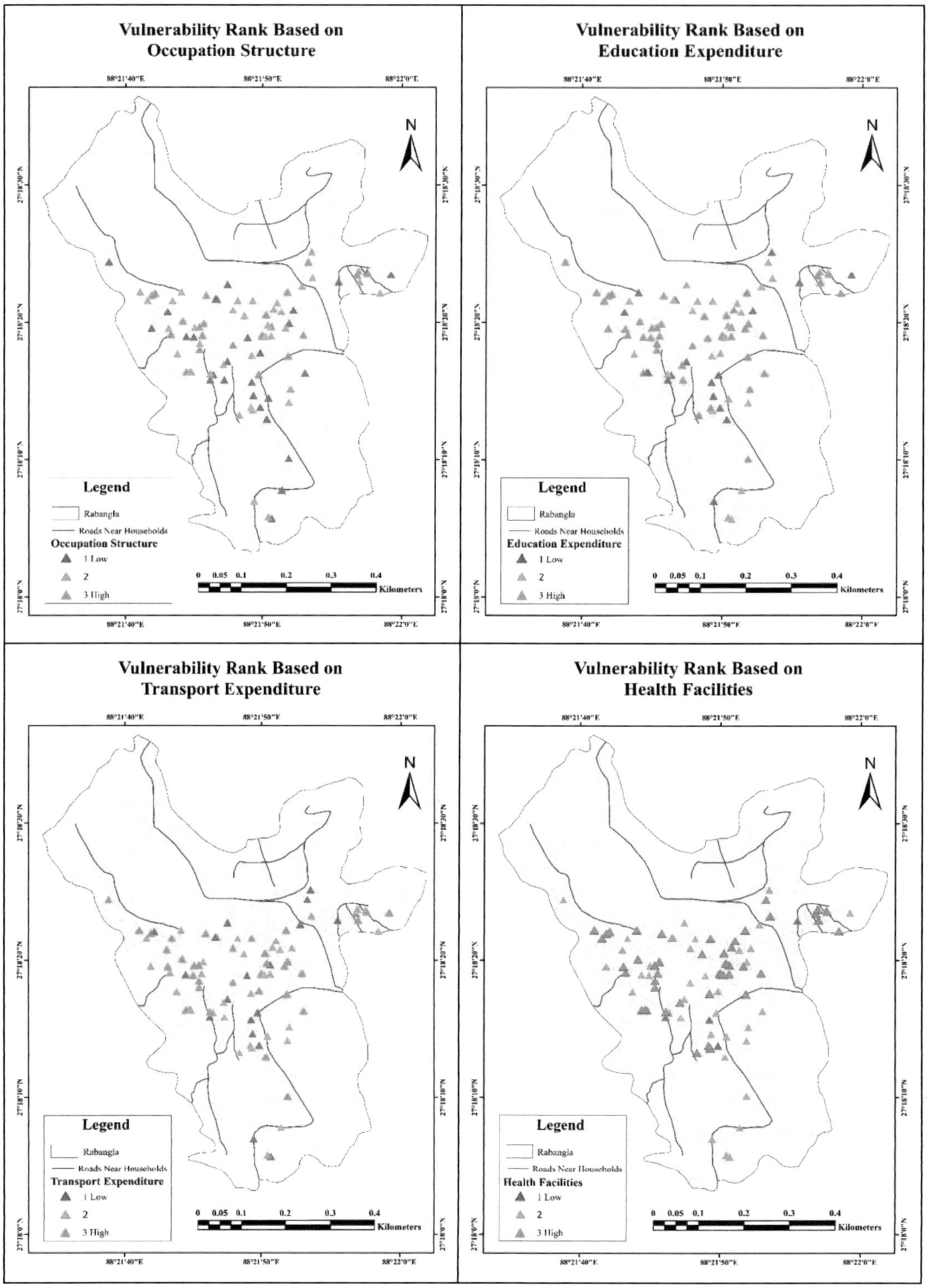


**Figure 4.** Vulnerability Rank based on occupation structure, educational expenditure, transport expenditure and health care facilities.

The primary data related to transport facilities have been analysed to understand the commonly used transport facility, frequency of the services, and expenditure on transport. The study shows that the people of Ravangla mostly use taxi services, i.e. almost 79 per cent of the total surveyed population. Auto and Rickshaw are very less in the town due to the topography of the region. The bus service is quite poor in this region. About 88 per cent of the surveyed population are facing the problem of infrequent bus service. The transport expenditure is average in Ravangla. Around 26 per cent of surveyed population pay rupees 2000 to 3000 per month on transport. They are economically moderately vulnerable (figure 4). Almost 23 per cent of surveyed population pay around 1000 to 2000 rupees per month on transport services. This group is economically highly vulnerable in terms of transport expenditure (figure 4). Around 22 per cent of the surveyed population pay more than 4000 rupees per month on transport services. This group is comparatively less vulnerable (figure. 4). Most of the people in Ravangla are facing problems with transportation for the narrow roads (i.e. almost 30 per cent of the surveyed population), and landslides (i.e. almost 26 per cent of the surveyed population). Other transport-related problems in the region include the bad condition of the roads, traffic congestion, and Low-frequency of public bus services. Based on transport expenditure 40 per cent of surveyed households are highly vulnerable (figure 4).

Health infrastructure is another important parameter to assess the economic condition of a region. Types of health institutes that are present near the households are hospitals, nursing homes, and health centres. Around 50 per cent and 20 per cent of people in the surveyed household go to hospitals and health centres for their treatment respectively. Almost 12 per cent of the total surveyed population visit unskilled doctors and 11 per cent used to visit Nursing homes for treatments. Due to the cold weather of the region, most of the surveyed population (i.e. 70 per cent) have a common cough, cold, fever, and related problems. Another common health issue in the hilly region is arthritis, with almost 18 per cent of people having this problem. Some other health issues are diabetes, blood pressure disorder, diarrhoea, and tuberculosis. Households, which have easy access to ambulance facilities, doctors, and treatment centres are comparatively less vulnerable. Moreover, the health facilities of the town are not satisfactory as most of the households have moderate to high vulnerability rank (figure 4).

Hosing position, housing condition, and household facilities are the other parameters that come under vulnerability ranking. The field investigation reveals that most of the surveyed households (i.e. around 61 per cent) are located on the hill slope, which makes them vulnerable (figure 5). Around 31 per cent of total households are located on the roadside which is comparatively less vulnerable.

According to the survey, the houses of the town are mostly pucca, which is 60 per cent of the total surveyed houses. The building material of pucca houses are mainly bricks, cement, iron rods, and modern tiles. Pucca houses are comparatively less vulnerable (figure 5). A small proportion of the houses are kutcha (i.e. around 25 per cent), which are mainly built with mud, wood, bamboo, and earthen tiles. These kutcha houses are more vulnerable (figure 5). Rest of the houses are Semi-Pucca houses (i.e. around 15 per cent). Semi-pucca houses are moderately vulnerable (figure 5).

The field investigation reveals that almost 27 per cent of the surveyed house has four rooms followed by 22 per cent and 20 per cent of the surveyed house have two rooms and three rooms respectively. Few numbers of houses have eight rooms (i.e. 5 per cent) and ten rooms (i.e. 3 per cent). The houses having less number of rooms are economically more vulnerable (figure 5). There are very few high-raised buildings in Ravangla. The region is mostly dominated by single-storeyed houses. About 62 per cent of the surveyed houses in the town are single-storeyed followed by the double-storeyed houses i.e. 22 per cent of the surveyed houses. Only 6 per cent, 7 per cent, and 2 per cent of the surveyed houses have three, four, and five stories respectively. There is only one high-raised house having 6 stories. The one-storeyed houses are economically more vulnerable (figure 5).

More than 50 per cent of the surveyed household ranked low in the vulnerability index (figure 5) based on the availability of household facilities as most of the households has basic facilities such as electricity, water supply, television, computer, and other necessary assets. Around 37 per cent of the total surveyed houses have a balcony, and 17 per cent of the houses have a garden. Almost 91 per cent of the houses have a television but few of them (i.e. 23 per cent) avail computer.

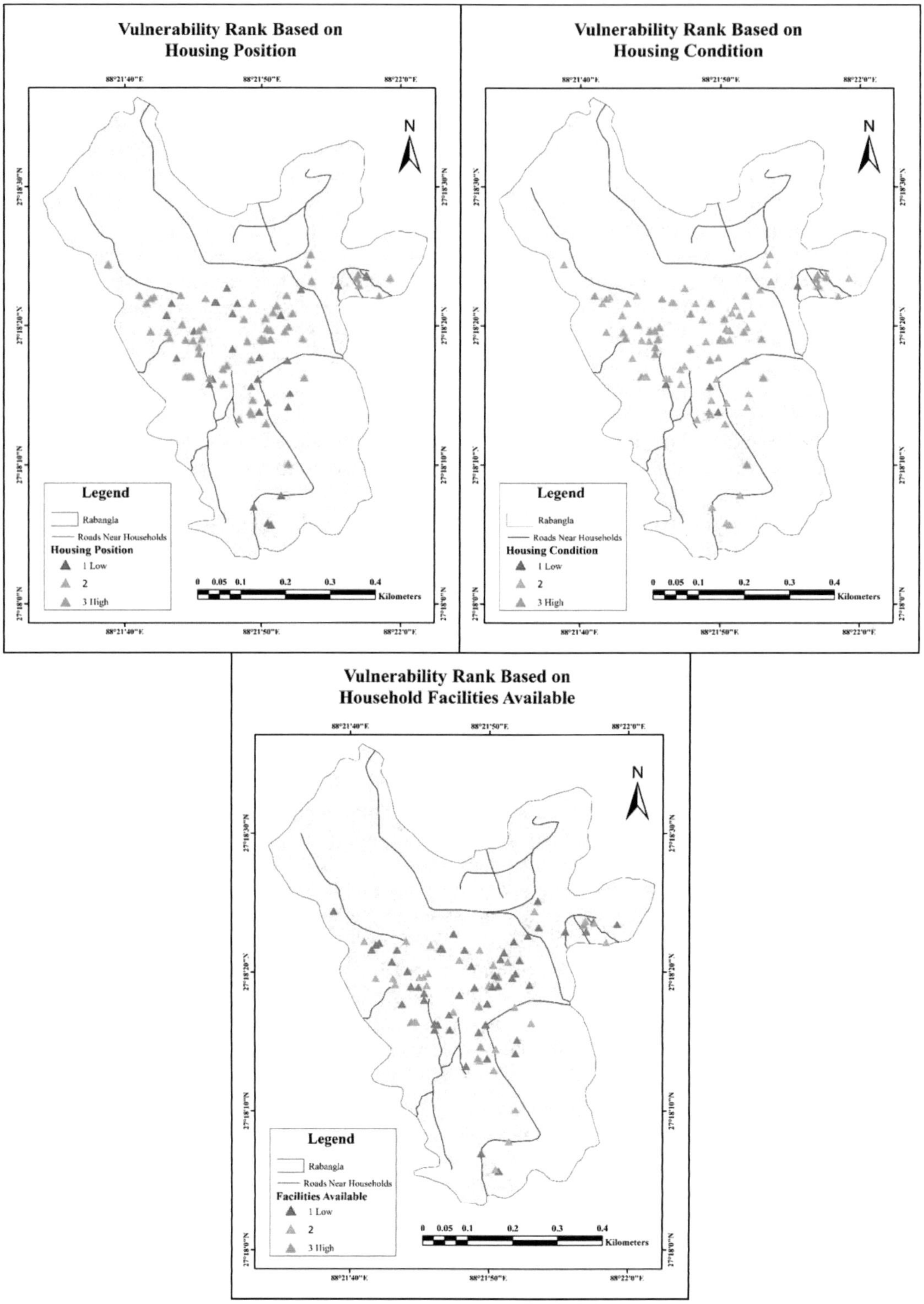


**Figure 5.** Vulnerability Rank based on housing position, housing condition, and household facilities.

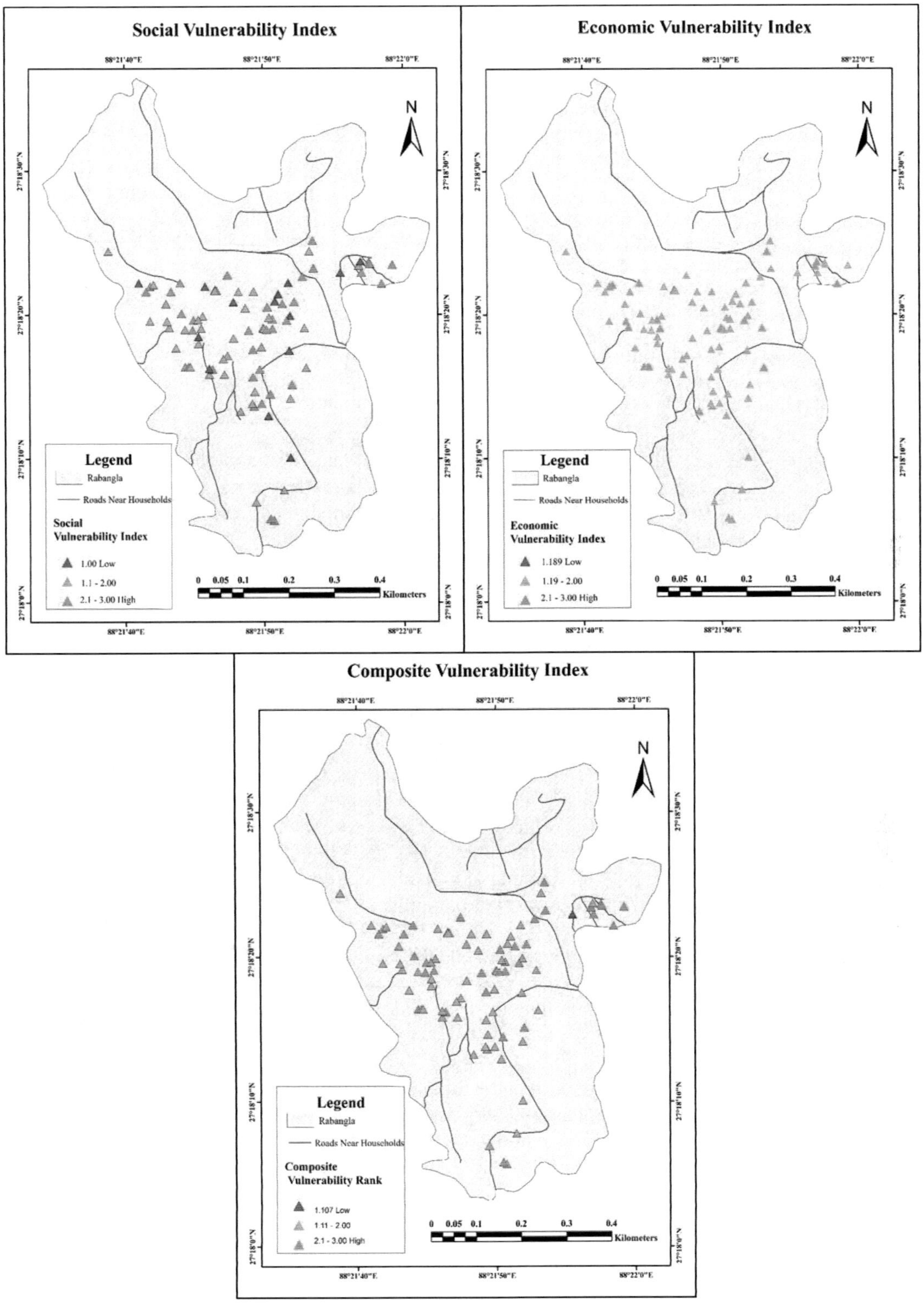


**Figure 6.** Community level Social, Economic and Composite Vulnerability.

Based on the social and economic indicators discussed in the previous section, social, economic and composite vulnerability maps are prepared. The social vulnerability Index shows that only 16 per cent of surveyed households have a low vulnerability rank, around 35 per cent have a high vulnerability rank, whereas, most of the households are socially moderately vulnerable (figure 6). In the economic vulnerability index, it is observed that there is no household under the low vulnerability rank. Around 77 per cent and 23 per cent of the surveyed households are moderately and highly vulnerable respectively. The composite vulnerability ranking represents the overall social and economic conditions of the households. The index shows that only one house has a low vulnerability rank, whereas, almost 66 per cent of surveyed households are moderately vulnerable and the rest of the households are highly vulnerable.

## 5. Conclusion

Conclusively it can be said that the community of Ravangla town is moderately vulnerable in terms of social and economic conditions. Every people have the basic facilities and amenities in their home to sustain life and have basic education and access to basic health care facilities. Everyone is engaged in different types of occupations, and none of them is too poor. The topography makes the households vulnerable but their condition is not too bad and for this reason, they come under the moderately vulnerable group. The Composite Vulnerability Rank shows numerous dimensions of vulnerabilities. It has also been found that most of the households in Ravangla are moderately vulnerable according to the composite vulnerability index. No specific zone is identified on the map of Ravangla town where the vulnerable houses are clustered. The whole map is covered with both moderately and highly vulnerable houses in a dispersed pattern. The low-vulnerable houses are found in the eastern part of the town. This study will help planners, policymakers, and administrators to take the necessary steps for mitigating socioeconomic problems and to reduce the vulnerability at the community level in the study area.

## References:

[1] Hung T and Yasuoka Y 2000 Integration and application of socio-economic and environmental data within GIS for development study in Thailand AARS, *GISdevelopment.net* **10**

[2] Tali BA, Khuroo AA, Nawchoo IA, and Ganie AH 2018 Prioritizing Conservation of Medicinal Flora in the Himalayan Biodiversity Prioritizing Conservation of Medicinal Flora in the Himalayan Biodiversity Hotspot: An Integrated Ecological and Socioeconomic Approach *Environmental Conservation* 1-8 https://doi.org/10.1017/S0376892918000425

[3] Qayum A, Arya R, Kumar P and Lynn AM 2015 Socio-economic, epidemiological and geographic features based on GIS-integrated mapping to identify malarial hotspots *Malaria journal* **14** 1-20

[4] UN-Habitat 2016 Urbanization and Development: Emerging Features. World Cities Report.

[5] Maleki R, Nooripoor M, Azadi H, and Lebailly P 2018 Vulnerability Assessment of Rural Households to Urmia Lake Drying (the Case of Shabestar Region) *Sustainability* **10** 1–20 https://doi.org/10.3390/su10061862

[6] Samanta B, Das S, and Hazra S 2017 Micro Level Vulnerability Assessment of a Community Living in Mousuni Island in the Indian Sundarban: An Integrated Study Employing Geoinformatics *Springer Remote Sensing/Photogrammetry Environment and Earth Observation* 195- 213 https://doi.org/10.1007/978-3-319-46010-9

[7] Senapati S and Gupta V 2017 Socio Economy vulnerability due to climate change: Derivating indicators for fishing communities in Mumbai *Marine Policy* **76** 90-97

[8] IPCC Climate Change 2007 The physical science basis. Cambridge: Fourth assessment report to the Intergovernmental panel on climate change *Agenda* **6** 333

[9] Sheridon SC and Dixon PG 2017 Spatiotemporal trends in human vulnerability and adaption to heat across the USA *Anthropocene* **20** 61-73

[10] Knowlton K, Lynn B, Goldberg RA, Rosenzweig C, Hogrefe C, Rosenthal JK and Kinney PL 2007 Projection heat-related mortality impacts under a changing climate in the New York City *American Journal of Public Health* **97** 2028-2034

[11] Rey G, Fouillet A, Bessemoulin P, Frayssinet P, Duffour A, Jougla E and Hemon D 2009 Heat exposure and Socio-economic vulnerability as synergistic factors in heat wave related mortality *European journal of epidemiology Springer science* **24** 495-502

[12] Ho HC, Knubdy A, Chi A, Aminipouri M and Lai DYF 2018 Spatiotemporal analysis of regional socio-economic vulnerability change associated with heat risk in Canada *Applied Geography Elsevier* **95** 61-70

[13] Malakar K and Mishra T 2017 Assessing Socio-economic vulnerability to climate change: a city-level index-based approach *Climate and Development* **9** 348-363 http://dx.doi.org/10.1080/17565529.2016.115444, 2015: 1-17

[14] Chakraborty S and Chakma N 2016 Economy and Social Development of Rural Sikkim *Space and Culture, India* **4** 61-79 https://spaceandculture.in/index.php/spaceandculture/article/view/198

[15] Sharma T and Rohatagi S 2013 Level of Rural Development In Sikkim: A case study http://14.139.206.50:8080/jspui/bitstream/1/4199/1/Level%20of%20rural%20development%20in%20Sikkim%20a%20case%20study%20of%20yuksam%20development%20block.pdf

[16] Sharma BK and Chettri L 2020 Cultural Aspects of Sikkim (India): A Review *International Journal of Arts and Social Science* **3** 180-186

[17] Das A 1999 Socio-economic Development In India *Development and Society* **28** 313-345

# The Quest for Quality Healthcare Services – What we can learn from a settlement-based case study?

**Sarani Roy[1, 2], Anu Rai[1]***

[1] Department of Geography, Adamas University, Kolkata, India.
[2]Cyber Swift, Kolkata, India

*Email: hianurai@gmail.com / anu.rai@adamasuniversity.ac.in

**Abstract**. Adopting the case study approach present study attempts to unwrap the status of the healthcare accessibility of the Berhampur Community Development (CD) block of Murshidabad. A network-based method integrated with the gravity model is used to identify under-serviced areas that do not have access to essential healthcare services at the settlement level. The attempt is made to map the significant socio-economic and demographic factors affecting the accessibility of quality healthcare services that can suggest the best possible solution to considering the heterogeneous needs of the population concerning healthcare services. The healthcare accessibility index (Ai) to the population demand is mapped to delineate the underserviced area. The Joint Hansen's Improved Gravity Model and Huff Model (probability model) is used with deprivation mapping. Following the human- rights approach of the good living standards, fourteen indicators were extracted from census data, and the indicators were standardized by applying a min-max normalization formula. The indices calculate to measure healthcare deprivation, and inaccessibility at the neighborhood scale indicate the sharp differences in the spatial context. The increasing coverage of healthcare facilities has yet to address the issue of accessibility as an interplay between facilities-connectivity matrix. The study used a connectivity matrix to analyze healthcare accessibility and found that network-based measurement of health accessibility is more suitable over straight-line distance. We found this work may help support policymaking and bridge the gap between the biased share of infrastructure coverage and health care disparity in the study area.

## 1. Introduction

The Lancet's (2018) recent study says that more than 2.4 million people died in India in 2016 due to conditions that could have been treated by healthcare. This number is the highest number of "amenable deaths" among the 137 low-income countries [1]. For rural India healthcare access is the bigger problem than quality, where 72% of urban medical practitioners are serving 28% of the country's population. The Sustainable Development Goals (SDGs) advocate offering desired accessibility to the health facilities for all sections of the population. In the context of human rights, it is also directed that "*health facilities, goods, and services must be within safe physical reach for all sections of the population, especially vulnerable or marginalized groups, such as ethnic minorities and indigenous populations, women, children, adolescents, older persons, persons with disabilities and persons with HIV/AIDS, including in rural areas*" [1]. The Sustainable Development Goals (SDGs) advocate offering desired accessibility to the health facilities for all sections of the population. In terms of the quality and accessibility of healthcare, India is ranked 145th out of 195 nations [1], however, the hospital industry

IOP Conf. Series: Earth and Environmental Science 1164 (2023) 012017 doi:10.1088/1755-1315/1164/1/012017

is rising at a CAGR of 16-17% to reach INR 8.6 trillion by FY2022 from INR 4 trillion of the current year [2]. The study seems to be conflating "access" with "coverage". The major goal of the study is to bridge the gap between understanding coverage and access to health care services by adopting a settlement-based study. We use the Berhampur CD Block as a case study to examine the demand for high-quality healthcare on a settlement-centric spatial scale while taking healthcare accessibility index into account. The Berhampur CD Block had a total population of 446,887, of which 75.55 % were in rural areas and 24.45% were in urban areas.

## 2. Literature Review

### *2.1. Access to health*

As per human rights context "health facilities, goods and services must be within safe physical reach for all sections of the population, especially vulnerable or marginalized groups' [1]. The provision of the appropriate service at the appropriate time and location is defined by Rogers (1999) as optimal access. [3]. Utilization has been defined by several scholars as realized access, and it is believed that demographic and health system variables determine utilization. Similarly, others highlighting the relation between the concepts of utilization and access with regards to facilitating or impeding the use of services by potential users [3]. Inequalities in geographic access to healthcare result from the configuration of facilities, population distribution, and transportation infrastructure. In a broader sense, accessibility is defined in four dimensions of access: availability, the geographical dimension that relates to the physical location, affordability, and acceptability. Accessibility in this context includes geographic distance between origin and destination and distance in terms of time spent on reaching the next location [4]. There is a difference between "having access" to health care and "gaining access" to health care. The former is related to the availability of health services and the latter refers to whether health demand can overcome financial, organizational, and socio-cultural barriers and utilize health services. Accessibility studies cover issues including 'affordability', 'physical accessibility, and 'acceptability' [5]. It represents the distinct existence of barriers, challenges, and impediments or of variables that are facilitators for the recipients of health care. It is the functional interaction between the population, medical facilities and resources. [6]. The classical (Euclidean) distance measure has been replaced in more recent accessibility studies with more realistic measurements like trip distance or time. In a Geographic Information System, approaches based on raster data and networks are frequently used to calculate journey times.

### *2.2. Accessibility, deprivation and population vulnerability*

Access to health care is a product of demand and supply factors including location, availability, cost, and appropriateness of services and demand factors, such as the burden of disease and knowledge, attitudes and skills, and self-care practices. Population heterogeneity and socio-economic diversity are crucial determinants. The healthcare-seeking episode is influenced by the physical and healthcare infrastructure available within the neighborhood. It is determined by the need of the healthcare seeker; personal, household, occupational, and latent perceptions of the users [7]. The identification and investigation of health inequities as well as the support for policy-making depend on indicators defining health phenomena [8]. In reality, accessibility research has given us useful tools to analyse how people can get to locations where they participate in routine activities within the context of the socioeconomic exclusion dimensions of mobility. Model-based estimates of distance travelled are used to calculate cumulative opportunities measure of accessibility [9]. Measures of deprivation are widely used method to identify localities or neighborhoods in which residents experience higher levels of 'poverty', or socio-economic 'need' and who, as a result, are more likely to suffer from poor health and social exclusion [10]. Where there is social isolation, unemployment, low-income levels, lack of available transport, and poor access to a range of services, health can suffer disproportionately. The higher the local level of deprivation, the greater the likelihood of the range of health and social care services that will be required to meet needs.

### *2.3. Health GIS and infrastructure modelling*

The application of Health-GIS has brought a major shift in the way of analyzing challenges of healthcare accessibility [11]. Many scholarly attempts have been made in accessing the healthcare infrastructure based spatial element of area and distance. Area-based measures describe areas like countries, towns, or physicians which have a properly defined boundary and the distance-based measures focus on the distance or travel time, or cost between the population and the health services [12]. Recently GIS application has been made for the evaluation of accessibility to mammography resources in Mississippi, USA based on an ideal driving route [13]. The use of GIS and remote sensing to manage, evaluate, study, and calculate the served and unserved population and area of supply and demand of health infrastructure and proposing healthcare model are widely acknowledged in literature [14].

## 3. Research Aim and Objectives

Adopting the case study of Berhampore Community Development (CD) block of Murshidabad present study is an approach:

- To unwrap the status of the healthcare accessibility of Berhampore.
- To identify the under-serviced area that does not have access to basic healthcare services at the settlement level.
- To map the major socio-economic and demographic factors affecting the accessibility of quality healthcare services.
- To suggest the best possible solution considering the heterogeneous need of the population in relation to healthcare services.

## 4. Methodology

The major goal of the study is to bridge the gap between in understanding coverage and access to health care services adopting a settlement-based study of Berhampur CD Block. The methodological section can be subcategorized under following sub-heads:

### *4.1. Study Area*

In the Berhampur subdivision of the Murshidabad district in the Indian state of West Bengal, there is a community development block called Berhampore that serves as an administrative division (Figure 1). Referring to Table 1, we see that with a total of 122 villages and ten census towns, the Berhampore CD Block has a total population of 446,887, of which 337,623 were rural, and 109,264 were urban [15]. There were 228,650 (51%) males and 218,237 (49%) females. Scheduled Castes numbered 76,935 (17.22%), and Scheduled Tribes numbered 10,809 (2.42%). The block is primarily dependent on the agricultural economy where most of the people are engaged in primary activities; hence the area (225.93 sq km) covered by agriculture is more than the other land use in the Berhampore CD block. The district is served by a total road length of 1219.50 km.

### *4.2. Database*

The major goal of the study is to bridge the gap between understanding coverage and access to health care services by adopting a settlement-based study. The secondary data like the LULC map of the Berhampore CD Block was downloaded from the District Land & Land Reforms Office, Murshidabad, Government of West Bengal, and different data related to the health facility, location, population, and other socio-economic data are collected from the Census of Murshidabad that includes the health center size (capacity), the number of people living in the block of the town, and the population density of this block.

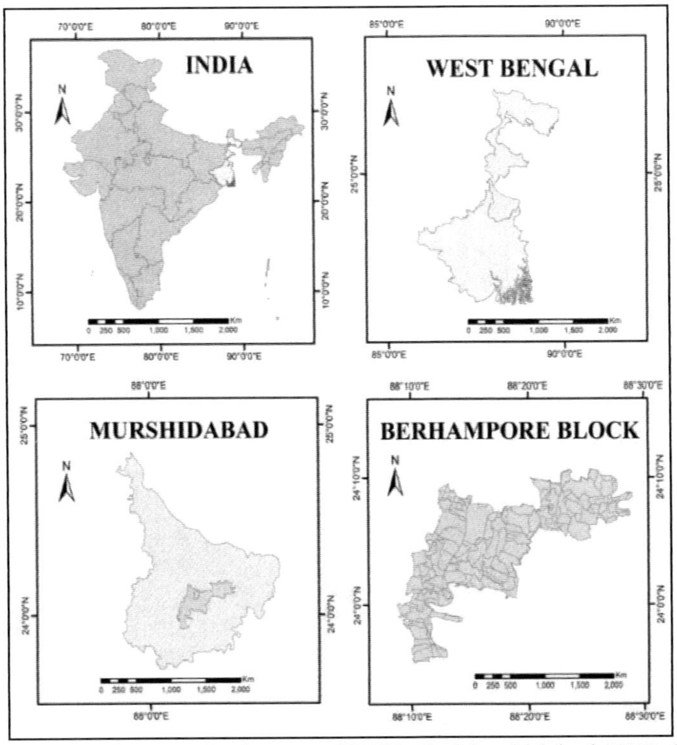


**Figure 1.** Study Area, Berhampur CD Block, Murshidabad, W.B. 2021

**Table 1.** Socio-Demographic Characteristics, Berhampur CD Block, Murshidabad, W.B. 2021

| Demography | | Number (%) |
|---|---|---|
| Population | Urban | 109,264 (24.45%) |
| | Rural | 337,623 (75.55%) |
| | Total | 446,887 |
| Population Density | | 2,300/$Km^2$ |
| Census Towns | | 9 |
| Villages | | 122 |
| Occupation (in percent) | Primary Sector | 52.81% |
| | Secondary Sector | 3.12% |
| | Tertiary Sector | 44.07% |
| Literacy (2011) (in percent) | Male | 53.31% |
| | Female | 99.96% |
| Religion (in percent) | Hindu | 85.67% |
| | Muslim | 53.86% |
| Health Statistics (2014) | Hospitals | 3 |
| | Primary Health Centre | 2 |
| | Private Hospitals | 18 |
| | Family Welfare Subcentres | 57 |
| | Indoor Patients treated | 6207 |
| | Outdoor Patients treated | 145229 |

Source: Official Website of District Magistrate Murshidabad, Government of West Bengal, India available at https://murshidabad.gov.in/ visited at 20th March 2020.

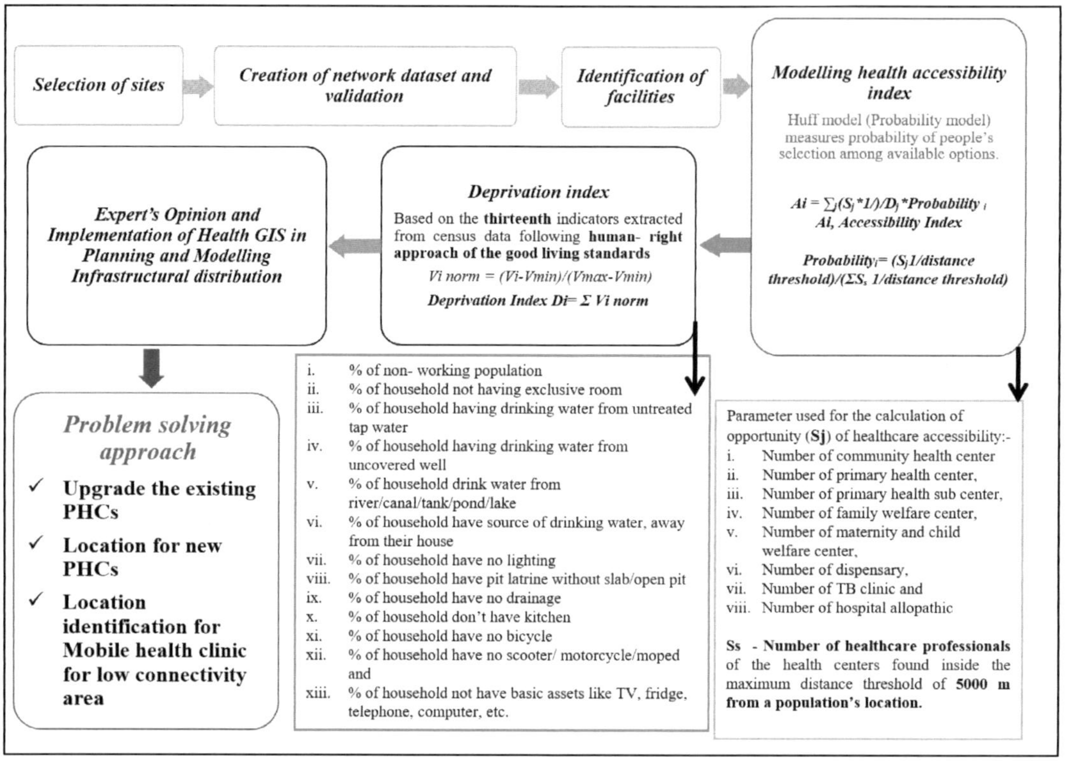


**Figure 2.** Methodology adopted

### *4.3. Selection of the study*

This study only includes the settlement boundary extracted from the LULC classification using Landsat 8 images for the year 2018 to fulfill the objectives of the study. Village boundaries which include various land use coverages like agricultural and barren lands, often-present blurred images of actual resource distribution and accessibility to population. Henceforth, this study adopts a settlement-based study and only includes the settlement boundaries irrespective of the administrative boundaries of the villages. The settlements were categorized into rural and urban types, and the names of the settlements have been verified from the toposheets collected from the Survey of India, Kolkata No. 78D/8, 79A/1, 78D/4, which were further ground verified from Google Earth.

### *4.4. Creation of network and facilities dataset*

The network dataset required for the analysis is extracted from Google Earth 2018. It includes national highways, state highways, and neighborhood connectors. Cart track, pack track, and other non-permanent paths are excluded from the study. Further, all primary healthcare facilities like primary health centers (PHCs), sub-centers (SCs), community health centers (CHCs), public health centers (urban), and major hospitals and clinics were extracted from Google Earth 2018. Locations of all facilities were verified using a GPS survey that was conducted from December 2018 to March 2019. The detailed infrastructure statistic including the number of beds, doctors, man-power, and other amenities required for the PHCs, SCs, and CHCs was collected from the District Census Handbook (Town and Village Amenities), 2011, Office of the Registrar General & Census Commissioner, Ministry of Home Affairs, Government of India.

### 4.5. Modelling health accessibility index

Three factors—healthcare capacity, population demand, and geographic impediment—are crucial when measuring and evaluating healthcare accessibility [16]. As to measure and evaluate healthcare accessibility, three factors are essential: healthcare capacity, population demand, and geographic impedance. Healthcare capacity is the supply of healthcare services. Additionally, it can be represented by using the number of specific facilities, physicians, or sickbeds. Population demand means the number of people who may need the services. Geographic impedance reflects the degree to which accessibility will be impacted by the "distance" between a service's location and the population it is intended to serve [17,18]. In 1959, Hansen proposed the first law of geography as a measure of accessibility. He considered all service facilities for the selection to calculate the opportunity-accumulated value.

$S_j / D\frac{\beta}{ij}$ is the gravity force from $j$ to $i$, where $S_j$ is the scale of $j$, $D_{ij}$ is the impedance between $i$ and $j$, and $\beta$ is the friction coefficient that describes travel impedance. The gravity model is expressed as following equation: $A_i = n \sum j=1$

$$A_{ij} = n \sum j=1 \; S_j / D\frac{\beta}{ij} \tag{1}$$

Further, Hansen (1959) integrated the concepts of the Huff Model and Gravity Model. According to the Huff model, a commercial facility's consumption rate should be proportionate to how well a choice was made [19]. This measure of probability solves the service area division problem for multiple stores and calculates the probability of selecting a commercial zone among consumers. This study adopts an integrated Huff-Gravity model to devise a healthcare accessibility index for the study area. The community health center, primary health center, primary health sub-center, family welfare center, maternity, dispensary, TB clinic, and allopathic hospital centers had been selected to measure the opportunity of the health care accessibility. The travel time from these facilities had been extracted using Google API. Origin- Destination (OD) Matrix-based on straight-line distance is computed based on the network and settlement centers to identify geographic impedance.

Integrated Huff and Gravity Model state that accessibility in a unit area $i$ (a census block) can be expressed in the following generalized measure of accessibility:

$$A_i = \sum_j \frac{S_j * C_{ij}}{D_j} \tag{2}$$

*Where, $S_j$ = the opportunity in the area j (Hospital/number of beds or healthcare professionals)*

*$C_{ij}$ = a function of travel cost from unit area i to area j (network distance/ travel time)*

*$D_j$ = population demand (Σ of population may access hospital)*

To select population, that to be included in the study, travel distances from each census block centroid (settlement/village/urban) to the nearest health service centres have been obtained using network straight-line distance in meter and average straight-line distance is estimated.

$$A_i = \sum_j \frac{S_j * \frac{1}{5000}}{D_j} \tag{3}$$

*Where, 1/5000 = maximum travel distance threshold*

Further, probability of selecting particular facility among other available facilities is calculated using following equation:

$$Probability_i = \frac{S_i \frac{1}{5000}}{\Sigma S_* \frac{1}{5000}} \tag{4}$$

*Where, Probability $_i$ = Probability of population in location 'i' visiting a primary healthcare services j; $S_s$ = number of healthcare professionals of the health centers available inside the maximum distance threshold of 5000 meter from the location of population.*

Henceforth, this study measures index of healthcare accessibility (*Ai*) as:

$$Ai = \sum_j \frac{S_j * \frac{1}{5000}}{D_j} * Probability_{\,i} \tag{5}$$

Parameter used for the calculation of opportunity (*Sj*) of healthcare accessibility:

*Number of community health centre*
*Number of primary health centre,*
*Number of primary health sub centre,*
*Number of family welfare centre,*
*Number of maternity and child welfare centre,*
*Number of dispensaries,*
*Number of TB clinic and*
*Number of hospital allopathic*

The index of healthcare accessibility was further normalised using min-max normalisation method.

### 4.6. Computing deprivation index

Assuming that access to quality healthcare is an outcome of an interplay between network connectivity, socio-economic and demographic factors, we compute the relative material deprivation index [20] to identify the most vulnerable population. Relative material deprivation measures the population's relative ability to access services or goods (educational/employment/community resources/housing). The House Listing and Household Amenities that is House Listing Primary Census Abstract, HLPCA, and the Primary Census Abstract (PCA) published by Census of India 2011 are utilized to extract the data of selected variables. Based on the thirteenth indicators extracted from census data following the human right approach of the good living standards, material deprivation index is computed as follows:

$$Vi\ norm = \frac{Vi - Vmin}{Vmax - Vmin} \tag{6}$$

*Vi norm- the value of 'i' indicator in a unit area i*
*V min- minimum value of 'i' indicator among all areas*
*V max- maximum value of 'i' indicator among all areas*

$$Deprivation\ Index\ or\ Di = \Sigma\ Vi\ norm \tag{7}$$

Variables selected for computing *Vi* are:

i. *Share of non- working population*
ii. *Share of households not having exclusive room*
iii. *Share of households having drinking water from untreated tap water*
iv. *Share of households having drinking water from uncovered well*
v. *Share of households drinking water from river/canal/tank/pond/lake*
vi. *Share of households having source of drinking water, away from their house*
vii. *Share of households have no lighting/electricity*
viii. *Share of households having pit latrine without slab/open pit*
ix. *Share of households having no drainage system*
x. *Share of households do not have kitchen*

xi. *Share of households have no bicycle*
xii. *Share of households have no scooter/ motorcycle/moped and*
xiii. *Share of households do not have basic assets like TV, fridge, telephone, computer, etc.*

Further, most vulnerable population is identified based on overlay analysis of healthcare accessibility index and deprivation score.

*4.7. Health Infrastructure Model Building for Berhampur CD Block*
We further conceptualise the application of GIS in proposing infrastructure model for Berhampur CD Block. Based on our field visits to major healthcare centres, which are remarkably conflating the notion of accessibility and quality, we propose three major recommendations to address the need and demand of quality healthcare access to Berhampur CD Block:

i. Upgrade existing infrastructure
ii. Connectivity improvement through mobile clinics
iii. Site selections for new facilities

Referring to previous studies, we also include proximity to public transit [21], distance to existing similar facilities [22], land cost, and proximity to targeted service recipients [23] as the major determinants of the site selection. The average weights were determined through an experts' survey, which included invitees from academia, block development office, hospital, and a representative from the patient communities. In total, responses of seven experts have been recorded, and weighted overlay analysis is performed on all thematic layers (land use and land cover, transport network, settlement with the deprived region, not accessible regions, unserved population, service area of health centers) using average weight score.

**Parameters used for the up gradation of existing PHC (mean weight in %):**
- *Near to the population more than 1000 population (20%)*
- *Near to the PHC with no doctor (15%)*
- *Near to the National and State highway (10%)*
- *Near to the available road network (10%)*
- *Near to deprived area (15%)*
- *Near to the not accessible area (10%)*
- *Near to population, not included within 5km of service area of any health center (10%)*
- *Away from existing PHCs having doctor (5%)*
- *On suitable land use land cover (5%)*

**Parameters used for connectivity improvement through mobile clinics (mean weight in %):**
- *Near to the population more than 500 person (20%)*
- *Near to deprived area (20%)*
- *Near to population, not included within 5km of service area of any PHCs (20%)*
- *Near to the inaccessible area (15%)*
- *On suitable land use land cover (10%)*
- *Away from existing PHCs (5%)*
- *Away from the National and State Highways (5%)*
- *Near the neighborhood streets (5%)*

**Parameters used for site selections for new facilities (mean weight in %):**
- *Near to the population more than 1000 population (20%)*
- *Near to the National and State Highway (10%)*

IOP Conf. Series: Earth and Environmental Science 1164 (2023) 012017 doi:10.1088/1755-1315/1164/1/012017

- *Near to the available major road network (15%)*
- *Near to deprived area (15%)*
- *Near to the inaccessible zones (10%)*
- *Near to population not included within 5km of service area of any health center (15%)*
- *Away from existing PHCs (5%)*
- *On suitable land use (10%)*

## 5. Results and Discussion

### *5.1.* Health Infrastructure at a glance

According to Indian Public Health Standards, at the block level, there must be at least one Primary Health Centre (PHCs) for every 30,000 population in plain area and 20,000 people in tribal, hilly, and backward regions (Table 2). A PHC acts as a referral unit for Sub-Centres (SCs). A Community Health Centre (CHC) is designated to operate at the district level per 100 000 population. Considering all the necessary facility standards, we record that Berhampore CD Block lacks PHCs (6), SCs (2), CHCs (2), and urban family welfare centre (3).

Referring to Table 3, we note that the Berhampore CD Block has five primary health centers that lacks in nursing and paramedical staffs; community health centers that lack doctors and nursing and paramedical staff, and the primary health sub-centers that lacks nursing, and paramedical staffs.

**Table 2**. Number of health centres in Berhampore CD Block in compliance to Indian Public Health Standards - IPHS (2012)

| Public health Centre (for rural) | Population Served for a Health Centre (IPHS Standards 2012) | Rural Population of Berhampur Cd Block | Number of public health centres | Population unserved | Required Number of health centres as per population size |
|---|---|---|---|---|---|
| Primary Health Centres (PHCs) | 30000 | | 05 | 186793 | 06 |
| Primary Health Sub Centres (SCs) | 5000 | 336793 | 56 | 56793 | 11 |
| Community Health Centres (CHCs) | 100000 | | 01 | 236793 | 02 |
| Public health Centre (for Urban) | | Total Population | | | |
| Urban Family Welfare Centre | 100000 | 641280 | 03 | 341280 | 03 |

Source: Census of India 2011.

**Table 3**. Human resource strength in compliance to IPHS (2012)

| Health Centers | Number | Total Doctor Strength (Required as per IPHS-2012 per center) | Total strength of nursing and paramedical staff (Required as per IPHS-2012 per center per center) |
|---|---|---|---|
| Community Health Centers | 01 | 6 (11) | 0 (21) |
| Primary Health Centre | 05 | 5 (1) | 6 (3) |
| Primary Health Sub Centre | 56 | 5 | 46 (2) |
| Maternity and Child Welfare Centers | 03 | 11 | 0 |
| Hospital Allopathic | 04 | 87 | 67 |
| Hospital Alternative Medicine | 04 | 6 | 0 |
| Dispensary | 09 | 3 | 8 |
| Family Welfare Centre | 03 | 2 | 1 |

### *5.2. Facilities Accessibility and Population Reach*

In their study, de Souza and David (2005) reported that the mean potential travel time to the nearest hospital was 14.2 minutes for the survey group of Britain for the life-threatening medical conditions [24]. For those with a positive outcome, it was 13.8 minutes, and for those with a negative result, it was 14.4 minutes.

The operational guidelines of *Ayushman Bharat* provide the framework for the Comprehensive Primary Health Care through Health and Wellness Centers. It prioritizes the up-gradation of sub-health centers where a community of about 3000 – 5000 population is not within reach of PHCs, SCs, CHCs, or a Sub-Divisional Hospital within 30 minutes travel time. To unwrap the status of the healthcare accessibility of Berhampore CD Block and to identify the under-serviced area that does not have access to basic healthcare services at the settlement level, we created a service area for each healthcare facility considering network connectivity and travel time.

During our field investigation, we have observed that most of the population uses motor transport to visit its nearest health centers for emergency medical purposes. Henceforth, the car travel time from each settlement centroid to healthcare facilities is extracted from Google Map Services. Further using the network analysis tool of Arc GIS 10.3 (Student Version), service areas are drawn for each population reach to show how accessibility varies with impedance (travel time in minutes).

*5.2.1. Facilities Accessibility and Service Area of CHCs.* Indian Public Health Standards (IPHS, 2012). has laid the minimum requirement of one community health center (CHC) per 1,20,000 population for plain areas. CHCs is a 30-bedded hospital designed to cater referral health care for cases from the Primary Health Centres and for the cases that need specialist care in Medicine, Obstetrics and Gynecology, Surgery, Paediatrics, Dental, and AYUSH. Our study records the presence of only one CHC in Berhampur CD Block, located at Harirampur and serves approximately 3,36,793, rural population compared to 1,20,000 population as per IPHS. It is approximately 3$^{rd}$ time more than its assigned norms. The overburdened CHC of Harirampur also lacks sufficient manpower, specialists, and paramedical staff against the minimum requirement of IPHS. The service area map generated (Figure

IOP Conf. Series: Earth and Environmental Science 1164 (2023) 012017 doi:10.1088/1755-1315/1164/1/012017

3) from the existing connectivity network depicts that most of the population of the block needs an average of 80 minutes and more travel time to reach a single CHC of the block.

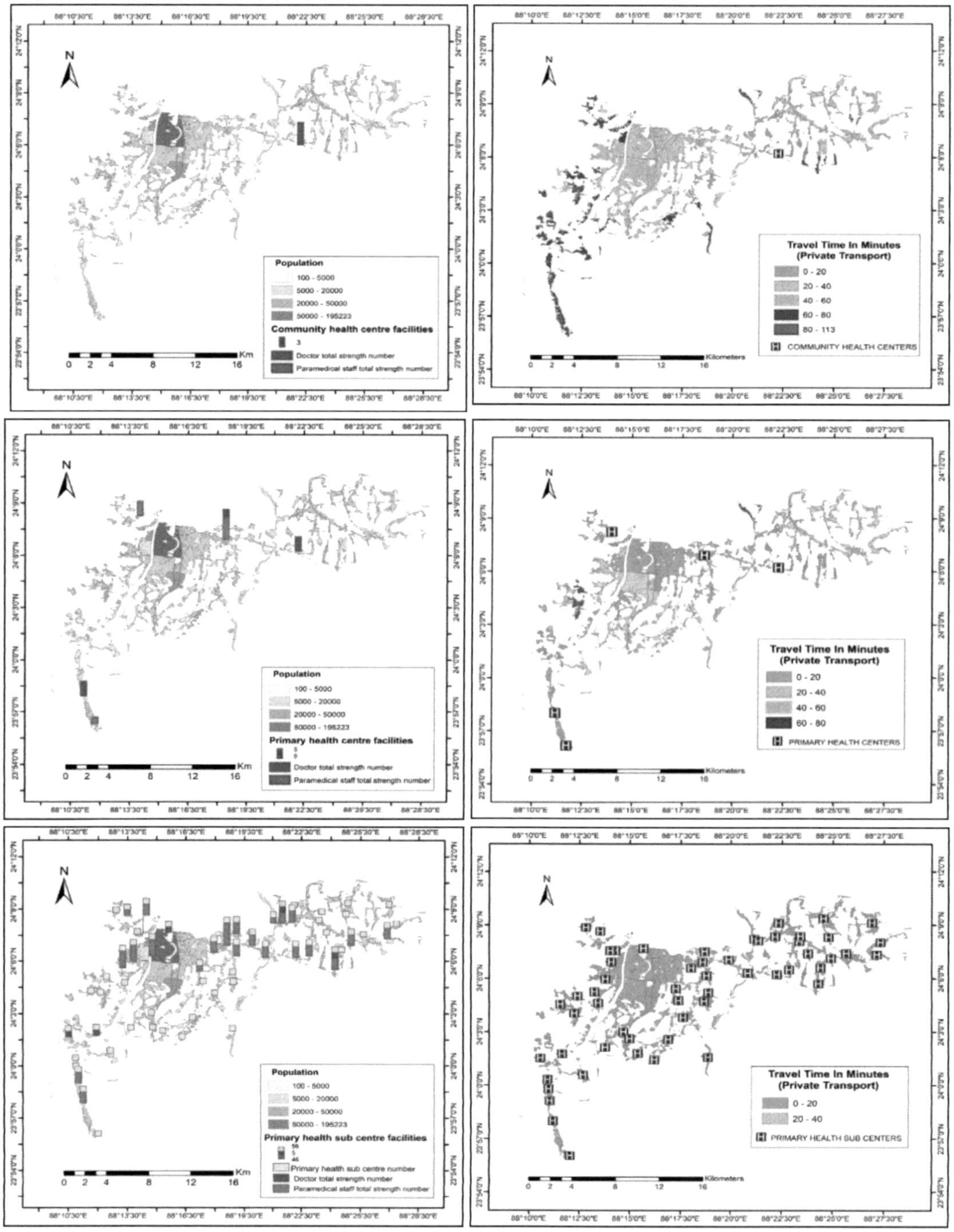


**Figure 3.** Facility Strength and Service Area of CHCs, PHCs and SCs

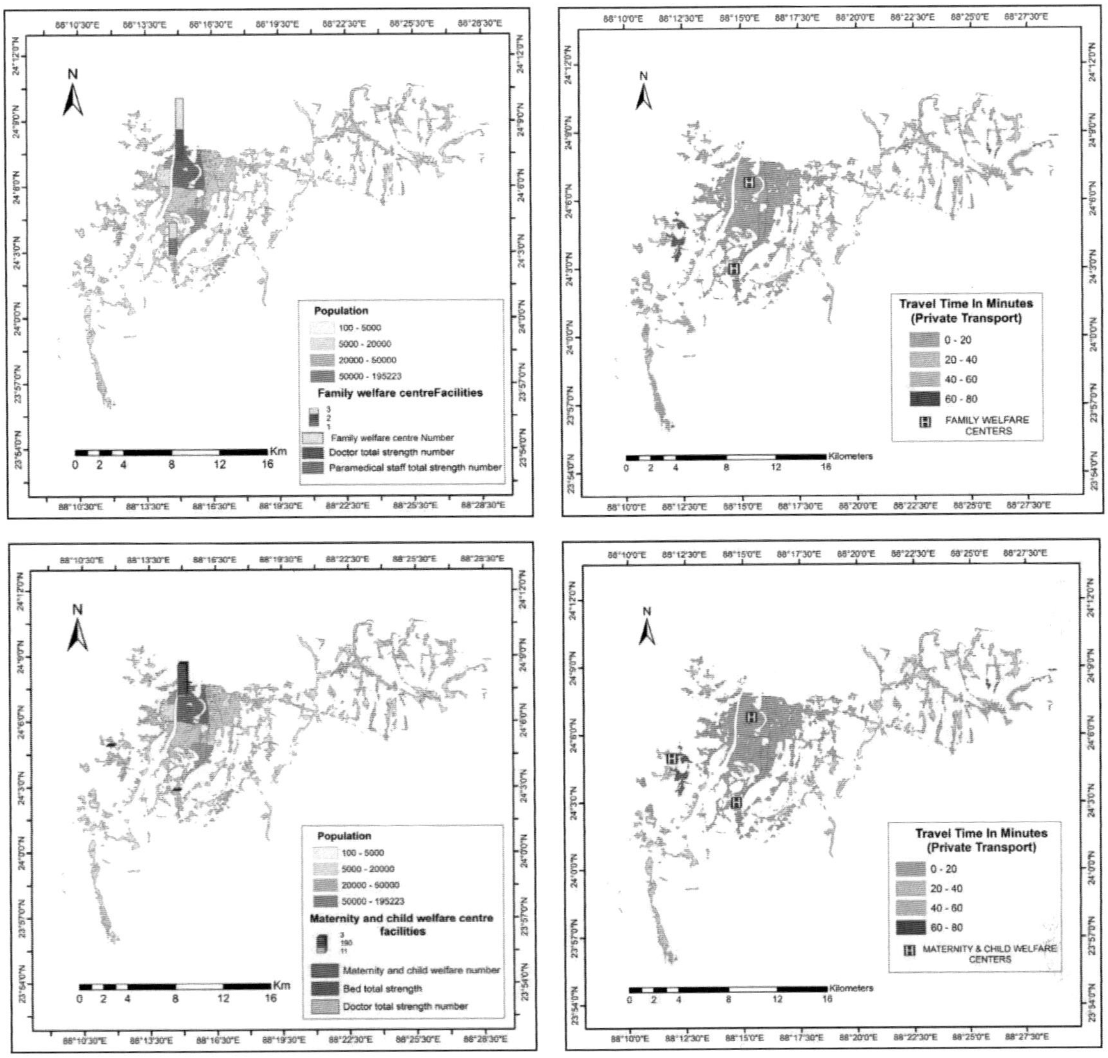


**Figure 4.** Facility Strength and Service Area of Family Welfare Centres and Maternity and Child Welfare Center

*5.2.2. Facilities Accessibility and Service Area of PHCs.* IPHS guidelines advocate providing a minimum of one primary health center (PHC) per 30, 000 rural population with a general physician and paramedical staff. PHCs are designed to act as a referral unit to a minimum of six sub-centers. Although, Berhampore CD Block records inadequate presence, where approximately, 3,36,793 rural population served by five PHCs only. At the settlement-based analysis, we found that 28 census villagers need to travel more than 40 minutes to reach the nearest PHC (figure 3).

*5.2.3. Facilities Accessibility and Service Area of SCs.* With a minimum standard of having one sub-center per 5,000 rural population, Berhampore CD Block has only 56 sub-centers per 3,36,793 rural population and lacks behind of another 11 centers to serve the basic medical requirement of these rural patients (figure 3). Within these 56 SCs, most of the centers have vacant positions of the medical practitioner, doctor, officer, or paramedical staff.

IOP Conf. Series: Earth and Environmental Science 1164 (2023) 012017 doi:10.1088/1755-1315/1164/1/012017

*5.2.4. Facilities Accessibility and Service Area of Family Welfare Centres and Maternity and Child Welfare Center.* In two-tier urban systems, Urban Health Centre (UHC) or Urban Family Welfare Centre (UFWC) for every 1,00,000 population is a norm laid by IPHS. In Berhampur CD Block, three centers are serving the need of 6,41,280 which is far away from the public health standards. Majority of the population spending more than 20 to 40 minutes of travel time to avail services at these UHCs (figure 4). At present, Berhampur CD Block has three Maternity and Child Welfare Centres (MCWCs) as shown in Figure 4. Surprisingly, two MCWCs which are established to serve the need of rural mothers struggling to have minimum facilities. These centers have no qualified practitioners. Berhampore (M) Maternity and Child Welfare Centers alone serve a total of 2,79,520 populations. To avail of the services here, the majority of the population travels more than 40 minutes which poses risks to lives in case of an emergency. The situation is worst for the Kandibandha Maternity and Child Welfare Center which is unable to serve the need of the local population due to its inaccessible location. Poor transportation connectivity, lack of basic infrastructure, and no qualified practitioners attributed to its poor functioning.

*5.2.5. Facilities Accessibility and Service Area of Allopathic Hospital.* Berhampore CD block has three hospitals (Allopathic) that serve the general and specialized needs of medical care to the local population. These hospitals are situated in Berhampore (M) and Haridasmati and cover a total population of 2,92,126 with approximately 20 minutes of travel time (figure 5).

*5.2.6. Facilities Accessibility and Service Area of Other Facilities.* Dispensaries are small facilities with minimal medical resources where a doctor provides primary care. Dispensary does not provide any hospitalization facilities however; it offers outpatient services. Berhampore CD Block has 67 dispensaries that serves all population within the travel time of 60 minutes. Four TB clinics that are available to serve TB patients. However, TB clinics located at Arazi Madhupur, Chhiruti, and Jadupur have no facilities. The average travel time for the villagers to access the clinic varies from 60 to 80 minutes (figure 6).

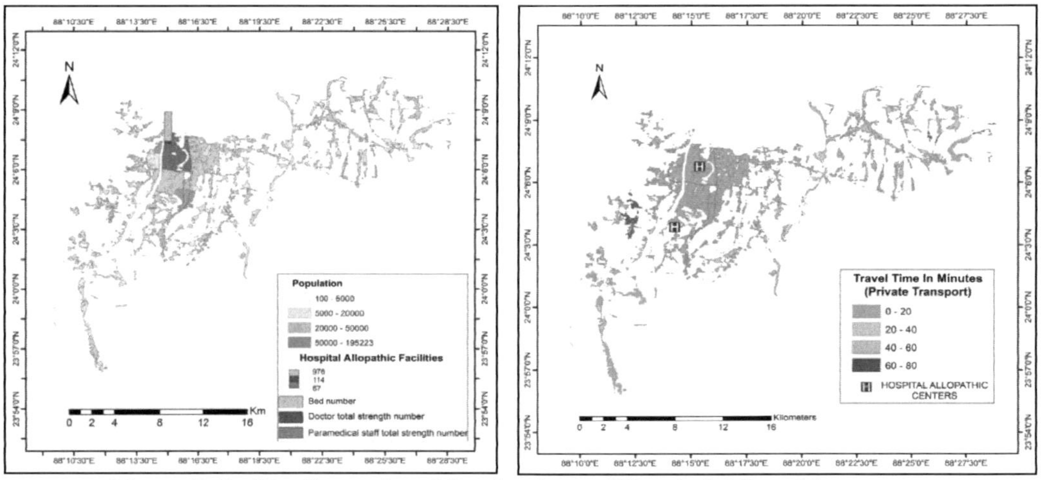


**Figure 5.** Facility Strength and Service Area of Hospital Allopathic

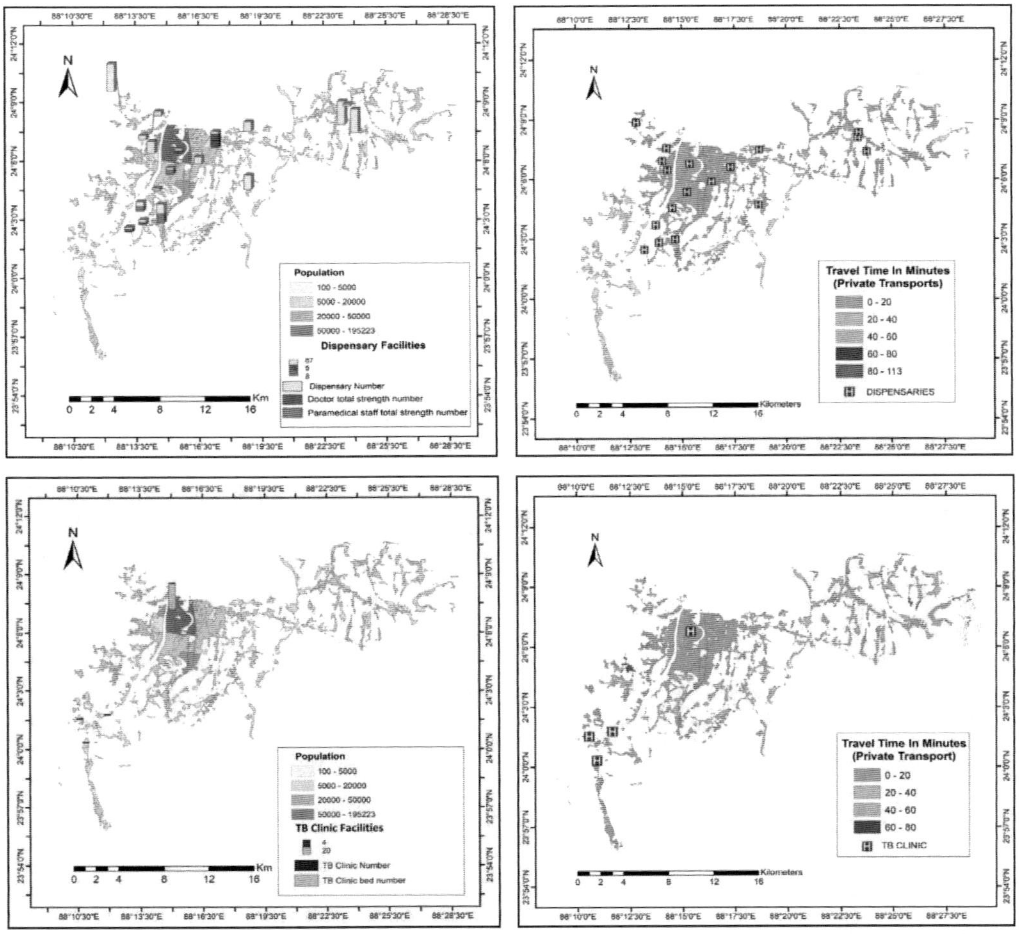


**Figure 6.** Facility Strength and Service Area of Dispensaries and TB Clinic

*5.2.7. Index of healthcare accessibility.* Based on Hansen's Improved Model of accessibility (Equation 1 to 5), healthcare accessibility index for each settlement have been computed. Further, the normalized values of accessibility index have been categorized into different classes with an assumption of having the value closer to '0' as inaccessible/very low accessible to regions to '1' that is highly accessible (Table 4). From the healthcare accessibility index, results exhibit (figure 7) that most of the settlement of the Berhampur CD Block are classified under no accessible (0 to 0.10) to least accessible zones (0.10 to 0.30).

**Table 4**. Classification of Accessibility Index

| Category of Accessibility Index | Accessibility |
|---|---|
| 0-0.10 | Not accessible |
| 0.11-0.30 | Least accessible |
| 0.31-0.60 | Moderate accessible |
| 0.61-0.90 | Moderately high accessible |
| 0.91-1.00 | High accessible |

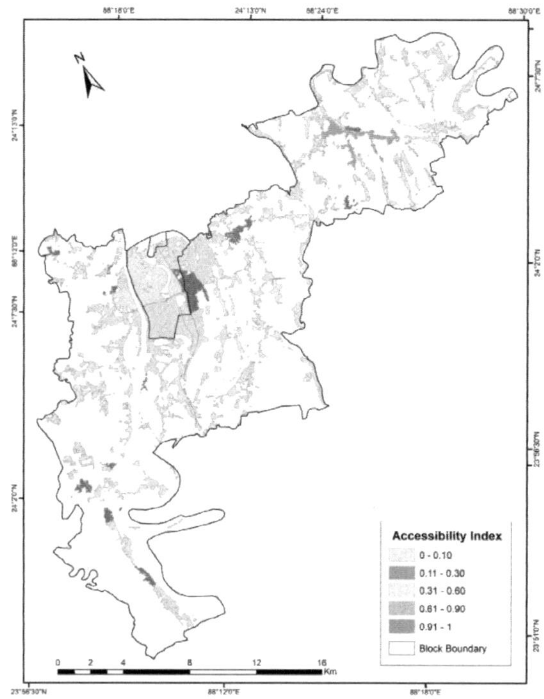


**Figure 7.** Accessibility Index Map based on Improved Gravity Model (Hansen and Huff, 2003)

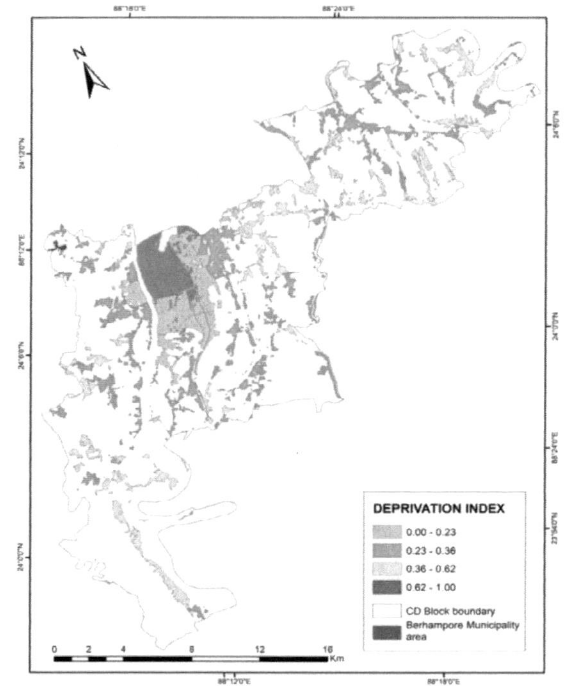


**Figure 8.** Results of Relative Material Deprivation Index, Berhampore CD Block, 2011

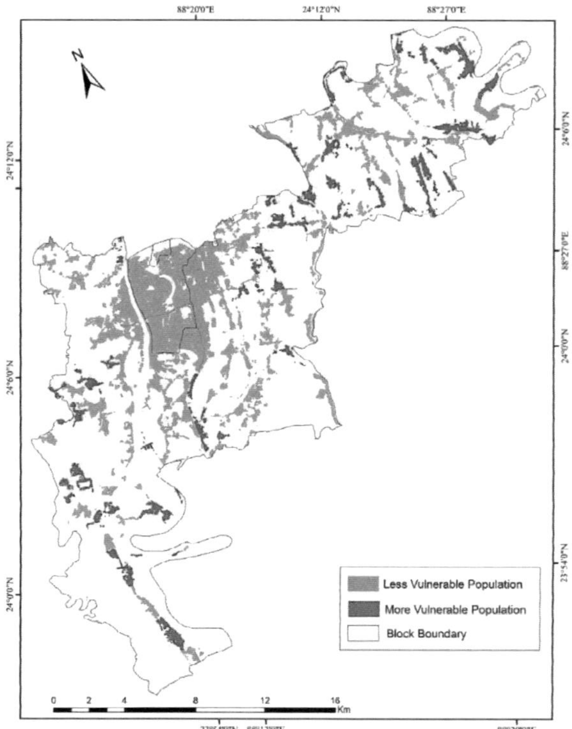


**Figure 9.** Vulnerable Population

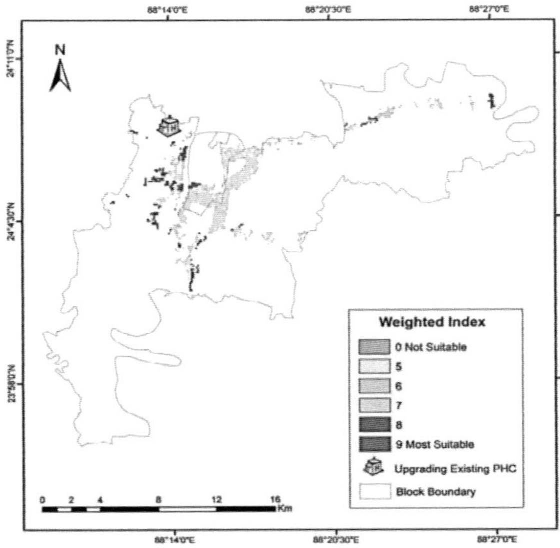


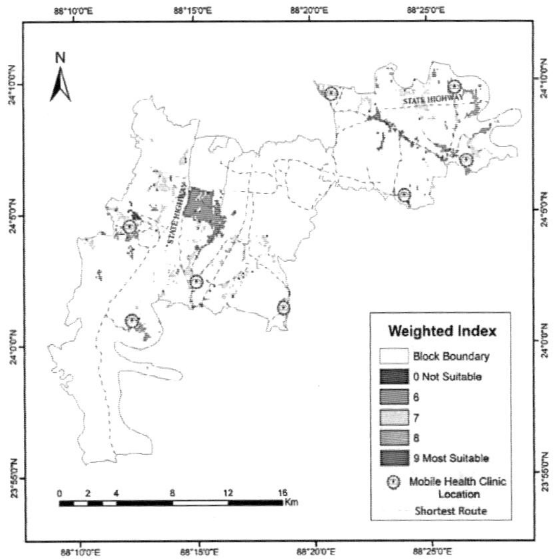


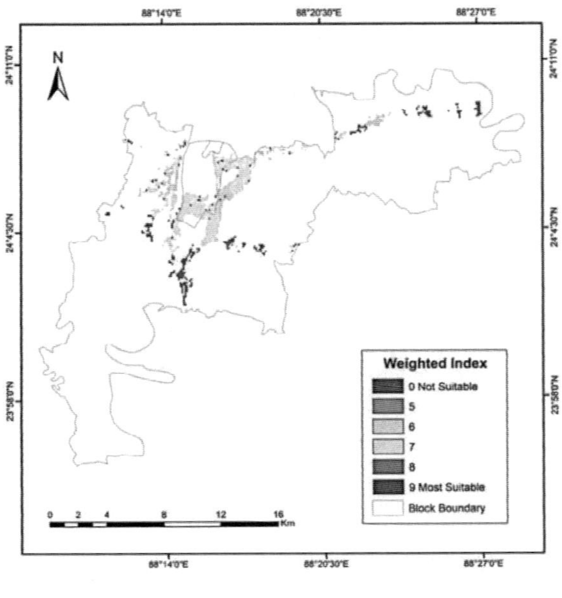


**Figure 10.** Location of existing PHC need up gradation – Bagmara Village (Top Left)

**Figure 11.** Suitable location and route for Mobile health clinic (Top Right)

**Figure 12.** Location identification for new PHCs (Bottom Left)

*5.2.8. Population heterogeneity and deprivation.* The result of the relative material deprivation index points towards poorer socio-economic status of the population. Most of the settlements record the deprivation index value of 0.36- 0.62. It covers 58 villages with 3,27,777 population (figure 8).

*5.2.9. Healthcare inaccessibility, deprivation and vulnerability.* The most vulnerable population is identified based on an overlay analysis of the healthcare accessibility index and deprivation score. The settlements surrounding Berhampore Municipality (M), where most of the villages have a population of more than 5000, but inadequate facilities to support the same. Through overlay analysis, where the deprivation index is high and the accessibility index is low, we computed the vulnerability index. We found that approximately, 1,19,429 population of Berhampore CD Blocks fall under the vulnerable category (figure 9).

### *5.3. Health Infrastructure Model for Berhampur CD Block*

This study has successfully shown the application of the GIS technique at the settlement-based level in identifying vulnerable populations immediately demands more healthcare attention. An issue of healthcare accessibility becomes more critical in the socio-economically deprived zones. Considering the present scenario of Berhampore CD Block, we further conceptualize the application of GIS in proposing an infrastructure model for Berhampur CD Block. Based on our field visits to major healthcare centers, which are remarkably conflating the notion of accessibility and quality, we propose three major recommendations to address the need and demand for quality healthcare access to Berhampur CD Block:

- **Upgrade existing infrastructure** - Based on the result of site suitability analysis, we found that the Primary health center situated at the Bagmara village needs up-gradation to serve the healthcare demand of the population (figure 10). This discovered PHC has zero doctor strength and the infrastructure also need more up-meet the demand of more than 50,000 populations within the reach of 20 min and more.
- **Connectivity improvement through mobile clinics** -The location on the Berhampore CD Block needs improvements in terms of connectivity. The shortest routes for the mobile clinic are determined by looking for the route with a fewer number of vertices and edges from the origin (Berhampur municipality) to destination. This acknowledges the relevance of mobile health clinics that represent a vital component of the healthcare system in serving vulnerable communities and promoting high-quality care at cheap cost (mobile health clinic). As we are dealing with the emergency purpose, so the shortest path indicates the path of minimal delay between two components. The census villages of Simulia, Selamatpur, Nutanpara, Baradaha, Gauripur, Charmahula, Bairgacchi, and Nischintapur that all are situated outside the service area of 20 min (private transport) of travel time from the existing PHC, found suitable for the establishment of the mobile clinic (figure 11).
- **Site selections for new facilities** - Based on the result we found that the most suitable location to build new PHC are on the census tract of Bairgachi, Kaya, Rajdharpara, Basudebkhali, Kalidanga, and Haridasmati. This entire census tract has a population of more than 3000, hence the demand is relatively more but the census tract has no primary health center to meet the demand. The resulted census tract is not coming under the service area of 20 min (private transport) of travel time from the existing primary health centers. The new PHC location can fulfill the need for the basic health of more than 5000 population (figure 12).

## 6. Conclusion

The study carried out here to measure healthcare deprivation and inaccessibility at a neighborhood scale indicates the sharp differences in the spatial context. The increasing coverage of healthcare facilities is yet to address the issue of accessibility as the interplay between facilities connectivity matrix. The study used a connectivity matrix to analyze healthcare accessibility and found that network-based measurement of health accessibility is more suitable over straight-line distances. We found this work useful, to support policymaking, and to bridge the gap between the biased share of infrastructure coverage and health care disparity in the study area. The application of GIS techniques in solving issues of geospatial health may solve the direct need for quality healthcare exclusively in the context of rural India. Due to time and financial constraints this study limits itself to here, however, it is further suggested here for the ground validation of identified suitable sites before implementation of any planning measures.

**References**

[1] Evans D B, Hsu J and Boerma T 2013 Universal health coverage and universal access *Bull. World Health Organ.* **91** 10–1.

[2] IBEF – Healthcare accessed on https://www.ibef.org/industry/healthcare-india

[3] Nykiforuk C F L 2008 *Exploring the utilization of geographic information systems in health promotion and public health.* Technical report #08-001. The Centre for Health Promotion Studies

[4] Levesque JF, H. M. (2013). Patient-centred access to health care: conceptualising access at the interface of health systems and populations. *Int J Equity Health*, 3-9.

[5] Martin Gulliford, J. F.-M. (2002). What does `access to health care' mean? J Health Serv Res Policy, 186-188.

[6] Rashid L. Bashshur, G. W. (1971). Some Ecological Differentials in the use of Medical Services. *Health Service Research*, 61-75

[7] Huerta Munoz U and Källestål C 2012 Geographical accessibility and spatial coverage modeling of the primary health care network in the Western Province of Rwanda *Int. J. Health Geogr.* 11 1–11.

[8] Cabrera-Barona P, Blaschke T and Gaona G 2018 Deprivation, Healthcare Accessibility and Satisfaction: Geographical Context and Scale Implications *Appl. Spat. Anal. Policy* **11** 313–32

[9] Dos Anjos Luis A and Cabral P 2016 Geographic accessibility to primary healthcare centers in Mozambique *Int. J. Equity Health* **15** 1–13

[10] J Withall, R. J. (2010). Who attends physical activity. *Health Education Journal*, 206-216.

[11] Ibrahim Ramzi A and Abdl-Latif El-Bedawi M 2019 Towards integration of remote sensing and GIS to manage primary health care centers *Appl. Comput. Informatics* **15** 109–13

[12] McLafferty S L 2003 GIS and health care *Annu. Rev. Public Health* **24** 25–42

[13] Gu W, Wang X and McGregor S E 2010 Optimization of preventive health care facility locations *Int. J. Health Geogr.* **9** 1–16

[14] Murad, A. A. (2007). Creating a GIS application for health services at Jeddah city. Computers in Biology and Medicine, 879-889.

[15] District Magistrate Murshidabad, Government of West Bengal, India available at https://murshidabad.gov.in/ visited at 20th March 2020.

[16] Wang, F. (2012). Measurement, Optimization, and Impact of Health Care Accessibility: A Methodological Review. *Ann. Assoc. Am. Geogr*, 1104-1112.

[17] Kleinman, J., & Makuc, D. (1983). Travel for ambulatory medical care. *Med. Care*, 543-557

[18] Luo, W. (2004). Using a GIS - based floating catchment method to assess areas with shortage of physicians. *Health & Place*, 1-11

[19] De Beule, M., Van den Poel, D., & Ven de Weghe, N. (2014). An extended Huff- model for robustly benchmarking and predicting retails network performance. *Applied Geography*, 80-89.

[20] Schuurman, Nadine, & Fielder, R. e. (2006). Defining Rational Hospital Catchments for Non-urban Areas Based on Travel-time. *International Journal of Health Geographics*, 1-11.

[21] Soltani, A., & Marandi, E. Z. (2011). Hospital Site Selection using Two- Stage Fuzzy Multi-Criteria Decision Making Process. *Journal of Urban and Environmental Engineering*, 32-43.

[22] Wu, C. R., Lin, C. T., & Chen, H. C. (2007). Optimal Selection of Location for Taiwanese Hospitals to Ensure A Competitive Advantage by Using the Analytic Hierarcy Process and Sensitivity Analysis. *Building and Environment*, 1431-1444.

[23] Vahidnia, M. H., Alesheikh, A. A., & Alimohammadi, A. (2008). Hospital Site Selection Using Fuzzy AHP and its Derivatives. *Journal of Environmental Management*, 3048-3056.

[24] Souza, V. C., & David P, S. (2005). Relationship between travel time to the nearest hospital and survival from ruptured abdominal aortic aneurysms: record linkage study. *Public Health*, 165-170.

IOP Conf. Series: Earth and Environmental Science 1164 (2023) 012017 doi:10.1088/1755-1315/1164/1/012017

Appendix 1. Map of Berhampore CD Block

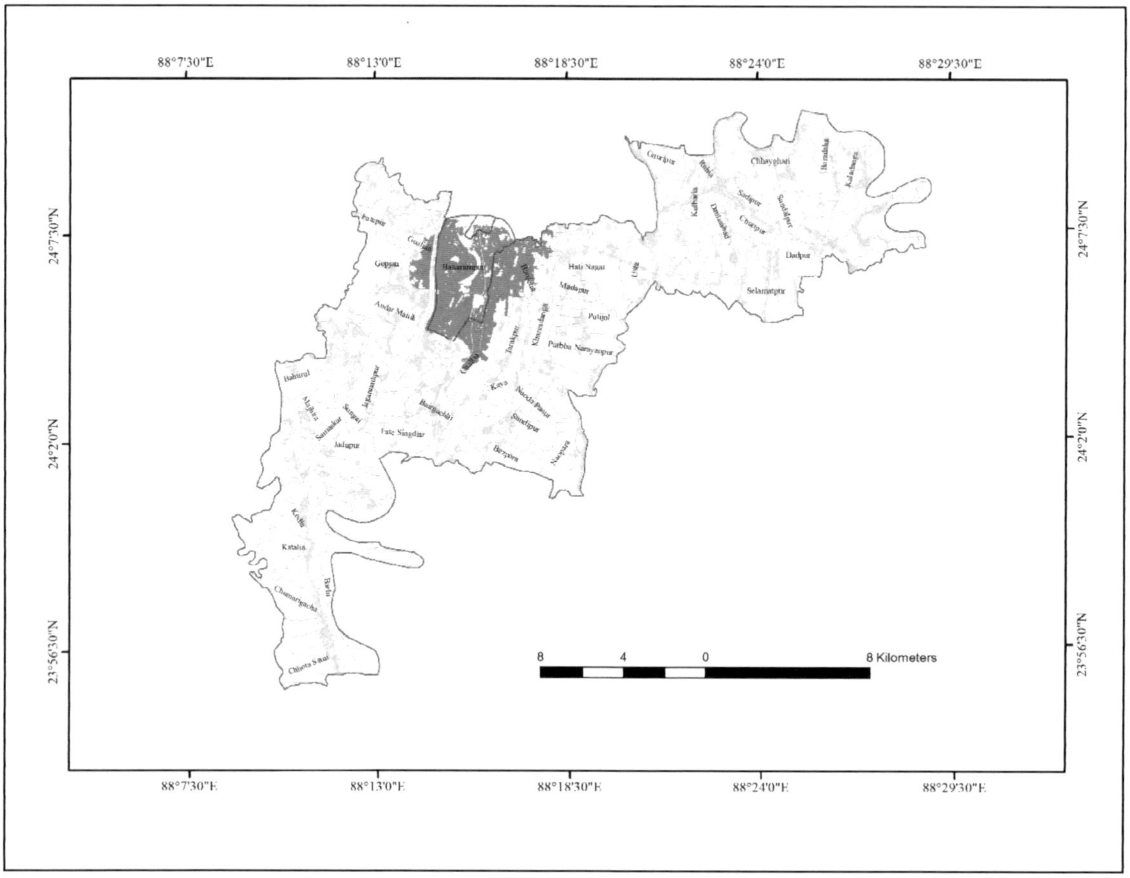

# AUTHOR INDEX

**Institute of Physics**
Dirac House, Temple Back
Bristol BS1 6BE UK

ISSN: 1755-1307
ISBN 978-1-7138-7285-6

CURRAN ASSOCIATES INC.
proceedings.com